SOTHIS PRESS

Sothis Press aims to completely reimagine the public domain book with an emphasis on universal access to captivatingly designed print editions. We believe that technology will enable every single book ever published to be perpetually accessible in beautiful print form. Sothis is beauty in ubiquity.

www.sothis.press
1 Waterhouse Square, London, ECIN 2ST

Practical Sheet and Plate Metal Work

For the Use of Boilermakers, Braziers, Coppersmiths, Ironworkers, Plumbers, Sheet Metalworkers, Tinsmiths, Whitesmiths, Zincworkers, and Others Who Require a Knowledge of the Working Up to Metals or Development of Surfaces

Evan An. Atkins

SOTHIS PRESS

ISBN 978-1-0368-0369-8

Practical Sheet and Plate Metal Work
is in the public domain.

This edition is published by
Sothis Press © 2023. The company
is registered in England and Wales.
The Company Number is 14900389.

PRACTICAL SHEET AND PLATE METAL WORK

PRACTICAL SHEET
AND
PLATE METAL WORK

FOR THE USE OF

BOILERMAKERS, BRAZIERS, COPPERSMITHS, IRONWORKERS, PLUMBERS, SHEET METALWORKERS, TINSMITHS, WHITESMITHS, ZINCWORKERS, and others who require a knowledge of :: :: the working up of metals or development of surfaces :: ::

BY

EVAN A. ATKINS, A.M.I.M.E.

Head of the Metal Trades Department, and Chief Lecturer on Practical Geometry, Municipal Technical School, Liverpool; Honours Silver Medallist in Metal Plate Work, City and Guilds of London Institute; Honours Medallist in Practical Geometry, Board of Education

SECOND EDITION, REVISED AND ENLARGED

WITH 450 ILLUSTRATIONS

LONDON

SIR ISAAC PITMAN & SONS, LTD., 1 AMEN CORNER, E.C.4

(INCORPORATING WHITTAKER & CO.)

BATH, MELBOURNE AND NEW YORK

Printed by Sir
& Sons, Ltd., I
Melbourne an

PREFACE

HAVING long felt the want of a book that dealt with sheet and plate metal work from a practical point of view, the author some years back decided to write one. In this he was encouraged by the editor of the *English Mechanic*, who invited him to contribute a series of articles on the subject to that paper. The present book, through the kind permission of the above-mentioned editor, is the outcome of those articles.

The distinct feature of the book is that the patterns are shown set out as they would be in the workshop, with hints as to allowances for joints, wiring and notches, bending up, use of tools, and so on.

Nearly all the patterns have been tested by the author himself or under his supervision.

It is hoped that by a careful study of the book the mechanic will become a more expert craftsman, and at the same time learn the art of cutting up metal economically.

The book, too, should serve as a work of reference for works managers and draughtsmen.

In conclusion, the author hopes that the work may lead to the more extended use of sheet metal in aiding the present "new art" tendency to evolve newer shapes of articles for household and general use.

PREFACE TO THE SECOND EDITION

THAT there has been a demand for such a book as this is shown by the very favourable way in which it has been received, and by the call for a second edition. The author has been particularly gratified by the many, both in this country and abroad, who have stated that the book has been of assistance to them. Special thanks are due to old colleagues both in Staffordshire and Lancashire for making the book known.

Opportunity has been taken to add to this edition a chapter on annealing, acetylene welding, etc.; a further chapter on miscellaneous patterns, and a fuller chapter on tools and machines.

Thanks are here expressed to Messrs. J. Rhodes and Sons, Messrs. R. Becker and Co., and the Acetylene Illuminating Co., for the loan of blocks.

E. A. A.

LIVERPOOL, 1912.

CONTENTS

CONTENTS

PRACTICAL SHEET
AND
PLATE METAL WORK

CHAPTER I.

INTRODUCTORY.

Every workman whose aim it is to become a proficient sheet or plate metal worker should at least have a fair knowledge of practical geometry, mensuration, and the properties of metals. Whilst no attempt has been made in the following pages to treat these subjects separately, yet their application has been shown and explained in all suitable cases.

It is impossible to become an expert in the striking out of patterns or templates except the basic principles are thoroughly grasped. The learning of pattern-cutting by attempting to remember the methods applicable in a number of articles is to be deprecated as it gives only a parrot-like kind of knowledge which invariably fails when dealing with an object whose shape is a little out of the ordinary run. Nearly all patterns come from the development of the surfaces of a few geometrical models, either singly or in combination, such as the cylinder, prism, cone, pyramid, and in the following chapters the objects have been grouped with this classification in view. To become a good pattern-cutter then it is essential that a careful study should be made of the methods followed in developing the surfaces of the solids above-named and their interpenetrations. From the above statement it will thus be seen that the first thing to do in the making of a pattern is to carefully examine the shape of the article for which the pattern is required, and having determined from what geometrical solid or portion of solids the surface is built up, then to develop the pattern by the method peculiar to those surfaces.

The only way to gain confidence in the marking out of patterns or templates for sheet and plate metal work is by continued practice, not only in drawing the patterns out on paper, but more particularly in cutting them out of thin sheet metal and bending into shape to test accuracy of work.

For work of double curvature, such as hollowed or raised articles, pipe bends, etc., it is particularly desirable that the beginner should experiment by the working up of parts of an object whose pattern has been set out to some definite scale before attempting to mark out a full-sized plate and shearing into shape. In this way by careful examination and measurement of the model plate any errors in the pattern may be detected and allowed for in marking out the full-sized plate.

Particular care must be taken in fixing the shape and size of notches, also in the allowances for wire, joints, and thickness of metal if an article is to be made accurately and without giving undue trouble in the making up. Patterns without proper allowance, it should be remembered, are useless.

Particular attention is called to the general method of "triangulation" used in the setting out of patterns, and every ambitious mechanic would do well to strive to thoroughly understand its principle as explained in Chapter XIII.

There is no particular reason why any one classified group of articles should be taken first, but it is generally found by experience that the setting out of patterns for simple pipe joints is easily followed by the beginner, hence these are dealt with first in the next chapter.

Before passing on to the setting out of patterns there is one important point that should be borne in mind, and that is whether a pattern is being made for a single article or many, or perhaps for a stock article. After some practice in pattern-cutting the smart workman should be able to mark out a pattern for a single job with a few lines, but where the pattern is to be used for many articles more lines should be used so as to ensure the greatest accuracy.

A good pattern, it should never be forgotten, means a saving of time in making up an article.

CHAPTER II.

ELBOWS FOR ROUND PIPES.

Pattern for Round Pipe Cut on the Slant.

It should be borne in mind that the most important point in the making of patterns is accuracy in determining the lines that are required for the pattern. It is always better to spend a little extra time in finding the correct length of these lines than to have an ill-fitting article, or to waste time in cutting or chiselling it into shape. If the pattern is for a stock article, then the greatest possible care should be exercised, so as to obtain a pattern as near perfection as possible; but, on the other hand, if it is required to set out a pattern for an odd job, the workman who has an ounce of common sense will know it is foolish to spend as much time in the setting out as will eat up the cost of the job.

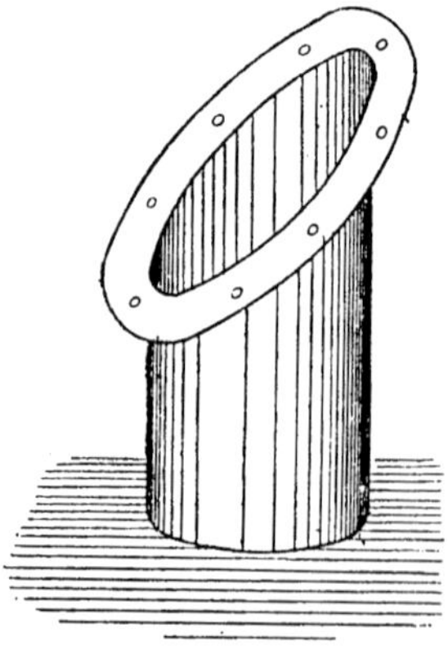

Fig 1.

Fig. 1 shows a sketch of the pipe, with flange fitted on the slant end. Generally, for those who have had but little practice at setting out, it is the best plan tc draw an elevation of the pipe or pipe-joint for which the patterns are required. First draw the centre line (Fig. 2), and then, as it were, "clothe" this with the pipe by marking half its diameter on each side, and draw lines parallel to the centre

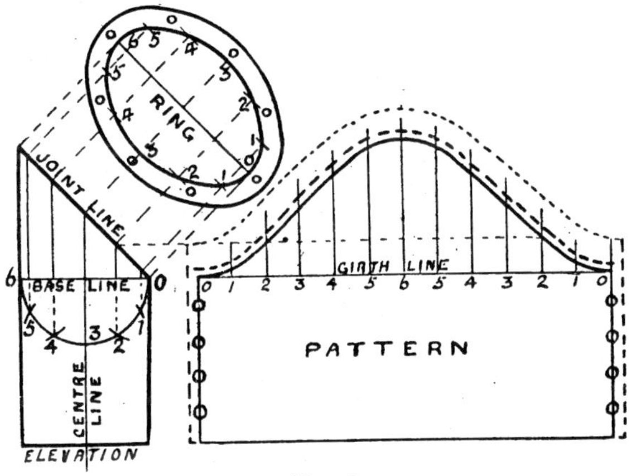

FIG. 2.

line, cutting them off to the required length. Now draw "base line," as shown, and on this describe a semicircle, and divide it into six equal parts by using the compasses at the same radius with 0, 3, and 6 as centres. Draw lines from "base line" to "joint line," passing through the points 1, 2, 3, etc., and parallel to the centre line, or square with the base line.

The pattern can now be developed by drawing a line, 0 0, equal in length to the circumference or girth of the pipe. This length can be obtained by carefully measuring along one of the six arcs 0 to 1 or 2 to 3, etc., into which the semicircle is divided, and setting it along the straight line twelve times. The arc can be measured by bending along it a strip of sheet metal or stiff paper, or a bit of thin wire; or it can be more accurately found by using the well-known rule for calculating the circumference of a circle:—"Multiply the diameter by 22, and divide by 7." Thus, in the present case, if the diameter of the pipe is $10\frac{1}{2}$ in., its circumference will be 33 in., and, dividing this by 12, the length of one of the arcs will be $2\frac{3}{4}$ in.

The simplest plan, however, and the one most often adopted in ordinary practice, is to take the lengths directly from the drawn semicircle. Lines perpendicular to 0 0 should be run up from each point, and numbered as shown, and their lengths cut off equal to the corresponding lines between "base" and "joint" lines in the elevation. In workshop practice, it is most convenient to take these lengths off with the compasses, and set them up the proper lines; but in developing a pattern on paper, the heights can be projected from elevation on the pattern, as shown with line cutting off point 2. The points marked can now be joined up with a free-flowing curve, and thus the *net* pattern is completed. To add the proper allowances for thickness of metal—laps, seams, joints, and wiring—is the most important part of the making of patterns, and this will be dealt with fully in subsequent chapters. In the present case, whatever is allowed for the side-riveted seam, half must be put on to each end of pattern. Thus, suppose the lap is $1\frac{1}{4}$ in., then $\frac{5}{8}$ in. will be the allowance for each end of pattern. It will be noticed that the centre lines for the rivet-holes are the end lines of the net pattern.

The thick dotted line at top represents the allowance for small flange for fastening ring to pipe, either by riveting, brazing, or soldering. The thin dotted line shows the allowance to be made if the whole flange is to be thrown off the pipe. Care must be exercised so as to get the allowance for flange the same width all along the pattern. This can be best done by setting the compasses at the required width, and drawing them along the curve at top of net pattern.

Attention is called to the method of numbering adopted. The figure 0 will in all similar cases be placed against the seam of pipe, and it will thus always come on the outside lines of net pattern.

The ring to form the flange can be set out from the elevation of pipe. The long diameter 0 6 will be equal in length to the joint line, and the intermediate points can also be taken from the same line. The widths at the different parts of the ring can be taken from the lines with the corresponding numbers on the semicircle in elevation. These points will now all be joined with a curve and the width of flange marked around. As the hole in the flange-ring is an ellipse there are many other ways that might be employed for marking it out—some shorter, some longer—and the best of these methods shall be shown as occasion demands.

Flanging.

A fair amount of skill is required to throw-off or stretch a flange properly. The first thing that should be done is to cut a gauge (Fig. 3) out of a bit of sheet brass, and with this mark the depth of the flange all round on the inside of the pipe.

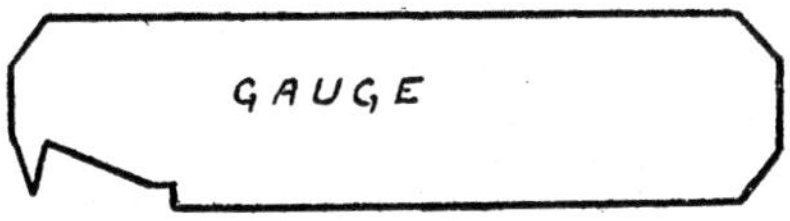

FIG. 3.

In stretching the flange on anvil, head-stake, or other tool it should be remembered that it is the outer edge of the flange that requires the greatest amount of hammering, as the length round the outside of flange will be greater than the inside by just about $6\frac{1}{4}$ times the width of the flange. If the pipe is made out of $\frac{1}{8}$ in. or thicker metal the flange will have to be turned over hot, and in this case the depth of flange should be marked on the plate when flat, with centre-punch marks.

In the flanging of plate metals there is no need to exercise quite so much care to avoid the splitting of the flange as there is with sheet metals, as there is a greater volume of metal to allow for drawing. Since the introduction of mild-steel plates of uniform structure, flanging operations can be carried out with a greater degree of certainty than in the old days, when iron of an indifferent quality had to be used. All the advice in the world, however, will not make a mechanic into a good flanger without plenty of practice.

If holes are required in the flange, no attempt should be made to put these in the sheet or plate before bending or flanging, as the flange is almost certain to break across the holes, and if, by good luck, it does not, it will be found that the holes are drawn out of shape.

In stretching, throwing-off, or flanging sheet metals, annealing plays an important part, so that as soon as an edge shows signs of becoming hard or brittle it should be at once got redhot and allowed to cool down.

Square Elbow for Round Pipe.

Possibly one of the commonest jobs an iron-plate worker is called upon to do is to make a square elbow for a round pipe. An elbow of this description may be required either for a stove pipe, a rain-water pipe, or a ventilating shaft. The pattern for it can be set out in a variety of ways, all giving the same result. One of these methods is shown in Fig. 4. This may be described as the general method, which is applicable to all kinds of pipe joints for circular non-tapering pipes. An elevation of the elbow is drawn in the usual way, and a semicircle described as shown. For the pattern the circumference of the pipe is set along the line 0 0, vertical lines are run up from each numbered point, and these cut off equal in length to the line with the same number running between base and joint lines in elevation. Before making any allowances for jointing, the method of fastening the pipes together should be decided. There are

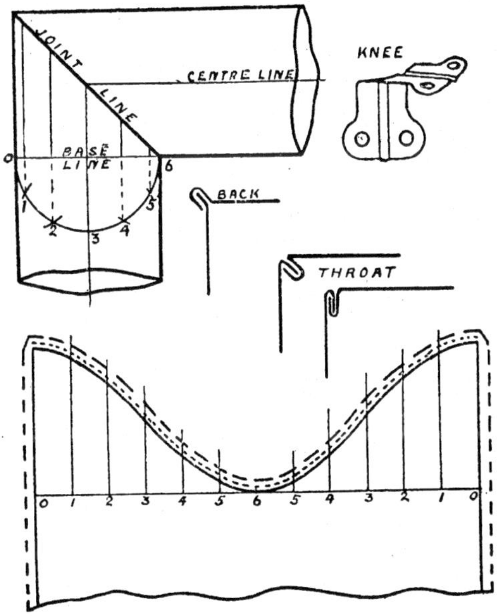

FIG. 4.

many ways in which the joint can be made, the method adopted depending upon the purposes for which the pipes are to be used. In Fig. 4 it is assumed that they are seamed together, the plan often followed in making elbows for stove pipes. A sketch of joint at back and two sketches of the joint at throat are shown. After the pipes are edged, or as it is called, paned together, it is usual to knock-up that part of the joint round the throat as shown in the bottom sketch. The four thicknesses of metal are of course hammered perfectly tight together. A knee is sometimes riveted in the throat of the elbow, which adds considerably to its strength.

For ordinary thicknesses of sheet iron, say 24 gauge, the allowance for the single throw-off may be 3-16 in., and for the double edge a little greater than $\frac{3}{8}$ in. These allowances are shown by the dotted lines on pattern. The side seam will be grooved, and it will be sufficient to allow $\frac{3}{8}$ in. on each side to cover for what is required for a $\frac{1}{4}$ in. groove. The way to make allowances for the different kinds of joints will be dealt with fully in subsequent chapters. Notches at 0 0 must not be cut too large, or the result will be a hole in the joint of the elbow. The object of the notches on pattern is to avoid having to stretch or throw-off the four thicknesses of sheet which form the groove, which if attempted would, in many cases, break the grooved seam. Besides this, if the groove would stand turning over, it would result in an unsightly lump on the joint seam. It is always the safest plan to cut a long notch, as shown in the pattern at 0 0 (Fig. 4).

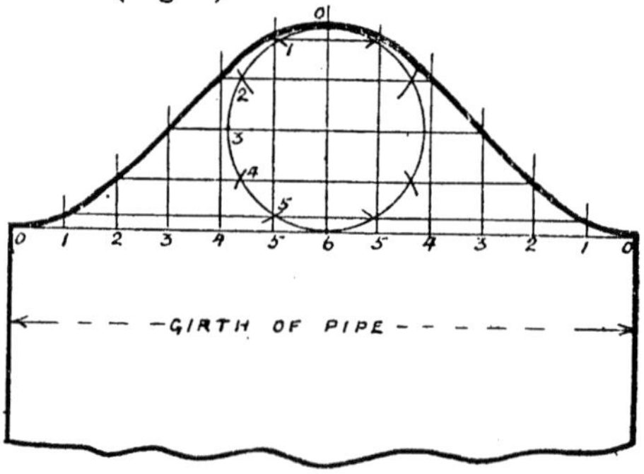

FIG. 5.

Without the sheet iron is of good quality, it is best to anneal around the edges for wide flange before attempting to throw it over. In fact the safest plan is to anneal twice, first before flanging, and then again after, before the edge is turned back. It might be taken as a sheet metal worker's maxim, "Never spoil a good job for the want of a little annealing."

A simpler method for marking out the pattern for a square elbow is shown in Fig. 5; but it must be distinctly borne in mind that this method applies to a square elbow only, and cannot be used for any other kind of elbow or bend. A circle equal in diameter to the pipe is described and divided into twelve equal parts, the girth line 0 0 being divided up in the same manner. Points on the curve are obtained by running construction lines up and across as shown.

Elbow with Slip-Joint.

A ready way of jointing the two pipes of an elbow together is to slip one inside the other, first having turned down the edge inside the throat, and then turn the edge at back over the inside pipe. The patterns for this kind of a joint are shown in Fig 6. The elevation is drawn in the usual way, and the lengths A D and 0 C made a little greater in length than the required lap. In setting out the pattern for the pipe with the outside lap, the lengths of lines are measured up to the line 0 D, and marked up on pattern on the corresponding lines. This will give the curve 0 D 0. In developing the pattern for the pipe with inside lap, lengths will be measured along to the line C A, and set up on pattern, and these will give the curve C A C. Hence, it will be seen that the curve for net pattern is 0 A 0, for bottom pipe 0 D 0, and for top pipe C A C. This is the way in which two pipes of exactly the same diameter can be jointed with a lap joint, one pipe fitting inside the other. The length of the curve 0 D 0 is, of course, greater than that of C A C, and the difference in the lengths of these two curves can be made anything we please by arranging the lengths of A D and 0 C in the elevation. The ellipse at the end of one pipe will be less in circumference than the ellipse at the end of the other;

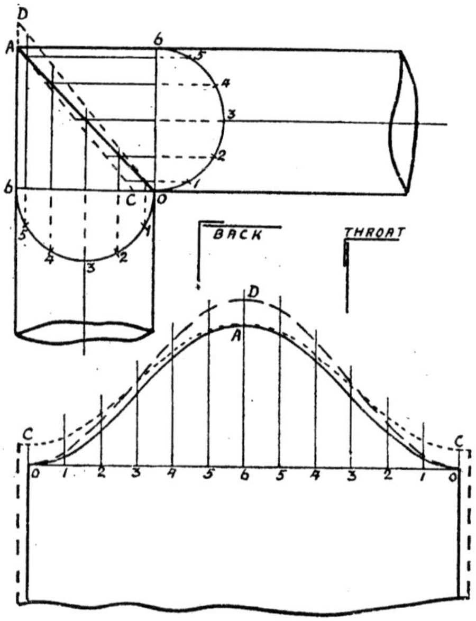

FIG. 6.

consequently, the smaller will go inside the larger. If no inside lap is required, as in the case of a galvanised sheet-iron rain-pipe elbow with soldered joint, then the length C 0 in top pipe will be made considerably shorter than in the figure. In every case the lengths that A D and C 0 are made will depend upon the thickness of the metal used.

Obtuse Elbow for Round Pipe.

The pattern for an obtuse elbow for a round pipe is shown in Fig. 7. The setting out of this pattern requires no additional explanation to that given for previous patterns. In drawing the elevation of pipe, however, care must be taken to set it out to the required shape of bend. In the workshop dimensions are generally given in one of

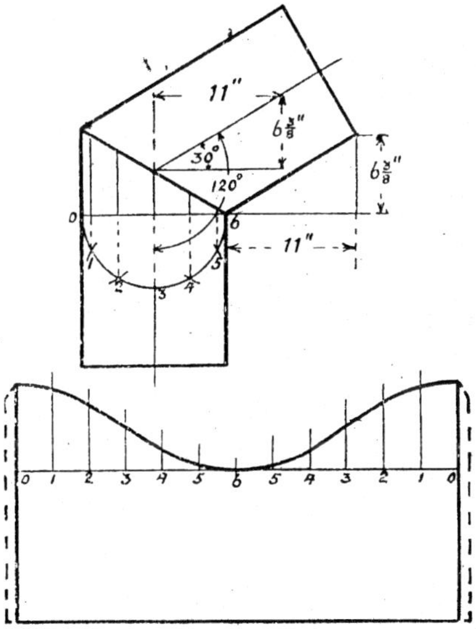

FIG. 7.

the three ways shown in the figure. The angle between the centre lines is sometimes given, which in this case would, of course, be 90° + 30° = 120°. Allowances for the side-seams only are shown in this pattern.

CHAPTER III.

TEE-PIECES FOR ROUND PIPES.

Square Tee-Piece.

IT is sometimes necessary for sheet and plate metal workers to make what is known as tee-pipes or elbows; the patterns, therefore, of a few examples in round pipes of this kind of work will be given. The pattern for a right-angle tee-piece, both pipes being of the same diameter, is shown set out in Fig. 8. An end elevation of the top pipe is drawn first, and the quarter-circle divided into three equal parts. Lines are now drawn through each point parallel to the centre line and down to the base line. The girth of pipe is set along 0 0, lines run up and cut off equal in length to the lines in elevation drawn from base line to corresponding number. The hole is marked out by making line 33 equal to half-

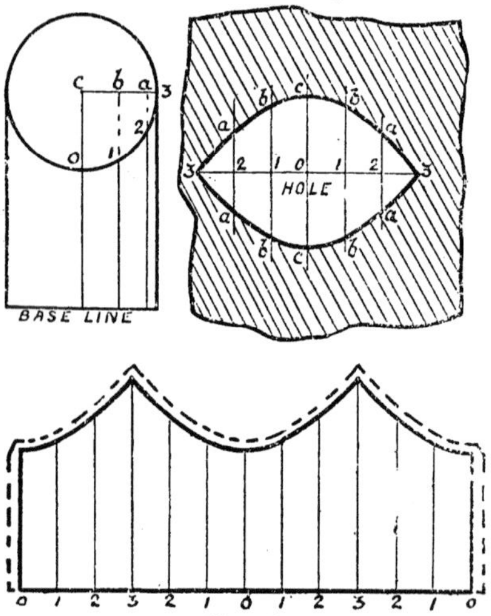

FIG. 8.

circumference of pipe, lines being drawn across through each of the five intermediate points, and cut off equal in length to the lines with the same number and letter in elevation. Thus *a* 2 in pattern will be the same length as line *a* 2 in quarter-circle. Care must be taken that the hole is marked in its proper position on the sheet or plate for top pipe. The line *c c* should be on the longitudinal centre line of plate.

The construction lines for obtaining the pattern by a more practically useful method are set out in Fig. 9. This is a most important case, and on account of the peculiar results obtained should be carefully studied. No elevation is needed, the view shown simply being drawn to exhibit the shape of the tee-pipe. A quarter-circle of same radius as pipes is set out

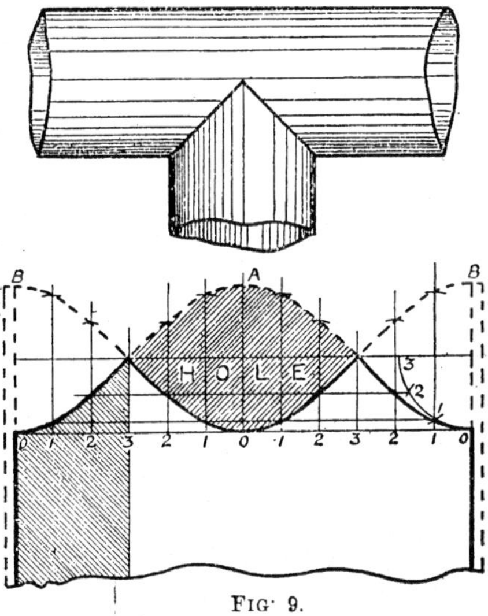

FIG. 9.

first, and then divided into three equal parts in the same manner as before-mentioned. Line 0 0 is drawn equal in length to the girth of the pipe, divided into twelve equal parts, and then numbered as on pattern. Through each point perpendicular lines are run up, and these are cut off the proper length by drawing lines through 1, 2, and 3 on quarter-circle parallel to line 0 0. Thus the point of intersection of line through 1 on quarter-circle with the line drawn up from 1 on girth line will be a point on curve of pattern. In the same way the other points will be obtained.

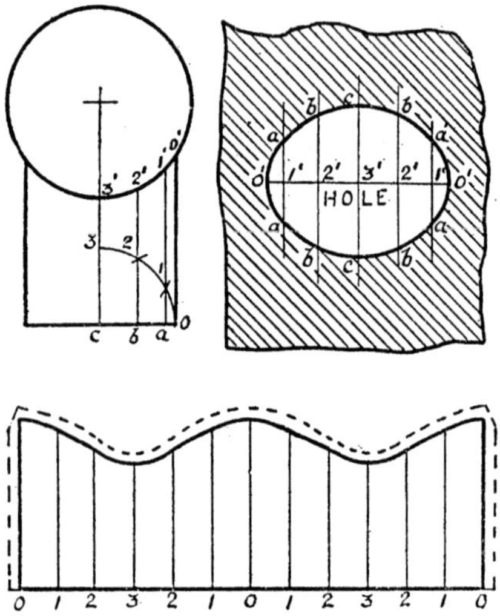

FIG. 10.

It will be noticed that the cut on pattern to form the joint is made up of four equal curves: hence in workshop practice all that is necessary to mark out is a template containing one of the curves, such as the shaded part shown at the left of pattern. This simple template can be used in a variety of ways. The pattern for the pipe can be set out by using it four times, marking the curve, and then reversing. The hole on top pipe can be drawn out in a similar manner, as will be seen by the four curves that form the hole at top of pattern. The template can also be used for setting out the patterns for a square elbow, the curve B 0 B showing the pattern for seam at back, and the curve 0 A 0 for seam at throat. Laps can be added on to the net patterns according to the method of joining adopted.

Tee-Piece with Unequal Pipes.

In the development of the patterns for tee-piece in which the branch pipe is smaller than the main (Fig. 10), the method pursued is the same as with Fig. 8. It will be observed that this pattern is also formed of four equal curves, and consequently in large work the setting-out of

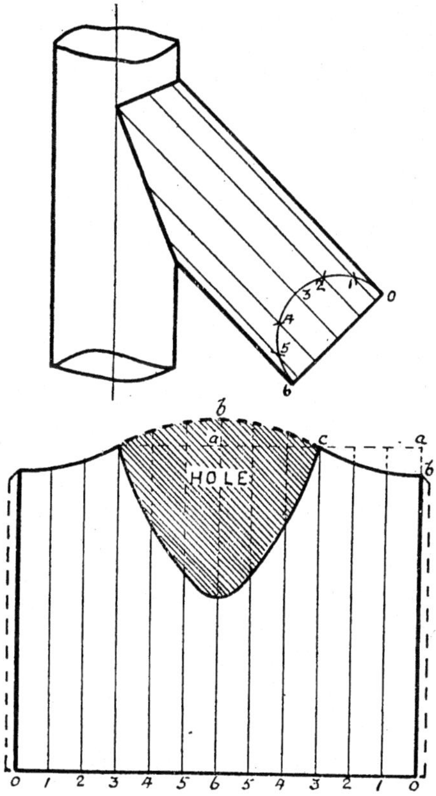

FIG. 11.

one-quarter of the pattern will be sufficient for practical purposes. In marking out the hole the lengths 0 1, etc., are taken from the corresponding lengths around the main pipe, and the widths at the same points from the quarter-circle on branch pipe. A test as to the accuracy of working can be applied when it is remembered that the girth around the hole should be the same as the length of curve on pattern.

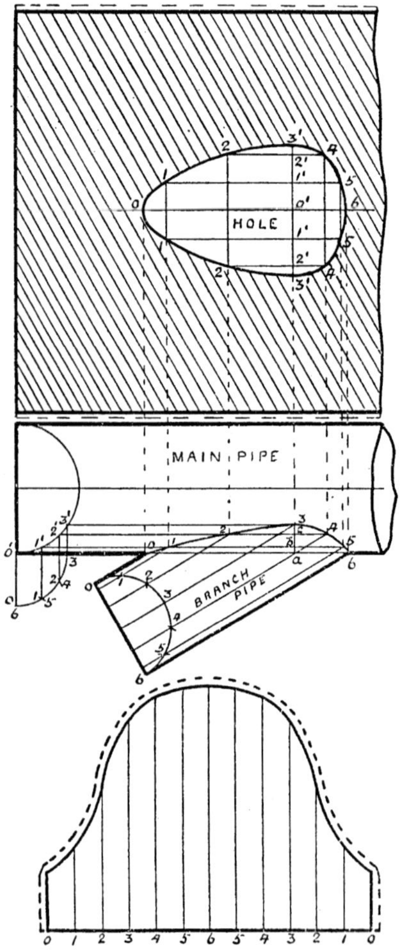

Fig. 12.

Oblique Tee-Pipe.

For an oblique tee-pipe (Fig. 11), in which both pipes are the same diameter, the elevation of the two pipes is set out to the required angle, and the pattern marked out in the usual manner. The shape of the hole can be obtained as in Fig. 8, or scribed directly from the pattern, the curve at top of hole being the same as the curve from *b* to *c* on pattern; and the heights *a b* being equal. The two halves of pattern are exactly the same, and after what has been said with regard to Fig. 9, the mechanic with an ingenious turn of mind will probably be able to see how the two curves on half of pattern can be used to set out an obtuse elbow and an acute elbow at the same angles at which the centre lines of the two pipes meet.

Oblique Tee-Piece for Unequal Pipes.

Where a junction of two pipes of unequal diameter is formed, as in Fig. 12, it will be necessary first of all to obtain an elevation of the joint line, or of points upon the same. This can be done by drawing a semicircle on the main pipe and on the bottom line of this pipe a quarter-circle of radius, equal to half the diameter of branch pipe. The quarter-circle is divided into three equal parts, and lines drawn up to cut the semicircle in 0′, 1′, 2′, and 3′. Lines are drawn up through these points parallel to the centre line of the top pipe, and where they intersect with the lines drawn through the corresponding points on the semicircle on branch pipe will give points on the joint curve. Great care should be exercised to obtain these points correctly, as the accuracy of the patterns depend upon the lines on the branch pipe being cut off to their proper lengths. In setting out the pattern for branch pipe the girth is, as usual, measured along 0 0, lines drawn up from all the points, and distances marked up these lines equal in length to the line with the same number on branch pipe in elevation.

To mark out the shape of the hole is somewhat more difficult than in the previous cases. Drop a perpendicular line from 3 on joint line to the bottom line of pipe. This in elevation is denoted by 3 *a*. To obtain points on curve of hole draw any line down the paper, and mark a point on it 0′. Set above and below this point the distance 0′ 1′, obtained by measuring along between 0′ 1′ on the semicircle on main pipe in elevation. In the same way set along the lengths 1′ 2′ and 2′ 3′. Lines at right angles to the line 3′ 3′ are now drawn through these points, and the corresponding lengths measured on each side of 3 *a* in the elevation marked along. Thus 0′ 0 will equal *a* 0 and 0′ 6 equal *a* 6, 1′ 1 will be the same as *b* 1 and 1′ 5 as *b* 5. In the same way, 2′ 2 and 2′ 4 will respectively equal *c* 2 and *c* 4. The points found will, of course, now be joined up with an even-flowing curve, and the shape of the hole is completed.

Facility in marking out the shapes of holes should be acquired by every sheet and plate metal worker, as it will save endless cutting, chiselling, and filing after the plate or sheet is bent into shape.

Nothing has been said so far as to any **allowance** that should be made for the thickness of sheet or plate; but this will be dealt with in later chapters.

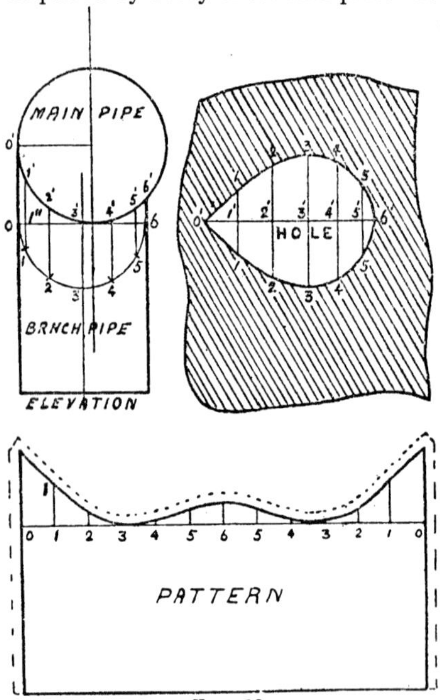

FIG. 13.

Offside Tee-Piece.

When a branch pipe which is smaller than, and square to, a main pipe, and is also required to fit flush on the back of the main pipe (say, to lie against a wall), then its pattern will be obtained as shown in Fig. 13.

The necessary lines for the hole and pattern are obtained by marking out an end elevation of the two pipes, as seen on Fig. 13. A line 0 6 to touch the main pipe is now drawn, and upon it a semicircle described, this latter being divided into six equal parts, and perpendiculars run up through the division points to meet the main pipe. The girth line 0 0 of the pattern is made equal in length to the circumference of the branch pipe, divided into twelve equal parts, and lines square to it run up through each division point. These perpendicular lines are now cut off equal to the same numbered line measured from the line 0 6 up to the main pipe circle in the elevation. Thus, for instance, the line 1, 1, on the pattern will be the same length as 1″ to 1′ in the elevation, and so with all the other heights.

The shape of hole in the main pipe can be marked out by drawing a line, 0′ 6′, made up of the lengths of the arcs 0′ 1′, 1′ 2′, etc., from the main pipe circle, drawing perpendiculars through each point, and cutting these off above and below the line 0′ 6′, equal to the similarly numbered line on the semicircle in the elevation. Thus, to give an example, the line 1′ 1 on the hole will be made the same length as the line 1″ 1 on the semicircle in the elevation, and so on for all the other lines.

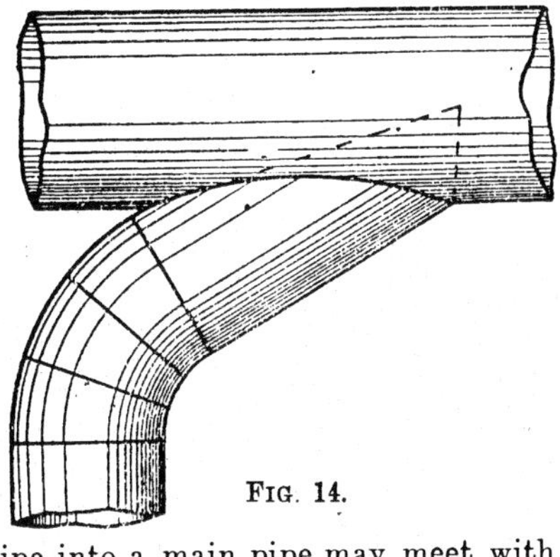

FIG. 14.

Offside Oblique Tee-Piece.

That the flow of a fluid from a branch pipe into a main pipe may meet with as little resistance as possible, a branch pipe may be required to join on to a main pipe, as in Fig. 14. Here it will be seen that the cut on the branch pipe where it joins the main pipe is somewhat peculiar, its shape at the back taking the form shown by the dotted line.

The patterns for the segments of the curved portion of the branch pipe can, of course, be set out, as in the former cases.

The striking out of the pattern for the branch pipe cut, and the hole in the main pipe, is shown in Fig. 15. Before a pattern can be made, an elevation of the intersecting line of branch and main pipes must first be obtained. This is done by describing a semicircle on the branch pipe in the side elevation, dividing into six parts, and running lines up parallel to the centre line as shown. These lines are cut off by drawing lines up from the points on the semicircle in end elevation until they meet the main pipe circle, then running along until they cut the same numbered line on the side view.

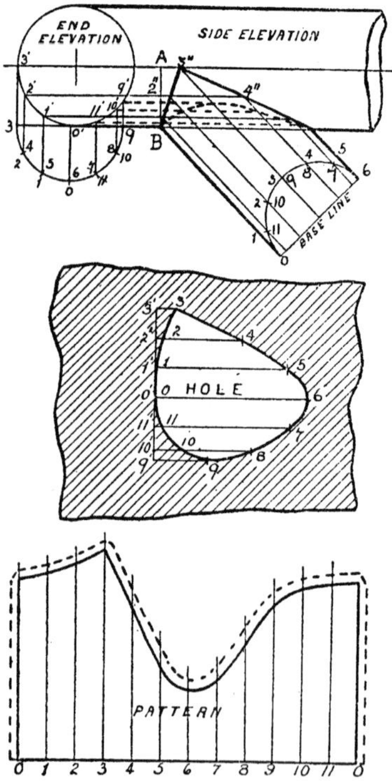

FIG. 15.

Thus, the line through point 10 on the semicircle in the end elevation gives point 10′ on the main pipe circle, the horizontal dotted line through this point then intersecting with the line drawn through point 10 on the semicircle in the side elevation, and so on for all the other points required for the elevation of the joint line. The pattern for the branch pipe is now marked out in the usual way by measuring lines from the base to the joint line, and setting these lengths up on the correspondingly numbered line on the pattern. It should be noticed that two lengths are measured off each line in this side elevation, except the two outer lines. Thus, to take one case, the height of line for position 8 on the pattern will be measured from the base up to the dotted curve, and that for position 4 up the same line to the point marked 4″, and so for each pair of lines.

To mark out the hole, a girth line, 3′ to 9′, is laid down, the parts of this being equal in length to the length of the correspondingly numbered arc on the main pipe circle. Through each of the division points lines square to the girth line are drawn. Now to get the lengths of these. Draw a line, A B, as shown in the elevation, and, using this as a base to measure from, measure the distance of the different points on the joint curve from this, and set along the corresponding line on the hole. Thus the line 2′ 4 on the hole will be the same length as 2″ 4″ on the elevation, and, again, 3′ 3 will be equal to A 3″, and so on for all the other lines. All the points on Fig. 15 have not been numbered, as this would probably have led to confusion; but the reader should find no difficulty in following the construction, as having obtained one set of points and lines, all the rest will follow the same rule.

In bending the plates, care must be taken that they are bent the proper way, so that the pipes will fit together correctly at the joint. This, of course, holds for all cases of tee-pipes in which the branch does not fit on the middle of the main pipe.

CHAPTER IV.

PIPE BENDS IN SEGMENTS.

In the two previous chapters we dealt with several examples of the striking out of patterns for circular pipe joints, we now extend the methods there shown to the cases of bends made up in segments.

Quarter-Bend for Round Pipes.

The exact shape of the bend is first set out, as in Fig. 16, and then divided up into any convenient number of segments. In the present case there are six. The smaller the segment the less work there will be in stretching and hollowing into shape, but at the same time it should be remembered that small segments mean a large number of joints; hence some reasonable mean should be chosen. A mid-line is drawn in for one of

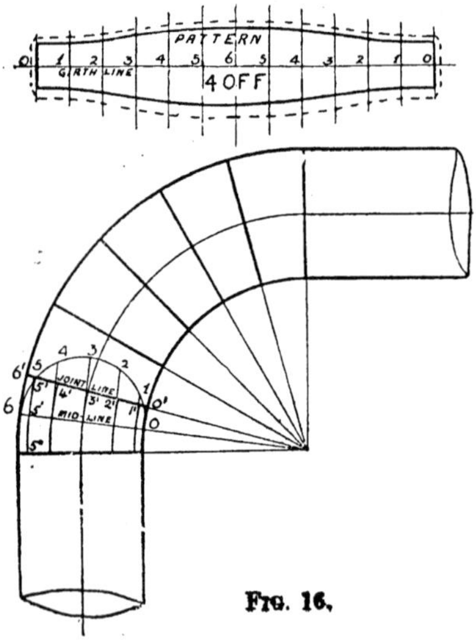

FIG. 16.

the segments, and a semicircle described on it as shown, this being divided into six equal parts, and perpendiculars dropped on to the mid-line.

For the pattern, a girth line is laid out equal in length to the circumference of the pipe, this being divided into twelve equal parts, and perpendiculars run up through each division point as shown. The compasses are now set respectively to the lengths of the perpendiculars between the mid-line and joint line and these transferred to the similarly numbered line on the pattern, and marked off both above and below the girth line. In each case these construction lines will be a shade too long, as will be seen by reference to the figure. To take one line only; instead of using the perpendicular $5' 5'$ for the pattern line, the length of the arc $5' 5''$ should have been used, and so with all the others. But manifestly the difference in length between the straight line and the arc is so small, that in nearly all practical work it is hardly worth taking into account. In all cases where there is no intention of working the segment into shape, by hollowing the back and stretching the throat, the straight line should be used.

The methods of fastening the segments together are various. They can be simply slipped over and soldered, either with or without sinking the seam, riveted together by having the segments alternately inside and outside, paned down or knocked up. The allowances on the pattern are shown for the latter two methods. If the segments are to be put together with one inside, the next outside, etc., then it will be necessary to take the thickness of the sheet metal into account, and have two patterns, the outside one being about seven times the thickness of the metal longer than the inside segment, as explained in Chapter XXXII.

In shaping the strips for small pipes, the usual plan is to hollow up the back part of the segment before bending the ends of the strip around to form the throat. Having hollowed the back part, the ends are then bent around, grooved or riveted up, and stretched to the required shape.

The kind of jointing and quality of work on the bend will, of course, depend upon what it is to be used for.

Large pipe bends that are made out of boiler-plates are constructed in a somewhat different fashion, the plates being arranged so as to break the joints. A sample of this kind of work is shown in Chapter XXXIII.

Quarter-Bend for Square Pipe.

A bend for a square or rectangular pipe can be made up much easier than for a round pipe. If the back and throat of the bend are flat, then of course the patterns for these parts will simply be straight sheets. The pattern for one of the cheeks will be obtained by setting out the side elevation of the bend, the outline of this giving the shape of the cheek pattern. If angle iron is used to connect the parts together, no allowance will be necessary; but if the sheet is flanged to form a lap, then of course an allowance for this will have to be made.

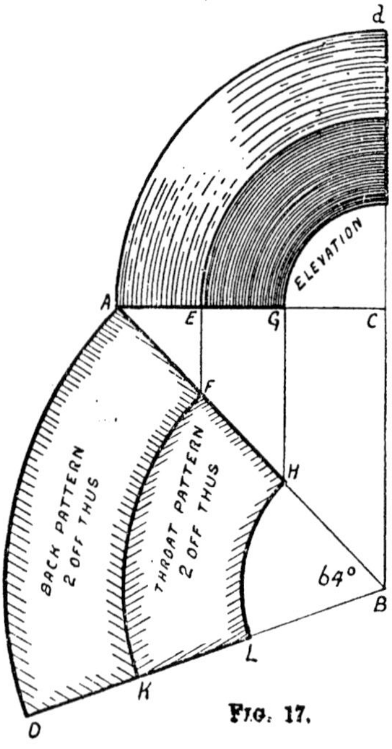

FIG. 17.

If the square pipe runs along diagonally, then the shape of a quarter-bend will be as shown in the elevation, Fig. 17. In setting this out, it should be remembered that A G is a diagonal of the square pipe, the length of a side being equal to A F.

The shape of the bend is built up in a somewhat peculiar manner, for we may consider each part as a portion of the surface of a cone (Chapter XII.). Remembering this, the development of the patterns becomes a very simple matter. Draw C B square to, and equal in length to, A C. Join A to B. Draw E F and G H square to A C. Then, with B as centre, and B A as radius, describe the arc A D, this being made equal in length to the back curve A *d*. The length of the arc A D may also be determined by calculating the angle A B D, and setting this out with a protractor, or otherwise.

$$\text{Angle } A\,B\,D = \frac{90^\circ \times A\,C}{A\,B}$$

In the present case A C = 22 in., and A B by calculation or construction will be nearly 31 in., therefore:

$$\text{Angle } A\,B\,D = \frac{90 \times 22}{31} = 64^\circ \text{ (nearly)}.$$

The arcs F K and H L are next described, and the two figures, as shown, will give the patterns for the back and throat-pieces respectively. It should be noticed that no direct measurement along the arcs F K and H L is necessary, as when A D is obtained, the others will be cut off proportionately. The back, side, and throat seams may be made by either lapping over and soldering, or riveting, or knocked up, the allowance on the pattern, of course, being such as to suit the kind of joint chosen.

The plates or sheets will be shaped by rolling or bending in the same way as all other conical work.

Double Bend for Round Pipe.

Where it is necessary to join together two lines of piping, so that the flow of liquid or gas passing through the pipe may be interfered with as little as possible, it is a good plan to make a connecting pipe of the form shown in Fig. 18. This shape of bend gives no abrupt break in the pipe, and maintains the full cross-sectional area throughout its length.

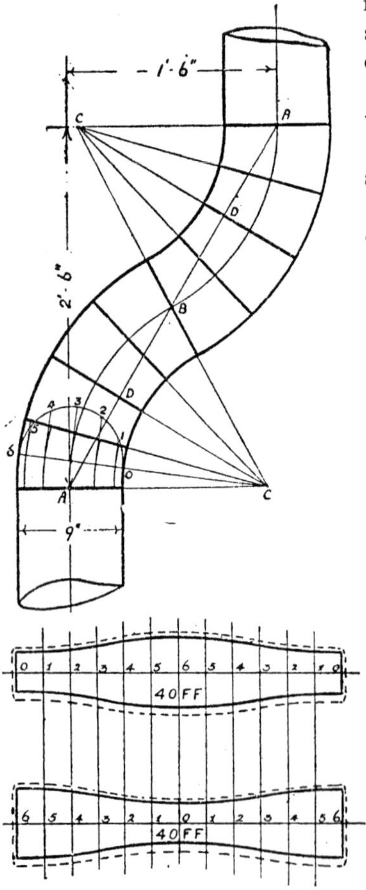

FIG. 18

It is most important that the exact shape of the bend should first be set out. This can be done by setting down the distance between the lines of pipes and the length of the bend (18 in. and 30 in. respectively in this case), thus obtaining the points **A A**. The line A A is now divided into four equal parts, and perpendiculars drawn through the end division points, D D, to meet A C in C. This gives the centres for the curves. If a line, C C, be now drawn, this will determine the points where the two curves of the pipe outline should join together. Each half of the bend is now divided into a convenient number of segments (in this case, four), and the pattern for one segment set out, as explained in connection with Fig. 16.

If for the sake of appearance it is required to run the seams in a line, or to alternate the longitudinal seams, then two patterns will be necessary, one giving the seam on the outer curve, and the other on the inner curve.

In jointing up pipes of this description, some care should be exercised, so as to get the bend without twist, and to the exact shape.

CHAPTER V.

TAPERED PIPE ELBOWS AND THREE-WAY PIECES.

In the former chapters we dealt with typical cases of pattern-cutting for cylindrical pipe-elbows, selecting such examples as would serve to illustrate the general principles involved. In conical pipe-work it would not be a difficult matter to pick out scores of apparently different forms of joints, but which, on examination would be found could nearly all be resolved into a few simple types. In this chapter we therefore propose to treat just one or two representative cases of conical pipe-jointing, and these should be sufficient to explain the general method that can be applied to all this class of work. The first example to be dealt with will be that of a

Cylindrical and Conical Pipe Elbow.

The centre lines of the pipes may be arranged to meet at any required angle; but, for the sake of simplification, a square elbow (Fig. 19) will be taken first.

In work of this character the important thing is to accurately set out a side elevation of the elbow, so as to obtain the correct position of the joint line. This is done by first drawing in the centre lines at the required angle, and then from their point of intersection describing a circle (shown dotted in Fig. 19) equal in diameter to the cylindrical pipe. The outside lines of the pipes are afterwards drawn to touch this circle, and where they intersect will give the ends of the joint line. Thus, in Fig. 19, the cone and cylinder intersect respectively in *a* and 6; hence the straight line *a* 6 will be the side elevation of the joint. It will save confusion to remember that this joint line does not pass through the point of intersection of the centre lines. The shape of the section made by the cut to form the junction of the two pipes will, of course, be elliptical, and by careful measurement it will be found that the size of the ellipses on the conical and cylindrical pipes will be exactly the same; hence the two pipes should fit together correctly. A cone base may be taken at any convenient position; but in the case of the square elbow it is, perhaps, best to produce the under side of the conical pipe until it meets the back of the straight pipe, and then use the line 0 6 as the cone-base. A semicircle is described as shown, divided into six equal parts, and lines drawn square to the cone-base, these being then joined up to the cone-apex *c*. From the points where the radial lines intersect the joint *a* 6, lines are run parallel to the cone-base on to the outside line of cone, thus obtaining the points 0′, 1′, 2′, etc. For the pattern, the compasses are first set to the distance *c* 6, and, with C as centre,

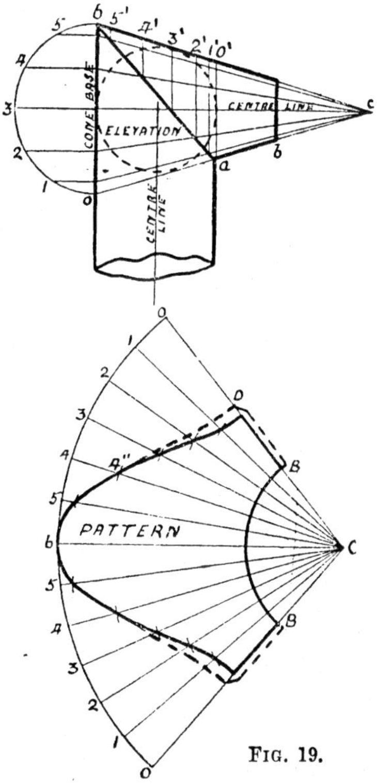

FIG. 19.

the arc 0 0 described, its length being obtained by stepping along the length of one of the arcs from the semicircle twelve times. To obtain points for the pattern curve, the compasses are respectively set to the lengths $c0'$, $c1'$, $c2'$, etc., these being marked from C along the correspondingly numbered lines on the pattern. Thus—to take one case only—the line C4″ on the pattern will be the same length as $c4'$ on the elevation. After marking all the points, they are joined up with an even curve. The cut for the other end of pattern is obtained by describing the curve B B from centre C with the radius *c b* from the elevation.

If the elbow is made of galvanised sheet iron, an allowance for jointing can be put on as shown by the dotted line 4″ D, the width of this depending upon the thickness of sheet metal used.

The cylindrical pipe pattern is not shown, as this will be struck out as before explained; but it should be noted that the allowance for jointing must be added to the back of pattern to correspond to that put on the throat portion of conical pipe.

In plate work more care will have to be taken to allow for thickness of metal in jointing. In setting out the elevation, the middle line of the metal thickness should form the outline of the figure. Suppose it is required to flange the tapered pipe over on to the cylindrical one; then the cone at the dotted circle portion should be made twice the thickness of the metal greater in diameter than the straight pipe. On the other hand, if the cylindrical pipe is to be flanged on to the conical part, then the former should be made two thicknesses in diameter greater.

Before proceeding to lay any lines down for a pattern or template, the arrangement of jointing should first be settled, as by a little forethought any method of connecting can be allowed for, and often much subsequent trouble avoided.

Swan-Neck or Offset.

The complete setting out for a swan-neck bend, made up of three conical pipes, is shown in Fig. 20. The double elbow might have been constructed partly conical and partly cylindrical, as in the last case, the same method for

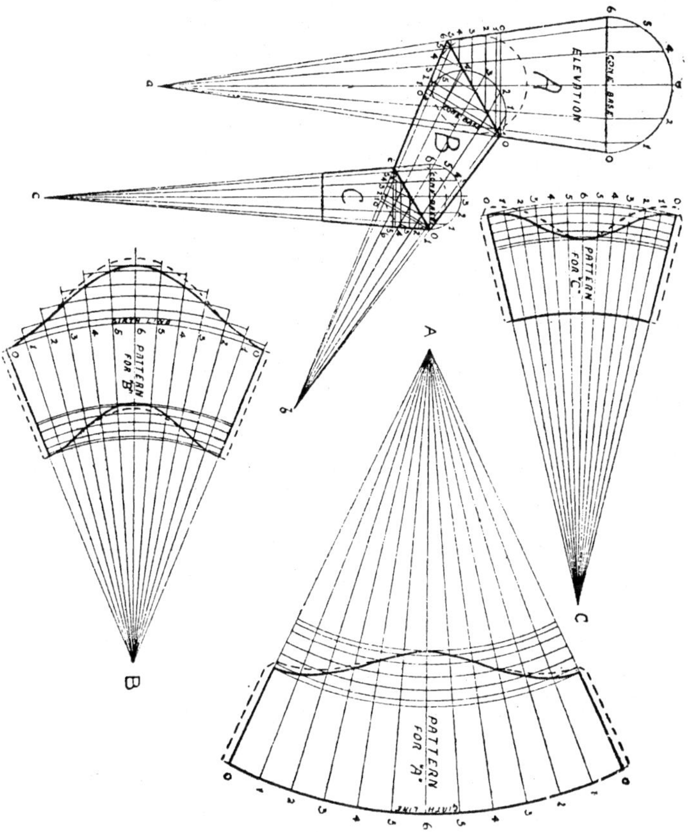

FIG. 20.

obtaining the joint line still holding good. There is no need for very much description in connection with this example after what has been said about the square elbow. The lines that form the cone-bases are indicated in the elevation, those for the parts "A" and "C" being similar to the last case, and that for "B" drawn to the right of the joint, and the radial lines produced through to meet it. It will be noticed that this latter arrangement brings the girth line of the pattern for piece "B" across the pattern instead of at the end, as in the other two patterns. The extras for jointing are added on the same principle as explained for the elbow. In fixing the parts together, it should be noted that "A" fits into "B," and the latter into "C."

Cylinder and Cone Breeches-Piece.

The forms and shapes of breeches-pieces are numerous. Those of the oblique cone order and coppersmith's kind are dealt with in Chapter XXIX.; but there are still many, that do not come under the above names, which can be formed of portions of cylindrical and conical pipes, or the latter alone. We shall now give two examples of this class of work—one regular in form, and the other irregular—and this should suffice for all practical purposes.

Regular Breeches-Piece.

An elevation of the above is shown in Fig. 21. The centre lines are first laid out at the required angle, and a circle described about their meeting-point of the same diameter as the cylindrical pipe. The ends of the conical pipes are then marked down in their proper positions and correct diameters. Lines are now drawn to touch the circle, and where required produced until they meet. The intersection points of these tangential lines will give points on the joint lines or joint lines produced. Thus, the line of connection, *d e*, between the pipes "A" and "C" drawn by joining *d* to *f*, and where this line cuts *g h* will give the point *e*. It should be observed that this latter point does not coincide with the centre of the circle.

The girth line of the pattern for the pipe "B" is

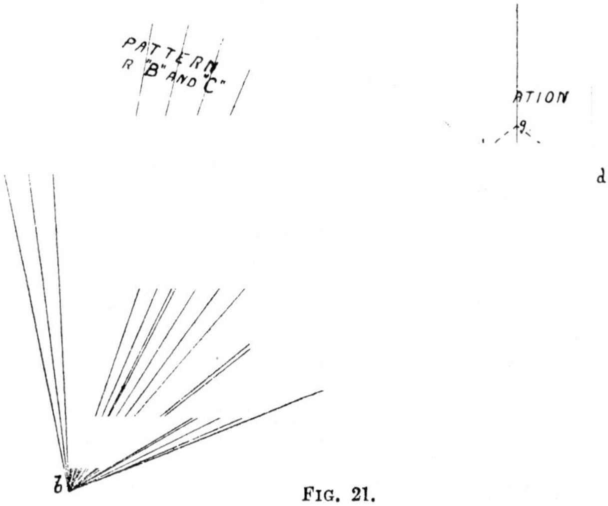

FIG. 21.

obtained by taking the end line of the pipe as a cone-base, and on this describing a semicircle, from which the girth line lengths can be measured and the radial lines drawn. Having projected the radial lines on to the outside line of the cone, the striking out of the pattern will be the same as in the former cases. There is, however, one little detail to which it is, perhaps, worth while calling attention. It will be noticed that the points E E do not lie on the regularly-spaced radial lines, but in between the lines passing through the points 3 and 4 on the girth line. To obtain the former points accurately, extra construction lines must be put in. To do this, join e to b, and from where the line crosses the cone-base run up a perpendicular to the semi-circle, so obtaining the point a. Now measure the arc 3 a, and set along the girth line from the point 3. Join b to a, and produce the line to meet the outside curve, which is swung around from e', in E.

The finding of the intermediate point has, in the above case, been explained at some length; and, as it is occa-

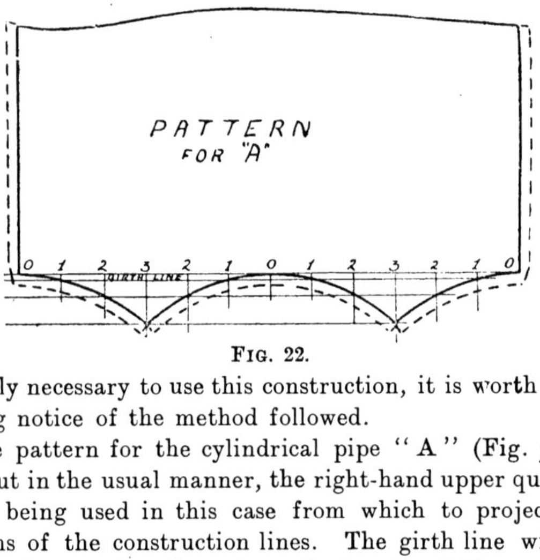

FIG. 22.

sionally necessary to use this construction, it is worth while taking notice of the method followed.

The pattern for the cylindrical pipe "A" (Fig. 22) is laid out in the usual manner, the right-hand upper quarter-circle being used in this case from which to project the lengths of the construction lines. The girth line will, of course, be equal to four times the length of the quarter-circle 0 to 3. The lengths of the cross lines are shown projected from the elevation. For a pane-down or knock-up joint a double lap is put upon the pattern for "A," a single lap for "B," and double lap along the middle part E E, and a single lap at the ends for "C."

Irregular Breeches-Piece.

The same principle as applied in the former cases can also be adopted as the method of construction for any kind of a three-way or other connecting-piece, built up wholly with conical pipes, or partly conical and partly cylindrical.

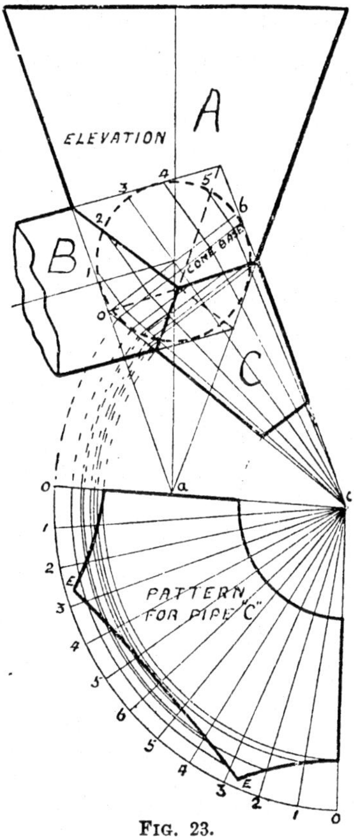

FIG. 23.

The elevation of an irregular breeches-piece, which is composed of two conical pipes and a cylindrical pipe, is shown in Fig. 23. The joint lines are obtained exactly the same as in the former cases. The only pattern set out is that for the conical pipe "C," as the others can be obtained in a similar manner. To complete the cone of which "C" is a part, the side lines are produced to meet in *c*, and the cone-base, drawn as shown, being made the same diameter as the dotted circle.

Half of the latter is used as the semicircle for obtaining the required construction lines, perpendiculars being drawn from the division-points down on to the cone-base. The pattern is struck out in the same manner as that shown for Fig. 21. The points E, E, on the pattern have been found without the use of intermediate construction lines, the curves through 1, 2, and 4, 3, being simply produced until they meet in E. No allowance has been attached to the pattern for jointing, as this can be put on according to requirements.

What has been said about allowing for thickness of metal in connection with Fig. 19 will apply equally well in the above class of work. For heavy plate work great care must be taken in this direction if work accurate to dimensions or neat joints is the desideratum.

Equal-angled Three-way Piece.

When three pipes of the same diameter fit together at equal angles, as in Fig. 24, the simplest way to obtain the elevation, and thus the pattern, is to draw a circle (shown dotted in the figure) equal in diameter to the pipes, and obtain the elevation of the joint line, as shown. The pattern can then be set out, as in other cases. If the joint is to be paned down, or knocked up, the allowance on pattern will be a double edge for middle part and a single edge on end parts, the pattern then serving for each branch.

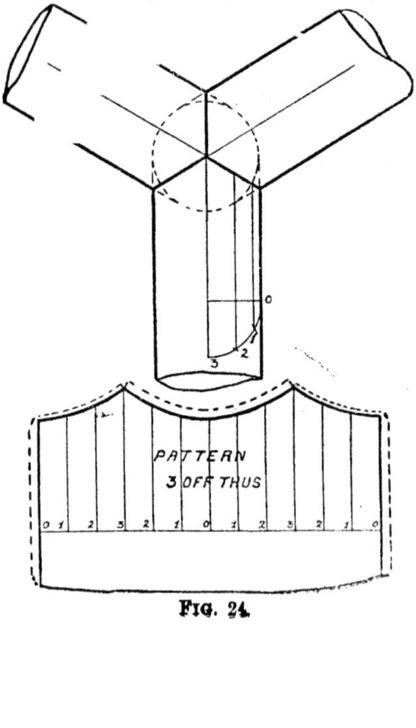

FIG. 24.

Unequal-angled Three-way Piece.

For this the elevation can be set out, as in the last case,

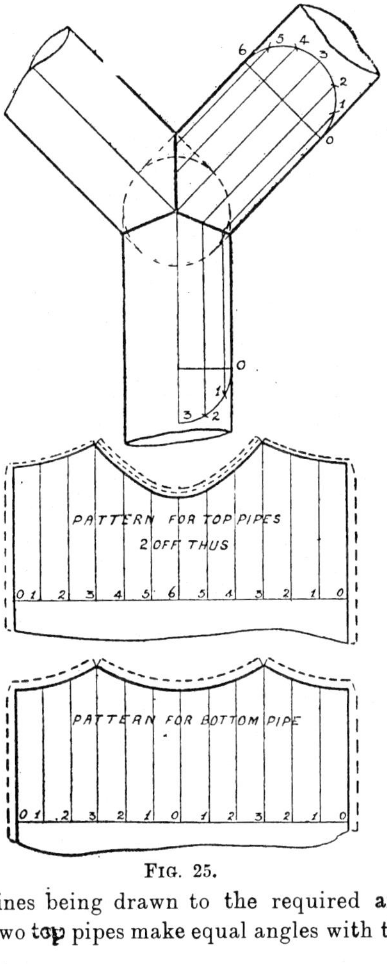

FIG. 25.

the centre lines being drawn to the required angles. In Fig. 25 the two top pipes make equal angles with the bottom pipe, consequently one pattern will do for the two pipes, the only difference being in the arrangement of the laps, the two dotted curves on the pattern showing respectively the laps for the two pipes.

If the three angles that the centre lines of the pipes make with each other are all unequal, then it will be necessary to have three distinct patterns, the setting out of these being similar to the cases already mentioned.

CHAPTER VI.

SQUARE PIPE ELBOWS AND TEE-PIECES.

In the making of square pipes for ventilating shafts and other purposes, it is often necessary to construct various kinds of elbows and tee-pipes. We shall, therefore, in this chapter, deal with the striking out of the patterns for a few representative cases.

Square Pipe Elbow.

A side elevation of the pipe (Fig. 26) is usually first set

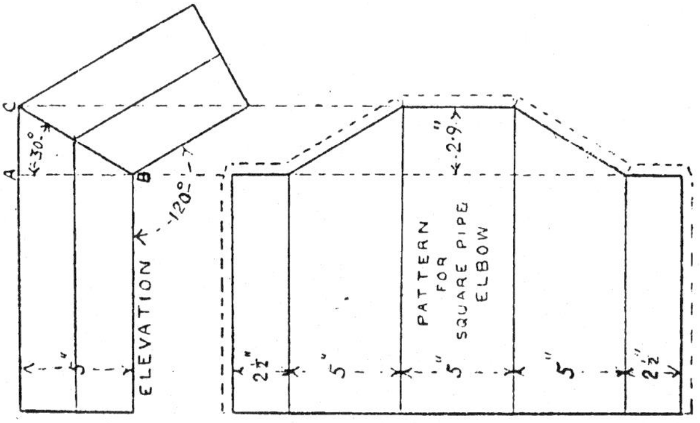

Fig. 26.

out, care being taken that the branches are put at the required angle. The object of this elevation is to obtain the difference between the height of the back and the throat of one of the branch pipes. This difference, which is marked A C on the figure, may be obtained by simply setting out the triangle A C B, the angle A B C having first been determined by the following rule: "To find the joint-line angle, deduct half the bend or elbow angle from 90°." Thus, in the present case the angle will be—

$$90 - \frac{120}{2} = 30°.$$

It will thus be seen that to get the length of line A C it is only necessary, in practice, to draw A B and the joint line B C at the proper angle. This dispenses with the elevation, which is always advisable, as in the setting out of patterns as little as possible in the way of plans and elevations should be drawn.

Those readers who have a smattering of mathematical knowledge, and can use tables, will be able to calculate the length of A C as follows:—

$$\begin{aligned} AC &= AB \times \tan. ABC \\ &= 5 \times \tan. 30° \\ &= 5 \times \cdot 58 = 2\cdot 9 \text{ in.} \end{aligned}$$

This height can be set directly on the pattern, and the same completely struck out without in any way using an elevation. It might be here remarked that what has been said in connection with obtaining the difference in height between the back and throat of a square pipe is also applicable to an elbow for a round pipe.

Referring again to Fig. 26, it will be noticed that the pattern is set out so that the seam will come down the centre of the throat when the sheet is bent up. This will necessitate three full widths of 5 in. each, and two half widths of $2\frac{1}{2}$ in., each being marked out to form the pipe girth. If the seam is to be in any other position, then the parts of the pattern must be arranged accordingly.

In Fig. 26 it will be observed that the length A C is projected by dotted lines on to the pattern. This should not be done when marking out on sheet metal, as it is most difficult to transfer lengths correctly in this way. It is done on the figure simply to better explain from where the length A C is obtained.

Allowances must be made on to the net pattern for the side seam, also for jointing the two arms together, either by riveting, soldering, or paning down and knocking over.

The pattern for the two arms may be set out in one piece, as shown in Fig. 27. The two side gaps are cut away, and after the sheet is bent up and seamed in the form of a square straight pipe, the elbow can be made by bending along the line D E and fastening together at the throat and sides. After the elbow is formed, it will be seen that the points F F come together.

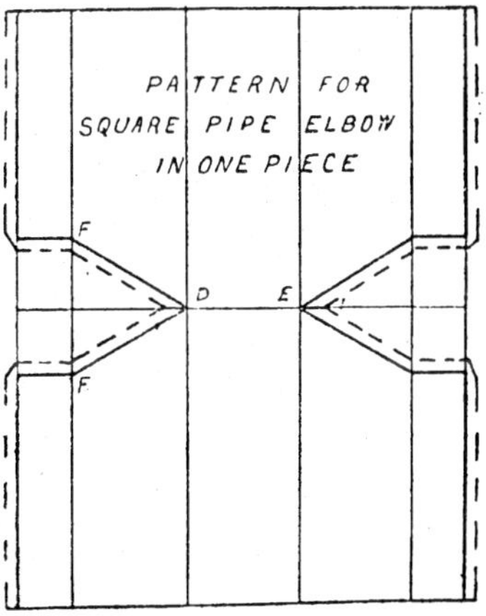

FIG. 27.

So that the elbow may have the correct offset, it is always a good plan to make a template for the required angle, and try this in the throat while the joint is being tacked.

For the special case of a square elbow it should be noted that the height of the back part of pattern above the throat portion will be equal to the diameter of the pipe.

Diagonal Square Pipe Elbow.

If the pipes run diagonally and an elbow is required, as shown in Fig. 28, it will be necessary to first obtain a diagonal of the pipe before the elevation can be drawn. This can be done by setting out the right-angle triangle A, B, D, the sides A D and D B being made equal in length to the diameter of the pipe. The diagonal A B can be calculated, and in the present case where the diameter of the pipe is 3 in. it will come out as follows:—

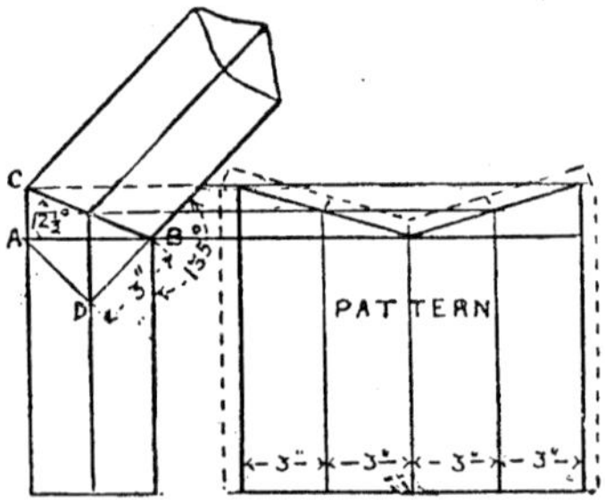

FIG. 28.

$$AB = 3\sqrt{2} = 4{\cdot}24 = 4\tfrac{1}{4} \text{ in.}$$

The height of A C can be found by either of the methods explained for the last elbow.

If the seam is down the back edge the girth of the pattern will be made up by four widths, each equal to the diameter of the pipe.

For elbows of this description it should be observed that the cut at top of pattern is simply two straight lines. The pattern for a

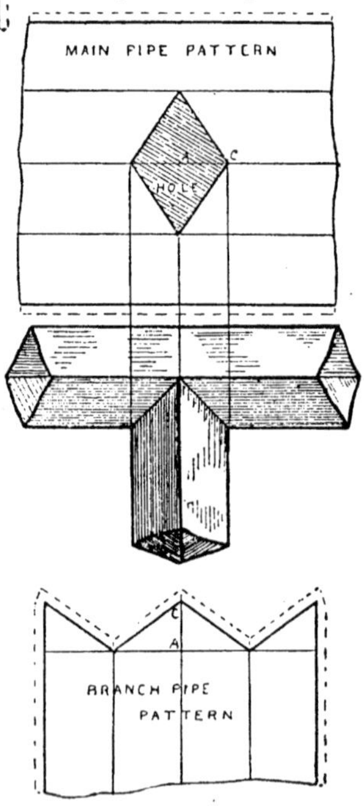

FIG. 29.

square elbow can be marked out in the same manner; but in this the height A C will be the same length as a diagonal of the pipe.

Diagonal Tee-Piece.

A tee-piece for a square pipe placed diagonally is shown in Fig. 29. The hole on the main pipe is easily marked out when we remember that its length must be equal to the two sides of the pipe and its width equal to the diagonal of the pipe. This is shown projected on the top portion of Fig. 29.

The pattern for the branch pipe is set out by taking the four sides of the pipe for its total girth, and the line A C equal to half the diagonal of the pipe or equal to the length of the line A C on the hole. It will be readily seen that the pattern can be marked out from the hole, or, which is perhaps better, the hole marked from the pattern, if this be made first.

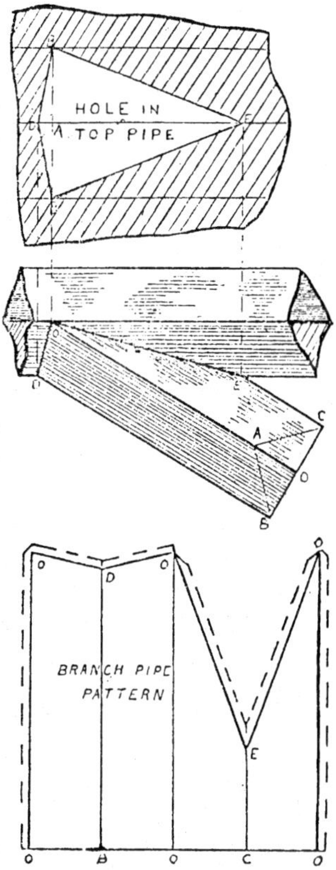

FIG. 30,

Oblique Tee-Piece.

When the two pipes are of the same diameter, very little more difficulty will be experienced with the patterns than in the last case. An elevation of the two pipes is first drawn (Fig. 30), making the required angle with each other, the lines A B and A C representing the diameters and the line B C the diagonal of the pipes.

The pattern for the branch pipe is set out by laying down four widths, each equal to A B, to make up the girth of pipe. To form the cut at top of pattern the lengths of lines are taken from the elevation. Thus lines B D and C E on the pattern are respectively equal to the lines with the same letters in the elevation. The three remaining lines O O are measured off the same length as the centre line O O on the pipe. The pattern is marked out so that the seam on the branch pipe will come on a side edge.

The shape of the hole can be determined from the pattern, the right-hand side being used to mark out that part of the hole, and the left-hand side the remaining part. The hole can, of course, be set out directly from the elevation, as seen in Fig. 30, the lettered lines of the hole corresponding in length to those with the same letters on the elevation.

Offside Oblique Tee-Piece.

When the branch pipe is smaller than the main, and the two pipes are required to lie flush against a wall, the setting out of the pattern and hole becomes somewhat complicated. We will take one typical case, which should be sufficient guide to cover most of the jobs that are likely to crop up in a practice of this character. In Fig. 31 it will be noticed that the two pipes fit together in a similar manner to those of Fig. 30. The branch pipe, however, being smaller than the main, will necessitate its being on one side, as seen in the end view.

Before it is possible to obtain the pattern for the branch pipe a proper elevation of the joint line must be found. This can be done by drawing an end view of the main pipe (Fig. 31), dropping a perpendicular 0′ to 0 and making 0 3 equal to the diagonal of the branch pipe; the lines 0 1 and 1 3 representing the sides of the small pipe. The base line is, of course, equal to the diagonal of the small pipe, and the lengths 0 2 and 0 4 on this being made the same as line 2′ 4 on the end view. Points on the joint line are obtained by projecting up from points 1 and 3 on to the square in end view, and then from these points running dotted lines along to meet the lines which are drawn from the points with these corresponding numbers on the base line. It will be seen that one dotted line cuts off the two points 1 and 5 on the joint line, and also that the bottom line of the top pipe determines the points 2 and 4.

The seam being up the corner of the branch pipe, four widths each equal to the side of the pipe will be set out for the girth of pattern. This width can be obtained from either 0 1 or 1 3 in the end view. Two intermediate lines 4 4 and 2 2 are required on the pattern, and for getting the correct position of these, the distances 5 4 and 2 1 on the bottom line of the pattern will be made the same as the lengths 5 4 or 2 1 on the end view. To cut off the lengths of lines on the pattern so

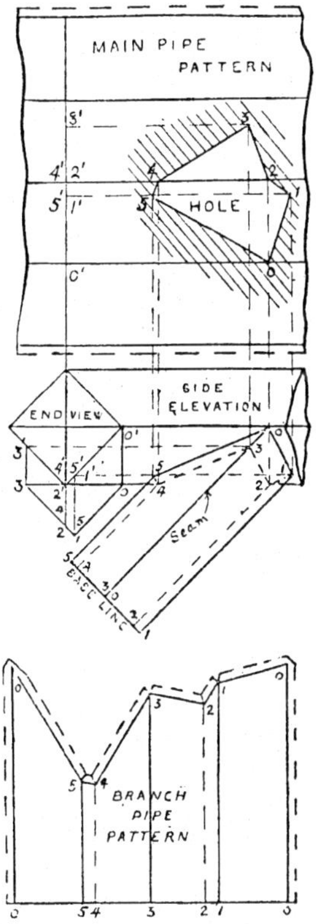

FIG. 31.

as to give the requisite shape to form the junction of branch and main pipes, the lengths 0 0, 1 1, 2 2, 3 3, etc., on the pattern will be made the same lengths as the lines on branch pipe in the elevation, having the same numbers.

To mark out the shape of hole in the top pipe, the girth is set out to represent the four unfolded sides of the pipe, the seam being along the top. The distances 0′ 1′, 1′ 2′, etc., are taken from the lines that are marked the same in the end view of main pipe. Now, referring to the hole, it will be seen that points on its outline are obtained by projecting up from corresponding points on the side elevation. It should be observed that this method of projection cannot conveniently be used in the workshop; but the reader will probably be better able to understand how the lengths are obtained by seeing them projected in this manner. In practice, the various lengths that are used to give the width of hole should be taken with the compasses directly from the elevation, and transferred to the pattern.

To test if the hole is the correct size and shape, its lines should be measured to see if they are of exactly the same length as the lines that are figured the same at the top of pattern. The laps allowed will, of course, be such as to suit the method of jointing adopted.

Particular notice should be given as to the way in which the plates are bent, as in work such as this, where the hole is not in the centre or the pipes fitting symmetrically, it should be borne in mind that out of the two ways in which the plates can be bent, one only of them is correct. Thus, in Fig. 31 the patterns have been so set out that if the edges of the plates are bent *up*, the two pipes will joint together properly.

Twisted Oblique Tee-Piece.

In Fig. 32 another representative example of square pipe-jointing is given. The two pipes are of the same diameter, the branch fitting on to the main obliquely, with two of its flat sides parallel to the edges of the main pipe. It will be seen that it is also arranged so that a side of the branch pipe and an edge of the top pipe come together, such that the two pipes will lie flat against a wall.

After going carefully over the last case, there will be little need to give much explanation in this. The important points to notice being that the lengths 2 3 and 4 5 on the bottom line of branch-pipe pattern are obtained from 2′ 3 in the end view, and also that the lengths 2′ 3′ or 4′ 5′ on the main-pipe pattern are measured from 2′ 3′ on the square, which is the end view of elevation on the main pipe. The lengths of lines for the branch-pipe pattern are, of course, measured in the usual way from the base line to the joint line, in this case (Fig. 32), as in the others, the numbered points on the pattern corresponding to those on the elevation of the joint line. The various dimensions to fix the shape of the hole are shown projected as before.

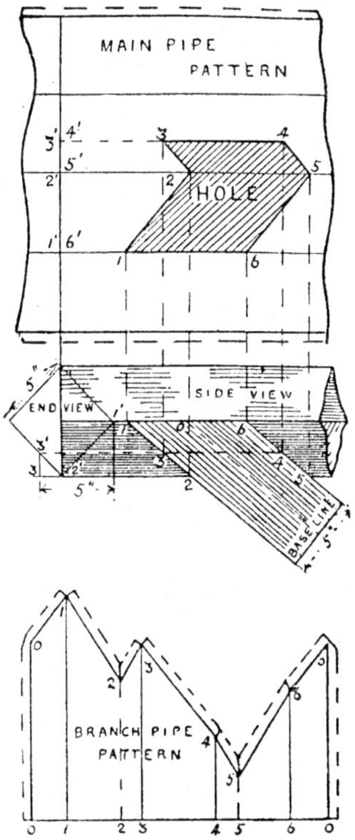

FIG. 32.

It will be seen that the seam for the branch pipe is arranged to come down the centre of the side.

The patterns have been set out so that if the sheets are bent up as in the last case the pipes will come together properly. The settings-out in connection with the last two elbows are good examples of the geometry of sheet metal work as applied to flat surfaces; and whilst in themselves have not a very extended application, yet serve to illustrate some of the methods that can be used in plain surface work.

CHAPTER VII.

RECTANGULAR PIPE ELBOWS AND TRANSFORMER PIECES.

In the previous chapter we dealt with the setting out of patterns for the various kinds of elbows used in connection with square pipe work. We now give a few examples that may be useful for rectangular pipes.

Square Elbows.

Fig. 33 shows a sketch of a square elbow, the broad sides of the pipe being at the back and throat. The elbow is

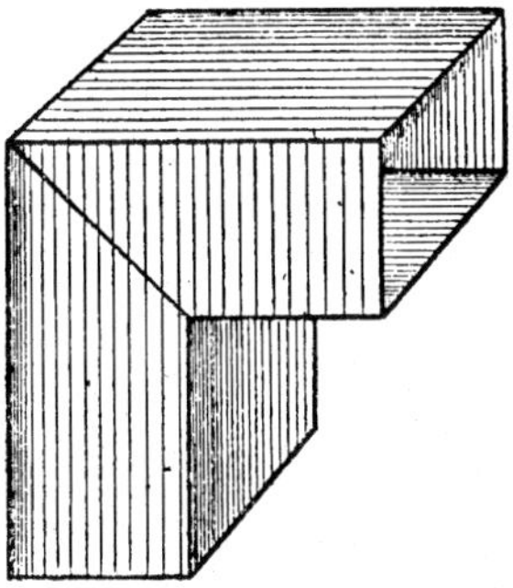

Fig. 33.

made up by two pieces of pipe, each being cut at 45°, and mitred as shown in the sketch. A pattern for one of the branches is shown set out in Fig. 34. A side elevation is first drawn, and before attempting to strike out the pattern, the position of the seam should be decided.

In Fig. 34 we have assumed the seam runs up the middle of the back and along the centre of the top. The girth line of the pattern is drawn, its total length being made up by the pipe dimensions, as marked on the figure. Thus,

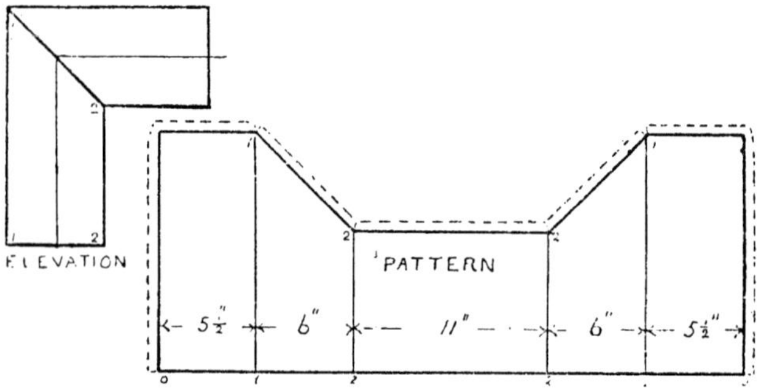

FIG. 34.

suppose the section of the pipe is 11 in. by 6 in., then these sizes will be used in obtaining the total length of the girth-line. The heights 11 and 22 on the pattern will be measured from the respective lines with the same number in the elevation. Allowance for seams must be added on to the sides of patterns for both arms of the elbow. If the joint is a simple lap, and riveted or soldered, it will be necessary to add laps on to the end of one pattern only.

The elbow, of course, could be constructed in the same manner as explained in Chapter VI., Fig. 27, or it may be formed of four pieces, two sides and back and throat, and jointed at the corners by knocking up. This latter method gives a very rigid form of pipe elbow, but has the disadvantage of costing more to make.

The sketch shown in Fig. 35 represents a similar kind of pipe worked into an elbow, with the broad sides forming the cheeks. In this again the seam is taken up the middle of the back. The size of pipe being 7 in. by 4 in., the length of pattern will be made up as seen by the dimensions in Fig. 36. The heights are shown projected; but these, in practice, would of course be taken directly from the elevation.

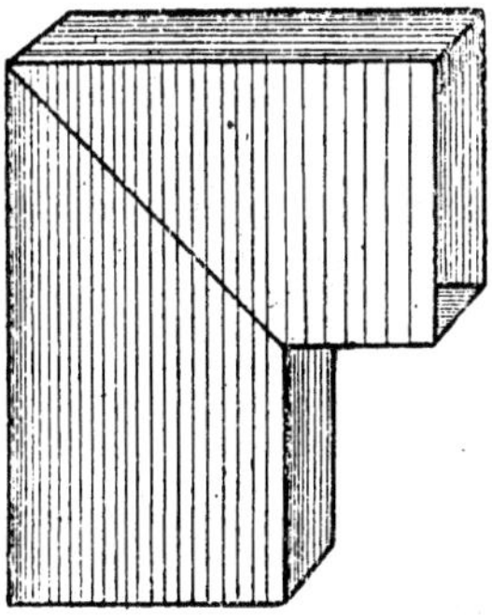

FIG. 35.

The heights of lines to form the cut, both in the case of square elbows and also offsets, can of course be calculated as explained in previous chapters. This would then do away with the necessity of drawing an elevation, the shape of pattern being marked directly on the plate.

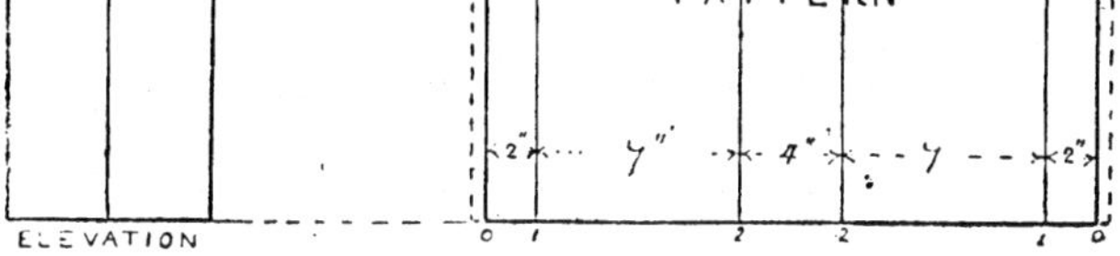

FIG. 36.

Twisted Connecting Pipe.

Some very peculiar jobs occasionally turn up in the way of connecting pieces. A simple but interesting example of this is shown in Fig. 37, in which two rectangular pipes are lying along the corner of a room, one fitting broadside, and

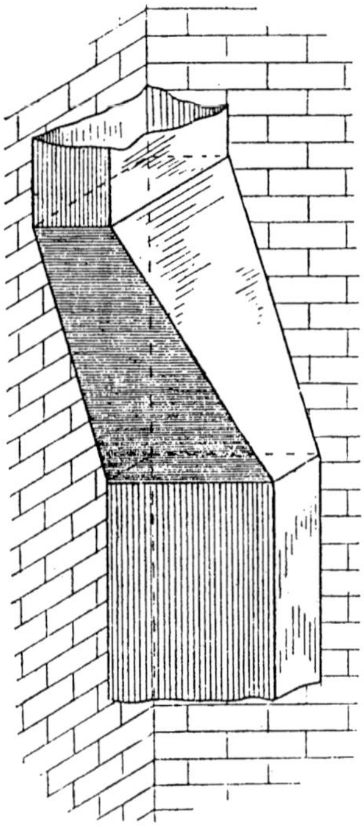

FIG. 37.

the other with the narrow side on the same wall. The problem is to make a connecting pipe to join together the ends of the pipes. The pattern for this can be set out on the plate or sheet from the dimensions; but it will, perhaps, add clearness to the description to have a plan and elevation before us, as shown in Fig. 38. In striking out the pattern all that is necessary to use will be the square and measure. Let us suppose that the seam is to run down the back corner. Draw the line A B; and make it equal in length to the depth of the connecting pipe. Run up perpendiculars A C and B D, cutting these off equal to the width and length of the pipe section respectively. Draw C E G and D F H square to C D, and cut off to the pipe dimensions, as shown. Draw H L and G K parallel to B D or A C, and set along them the two dimensions of pipe, as seen on the figure.

In large work, where it might be awkward to draw the lines parallel, as explained above, a simple method, giving the same result, would be to describe a semicircle on the line G H, and then with centre G and radius equal to 18 − 6 = 12 in. mark the point N. The remainder of the construction can then be completed without trouble. Allowances to cover the particular method of jointing adopted must be added to the net pattern.

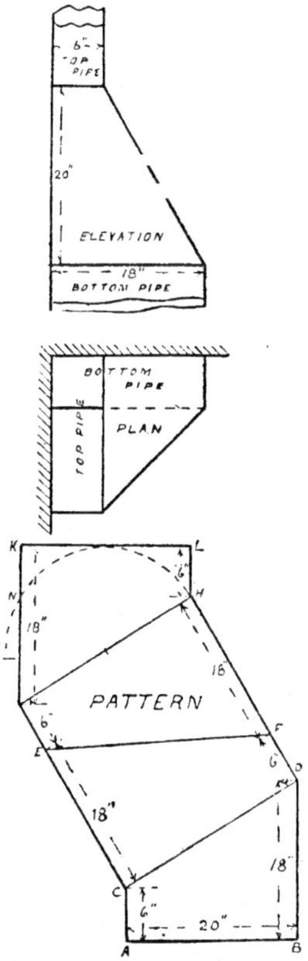

FIG. 38.

In shaping the plate, care must be taken to bend it in the right direction. In Fig. 38, if the ends of the plate are bent *up*, the connecting pipe will come into the correct shape and fit into position as seen in Fig. 37.

Pipe-end Ornament.

Multitudes of designs can be adopted to ornament the outlet or inlet end of a length of pipe, the method followed in setting out the shape of sheet or plate to form the cut being practically the same in each case. One simple design is shown in Fig. 39, in which the end of the pipe is flayed out and a bead turned on the edge of the sheet.

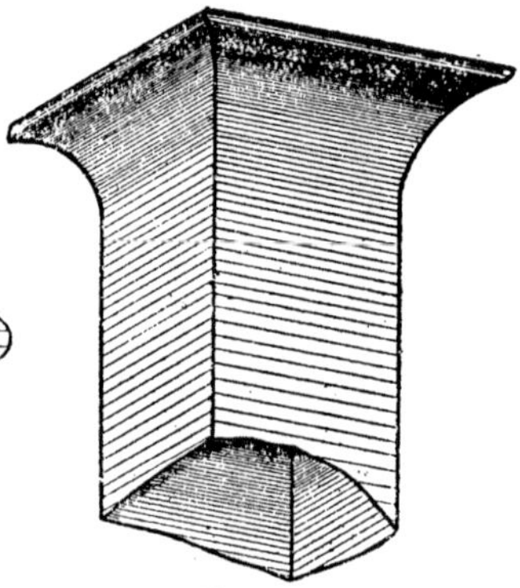

FIG. 39.

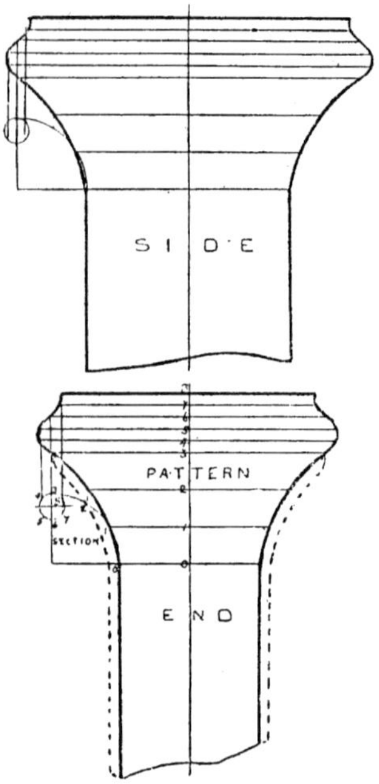

FIG. 40.

The setting-out of the patterns can be followed by reference to Fig. 40. The exact shape of the section of the end of pipe is first set out as shown on the end pattern. The quarter-circle is divided into three equal parts, and the small circle into six equal parts. The section girth is set along the centre line of the pattern, the length 0 to 1, 1 to 2, 2 to 3, etc., being the same length as the arcs on the section. Lines are drawn through these points square to the centre line of pattern, and these cut off the required length by projecting up from the corresponding points on the section as shown. A free curve is then drawn through the points, and the net pattern is complete.

Laps for riveting, soldering, or whatever form of seam is used, will of course have to be allowed for.

Where a bead is put on, as in this case, it will be an advantage to make it separate from the sheet, leave it slightly open, and slip on as with split-tube.

If required to fit together properly and to look well when finished, this kind of work will need setting out very accurately and making up as neatly as possible.

FIG. 41.

Bending Bench.

In a shop where pipe work is done, and there is no press for bending, or rolls for curving, long lengths of pipes or troughing, one of the most useful arrangements to have is a bench fitted up for bending as shown in Fig. 41. The bench must be a fairly strong one, and be rigidly fixed to the wall. Along its front edge should be firmly screwed a bar of stout angle-iron, say $2\frac{1}{2}$ in. or 3 in. Two heavy screw-cramps with large fly-nuts, to fasten to the bench as seen in sketch, will be required. These should be sufficiently long to have two or three holes in the flat part that fits under the bench, so as to be adjustable by drawing in or out.

The sketch shows the bending of a rectangular pipe. A bar of flat iron about 3 in. by 1½ in. is resting on the cramps, and when the sheet is inserted between this and the edge of bench, the cramps are screwed up and the sheet thus firmly held. It is now pulled over, and a sharp edge formed by beating down with a mallet or dresser. Each corner is thus treated in this way, the pipe being then grooved or riveted up. In stronger sheet a batten of wood is sometimes used by drawing along the edge and beating down with a heavy mallet or hammer. This avoids hammer or mallet-marks on the sheet.

In round pipes or half-round gutters it will, of course, be necessary to have a round mandrel to beat the sheet over, and to assist the leverage two battens of timber are nailed together in the form of a cross, and used as in the sketch. Two large eye-bolts are fixed near the edge of bench, and through these a bar or mandrel passed and secured. This bar simply acts as fulcrum, under which the end of the wooden cross is placed, and so enabling pressure to be put upon the edge of sheet.

In heavy work it will make the mandrel more solid to place props under each screw-cramp.

With different-shaped mandrels and some scheming and dodging quite a variety of work can be done on a bench of this description. With a strong bench and a stout mandrel, work up to ⅛ in. plate can be done in this way, and for short lengths up to 3-16 in. thick.

In fixing up a bench of this kind it is advisable to see that no leg is placed in between the two screw-cramps, as in some jobs where it is necessary to bend the plate under the bench a leg would be in the way.

CHAPTER VIII.

HOODS.

AN iron plate worker, whitesmith, or blacksmith may, some time or other, want to make a hood for a smithy hearth or some other purpose. We will therefore describe the setting-out for one or two typical cases.

Hoods are made in a variety of forms, depending upon the size, position, and shape of the hearth or other object to be covered. A common kind of hood, and the ones that we shall deal with in this article, are those that fit against a wall. A hood of this description, with flat front and sides, is shown in Fig. 42. It is constructed in three pieces, two sides and front. To make the describing of the setting-out plainer, it will be as well to fix some dimensions to the hood. Suppose them to be as follows:—Height 4 ft. 9 in., width 3 ft., depth 2 ft. 6 in., and turn down in front 2 in. The side can be marked out as shown in

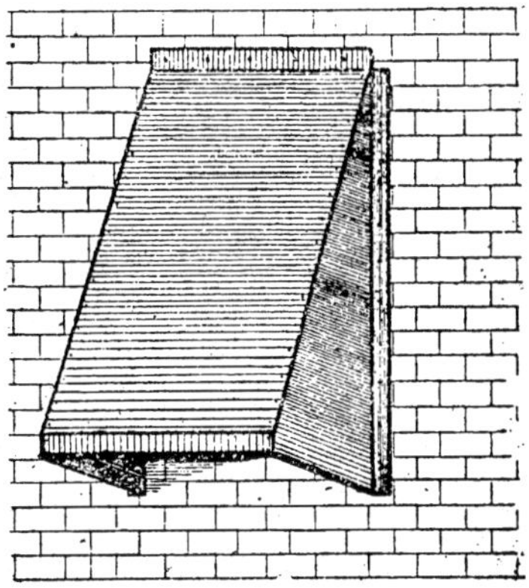

FIG. 42.

Fig. 43. Two lines are drawn square to each other, the one being made 4 ft. 9 in. and the other 2 ft. 6 in. A line 2 in. long is now drawn square to the end of the 2 ft. 6 in. line, and the end of this joined to the end of the height line. The slant line obtained will, of course, give the length of the front plate.

From the well-known property of the right-angle triangle:—"The square of the hypothenuse is equal to the sum of the the squares of the two sides," the slant height, or length of front, can be calculated thus:—

$$(55)^2 + (30)^2 = 3{,}925.$$

Extracting the square root:—

$$\sqrt{3{,}925} = 62{\cdot}65 = 62\tfrac{2}{3} \text{ in. (nearly)}.$$

Whilst in this particular job it is most convenient to obtain the length of the front from the side, yet there are in practice many cases, as we shall see, where this kind of calculation is most useful. A flange for attaching the hood to the wall must be allowed on the side pattern, and also on the top of front plate; flanges must also be left on the slant line of side for fastening front and side plates together. From the inspection of Fig. 42 it will be seen that the bottom of the hood is wired; hence it will be necessary to make an allowance for wiring on lower edges of sides and front. The amount of this allowance will of course depend on the size of the wire to be inserted, and also on the thickness of the plate used.

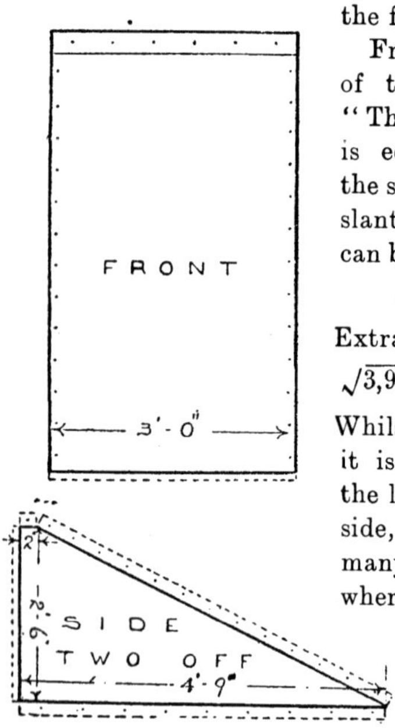

FIG. 43.

The following is the general rule:—

Allowance for Wiring.

"Add twice the diameter of wire to four times the thickness of metal." A careful study of Fig. 44, and the measurement of the length of centre line of metal will show the above rule to be as near correct as possible. Suppose that $\frac{1}{4}$ in. wire be used in the hood, and the sheet iron to be 1-16 in. thick, then the allowance to be added on to net pattern for wire will be—

$$2 \times \tfrac{1}{4} + 4 \times \tfrac{1}{16} = \tfrac{3}{4} \text{ in.}$$

For the flanges and wire edges to come into their proper positions the corners should be carefully notched, as shown on the pattern for sides. Holes for riveting can be marked and punched in the flange on side pattern, and this used for marking the holes on front plate.

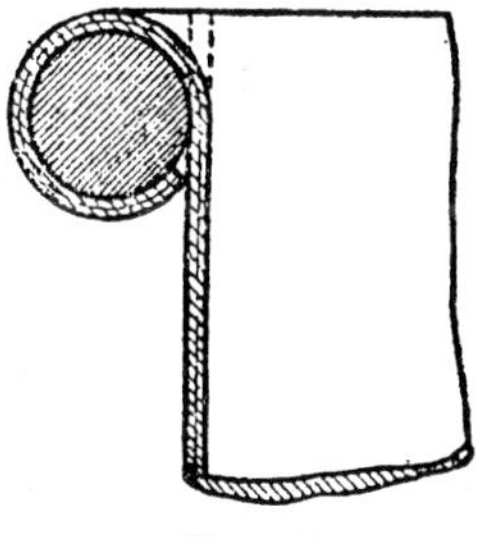

FIG. 44.

Sometimes angle-iron is used to joint the front and side plates together, and in this case no lap for riveting will, of course, be necessary. Again, sometimes an angle-iron frame is riveted around the bottom, and the two ends let into the wall, and when the hood is constructed in this way no allowance will be needed for wiring. Whilst speaking of wiring it should be remembered that although the above rule for wire allowance on pattern is strictly true for straight wiring, it is not exactly so for the edges of round tapered articles. If the large end of a circular article is to be wired, the calculated allowance will be slightly too much, and in the case of wiring the small end the allowance will be a little too small. In fastening the hood to wall, it is a good plan to bolt a bar of flat iron over the flange at top of hood, as this will materially assist in keeping the hood tight against the wall.

A method of marking a small hood in one piece is shown in Fig. 45. If the height and depth are given, one side can be marked out as at A, and the front part B set on this, the remaining side C being described on B by taking the height and depth and marking respectively from the top and bottom of the edge of B. If one only of the dimensions, either height or depth, be obtainable, then the sides can be marked on the front by describing a semicircle, as shown, and marking across with the given dimension from an end of the diameter.

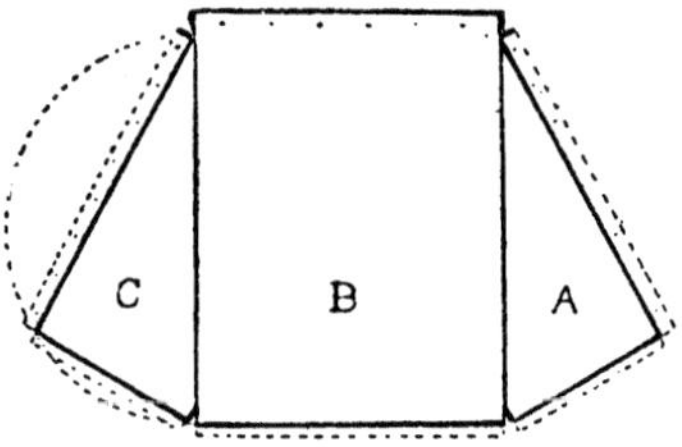

FIG. 45.

The use of the semicircle has an extensive application in sheet and plate metal work, as every pair of lines drawn from a point in the circumference to the ends of the diameter contain a right angle, or are square to each other. This property of the circle can often be taken advantage of in adding sides or ends on to a pattern.

A more expensive hood, both in labour and material, is of the kind required to cover a semi-circular hearth with flat sides, as shown in Fig. 46. The bottom of the hood is, of course,

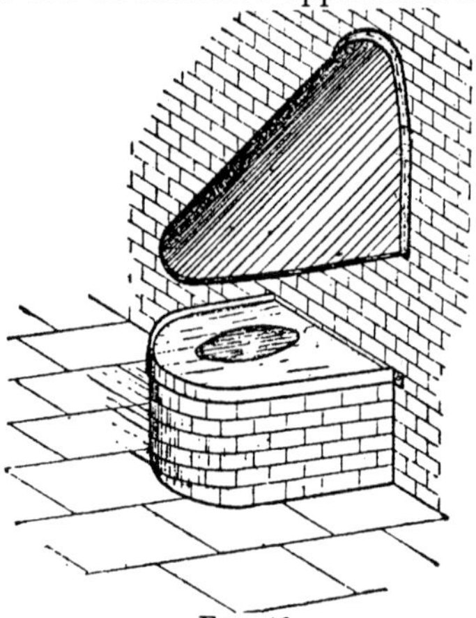
FIG. 46.

the same shape as the top of the hearth. To set out the plate in the flat required for the hood, the method illustrated in Fig. 47 can be followed. A side elevation of the hood is first drawn, and a quarter-circle described on the bottom equal in radius to half the width of hood. This quarter-circle is divided into three equal parts, and lines drawn up through each point square to the bottom line of hood giving points C, E, and G; and through these points lines are drawn parallel to the line A B. A base line to measure from is now fixed; and this may be drawn in any position square to the front of hood. The most convenient position, however, is when the base line is drawn to pass through a corner of the hood, as in the figure. Now measure the lengths of lines on quarter-circle, and set these distances on corresponding lines above the base line. Thus, H 3 = C 3^1, K 2 = E 2^1, and L 1 = G 1^1. The line M O will, of course, be made

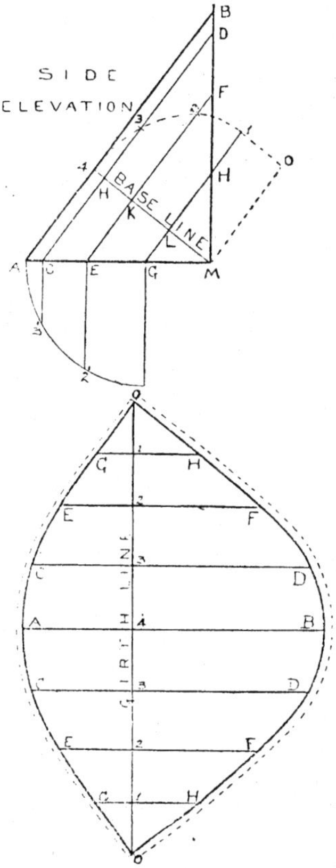

FIG. 47.

equal in length to G 1^1. Join the new-found points with a fair curve, and the length of this will give half the girth of hood. The shape of the curve from 1 to 4, it might be remarked, is a quarter of an ellipse, and this will give the form to which the hood should be shaped on each side of the centre line of front of hood.

In setting out the pattern, first draw a line, and along this mark the lengths 0 to 1, 1 to 2, etc., as obtained from the quarter-ellipse in elevation. Draw lines across, through these points, square to the girth line, and their lengths can be cut off by obtaining the lengths of similar lines in elevation. Thus, G 1 on the pattern equals G L in elevation, and H 1 equals H L. In the same way, E 2 and F 2 will be respectively equal to E K and F K, the other lengths being measured and set off in the same manner. All the points are now joined up with an even flowing curve; best drawn by bending a piece of hoop-iron round through the points, or a strip of wood and marking along.

If a flange is to be thrown off to fit against the wall, this must be allowed for on the pattern, and also if the bottom edge is to be wired allowance must be made for this. Care must be taken to notch the corners properly, so that the flange and wire edge can conveniently be turned over. The wiring around the bottom can be done either before or after shaping the plate, the flange for back being thrown off after the plate is bent.

It should be remembered that the bottom is semicircular in form, with the part from G to M straight, so that in bending it can be shaped to this; or if a very accurate job is required a template can be cut out of sheet iron, or a piece of strong wire bent to the shape of the curve 0 to 4, and this can be used as a gauge in bending. If the hood is large it will be made up in two or more pieces to avoid waste of material. Instead of a flange at back, angle-iron may be attached, and in place of wiring around bottom, cope-iron can be riveted on.

CHAPTER IX.

FLAT-SIDED TAPERED ARTICLES.

THERE are so many different kinds of the above class of work that there is some difficulty in making a selection of typical examples sufficiently broad to cover the general run of this character of work. In giving some representative cases it will perhaps be best to commence with a

Square, Equal Tapering Cap.

This form of cap, or bonnet, is what is known in geometry as a square pyramid (Fig. 48), each face being triangular in shape and equal-sided.

To readily set out the pattern for a pyramid (having

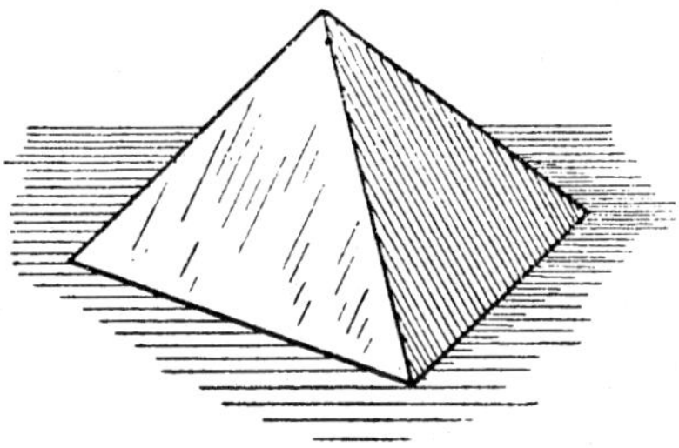

FIG. 48.

either four or more sides) the true length of one of the edges should first be obtained, and then used as the radius for the pattern circle. To do this a half-elevation is drawn (Fig. 49) by marking up the height, *b t*, and half the base, *b a*. The line *a c* is now drawn square to *a t*, and cut off equal to *a b*; then *t c* will give the true lengths of the edges, or the radius of the pattern circle. It might also be remembered that the triangle *a t c* gives the half-shape of one of the pyramid faces, or pattern triangles. After describing the arc C C C, to the radius *t c*, the compasses should be set to the length of the cap-base (twice *b a*), and this distance used to step around the arc. It will be seen that five lengths have been marked around; the last two being halved and joined up to T to form a seam line. It is always a good plan in pyramid work of this character, when the seam is required to come up the middle of a side, to set along the arc one more length than the number of sides that the cap has, and cut away half of each end triangle; this method insuring the correct position of the joints.

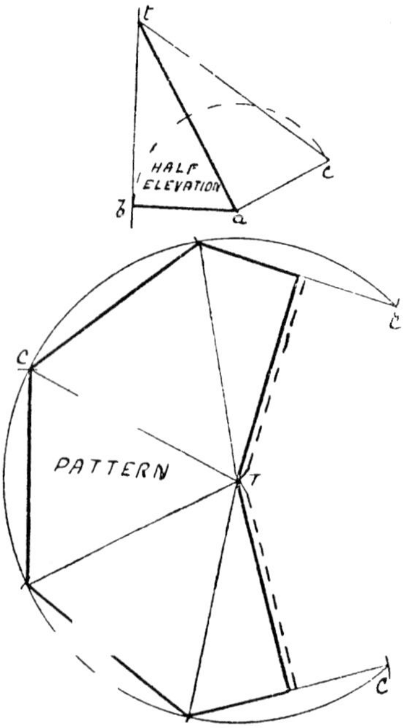

FIG. 49.

The above is also exemplified in Fig. 50, where the pattern for an

Equal Tapering Square Article

is shown set out. Here also a half-side elevation is drawn. The apex of the pyramid, of which this article is a frustum, can be found by producing *a d* to meet the centre line in *t*.

The pattern for the complete pyramid is first struck out, as previously explained. The line *d f* is drawn square to *a t*, and the compasses set to the length *t f*; the arc which passes through F F, on the pattern, then being described.

Large articles, like hoppers and hoods, would, of course, be made up in parts, with joints running down the corners; or, in the case of very large jobs, perhaps two or three plates for each side. No difficulty, however, should be met with in these cases, when it is remembered that the shape of one side of the hopper will be E F F E on the pattern, or, for half a side, the figure *a d f e* on the elevation.

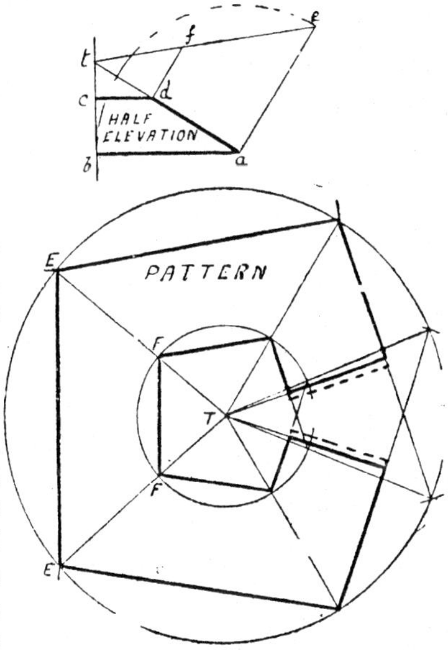

Fig. 50.

Equal Tapering Rectangular Article.

The plan of a rectangular hopper or hood is shown in Fig. 51, together with the necessary construction lines required for the patterns and the corner angles.

In setting out the plates the first thing required is the length down the slope of the hopper, and this can be found by marking off *a b* equal to the depth of the hopper, and drawing the line *a c* square to it; then *c b* will be the required length down the slope. This length will, of course, give the width of the plate, which, to avoid setting out, might be calculated thus:—

$$\text{Width of plate} = \sqrt{(\text{height})^2 + (\text{overhang})^2}$$

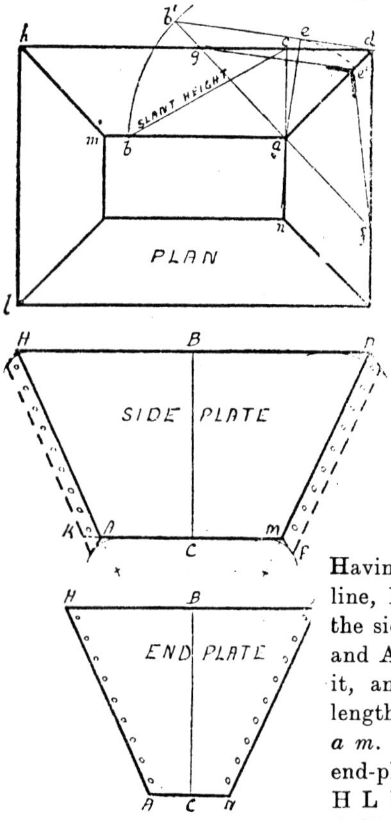

FIG. 51.

To illustrate the above by an example: Suppose the top of the hopper is 15 ft. by 11 ft., the bottom 7 ft. by 3 ft., and the depth 9 ft., then the overhang will be—

$$\frac{15 - 7}{2} = \frac{8}{2} = 4 \text{ ft.}$$

The width of the plate will equal—

$$\sqrt{(9)^2 + (4)^2} = \sqrt{97} = 9 \text{ ft. } 10\tfrac{1}{8} \text{ in.}$$

Having set down the centre line, B C, to this length, for the side-plate, the lines H D and A M are drawn square to it, and marked off equal in length respectively to *h d* and *a m*. In the same way, the end-plate can be marked out, H L being equal to *h l*, and A N to *a n*.

If flanges are to be turned on the side-plates, then laps must be left on, as shown. The notches on the laps can be formed by fixing the leg of the compasses at any points on D M and D M produced, and drawing arcs of circles to touch H D and A M; then drawing lines to touch these arcs, as at P M and the other corners. Without it is an exceedingly particular job, notches are not left on the bottom corners of the plates, on account of the difficulty of working and waste of material, but are sheared straight along, as shown by the line A K.

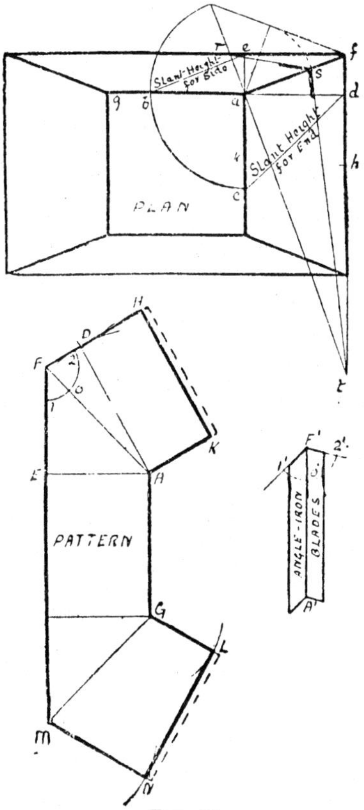

FIG. 52.

If the plates are not to be connected by flanges, but with corner angles, then it will be necessary to make a template, showing the rake of the angle-iron. The construction for this is shown on the plan in Fig. 51. Draw $f g$ square to d a, cutting off a b' equal to a b. Join b' to d, and draw a e square to b' d. Make a e' equal to a e, then the angle g e' f will be the required rake that the corner angle-iron must be set to.

Another construction for the same kind of thing is shown in connection with Fig. 55.

Oblong Hopper of Unequal Overhang.

The marking out of the plates for the above will be similar to the former case; the only difference being in having separate slant-heights for the sides and ends. To obtain these the lengths *a b* and *a c* (Fig. 52) are each made equal to the vertical depth of the hopper; then *c d* and *b e* will give the lengths of the respective middle lines of the plates. The rest of the construction is as before.

If templates are required for the blades of the corner-angles, then these can be marked directly from the plate patterns, or their end-cuts transferred with the compasses. Thus, taking F as centre, describe an arc 1 0 2 to any radius; then, with the *same* radius, and F′ as centre, mark the arc, 1′ 0′ 2′; cut off the arcs 0′ 1′ to equal 0 1, and 0′ 2′ to equal 0 2; join F′ to 1′ and 2′; so obtaining the rake for the ends of the angle-iron blades. The length F′ A′ will equal F A, and the cut on the bottom be parallel to the top.

The corner-angle *r s t* is shown marked out on the plan, the method of construction being exactly the same as in Fig. 51.

If it is desired to make a pattern for a smaller article, and to have the seams down the middle of the ends, then the marking out for this will be as shown in Fig. 52. Line A G = *a g*, A E = *b e*, and E F = *e f*, so giving the pattern for a side. The half-end must now be added thus: Set the compasses to *c d*, and with centre A, describe the arc passing through D; then, with the centre F, and radius *f d*, cut the former arc, so fixing the point D. Join F to D, and produce, making F H equal to *f h*, from the plan. Now draw A K parallel to F H, or square to A D, and cut off equal to *a k*. Another method, which is, perhaps, somewhat more convenient for workshop purposes, is shown on the other end of pattern. Set the compasses to *a k*, and, with G as centre, describe an arc (seen passing through L); then, with M as centre and *h f* as radius, draw the arc which passes through N. Draw a line to touch the two arcs, and on it drop perpendiculars from G and M.

Unequal Tapering Square Hood.

In the same manner as some circular tapering articles are formed as portions of round oblique cones (Chapter XVIII.), so we may have square objects coming out as parts of square oblique pyramids. Fig. 53 is an illustration of the latter. The top and bottom of the hood are square, and also parallel; but it will be seen from the plan that their centres do not come on the same vertical lines. If the corner lines in plan are produced, they will meet in the common point *t*, this being the plan of the apex of the oblique pyramid of which the hood is a frustum.

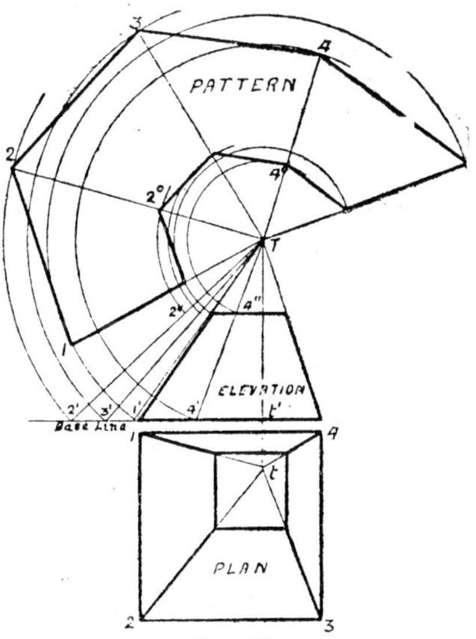

Fig. 53.

To set out the pattern: Produce the side lines in the elevation, and obtain T, the apex of the cone (this should come exactly over *t* in the plan). From *t'* mark the lengths *t* 1, *t* 2, *t* 3, and *t* 4, along the base line, so fixing the points 1′, 2′, 3′, and 4′. Join these latter to T, and then draw the arcs to the radii T 1′, T 2′, etc., as shown. Open out the compasses to the length of the side of the square, say 1 to 2, and, commencing at 1 (on the arc drawn through 1′), step around from curve to curve the points 2, 3, 4, and 1.

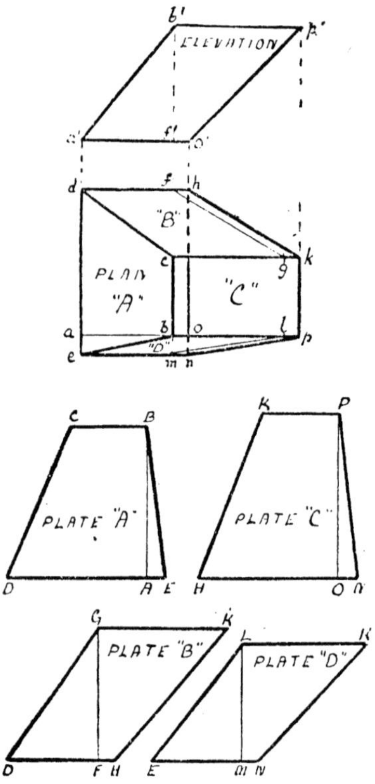

FIG. 54.

Join these up to each other and T. This figure would give the pattern for the complete oblique pyramid. We now want to cut away the part of pattern that corresponds to the top of the pyramid. Take T again as centre, and as radii the distances down to where the respective lines cross the top line or top line produced, and swing around on to the corresponding lines. Thus T 2° will be equal to T 2″, T 4° to T 4″, and so with the other lines.

If it is a large hood, then the plates can be set out separately, as in the former cases.

No allowances for jointing have been put on the above pattern, as this will depend upon the size of hood and the number of plates into which the complete pattern is divided.

Whilst the construction, as shown above, is for a square

oblique pyramid, it should be borne in mind that the same principle will apply to any other shaped article that comes out as a frustum of an oblique pyramid.

Irregular-shaped Oblong Uptake.

The plan and elevation of an irregular-shaped funnel or uptake is shown in Fig. 54. The laying out of the plate shapes follow the general principles as explained in connection with Fig. 52. First mark $c\,g$ and $b\,l$ each equal to the depth $f'\,b'$. Referring to plate "A," the line A B = $a'\,b'$, A D = $a\,d$, A E = $a\,e$, and C B = $c\,b$. In plate "B" the line F G = slant height $f\,g$, F D = $f\,d$, F H = $f\,h$, and G K = $c\,k$. For plate "C," O P = $o'\,p'$, O H = $o\,h$, O N = $o\,n$, and P K = $p\,k$. Then, on plate "D," L M = $l\,m$, M E = $m\,e$, M N = $m\,n$, and L K = $p\,b$. After what has been said in connection with the former examples, no further particulars than the above should be necessary to set out the four plates.

Overhanging Oblong Shoot or Hopper.

This example (Fig. 55) has been chosen to show the laying out of the plate patterns by turning back from the plan; also, and principally, to illustrate the obtaining of the corner angle-iron rakes by a second method more adaptable for workshop use, in many jobs, than that shown in Figs. 51 and 52.

The patterns are shown laid out in thick dotted lines, the lengths for their widths being obtained as before. Thus, A B = $a\,b'$, and $b\,c$ = $b'\,c'$, from which D C = D c. So with the other pair of plates.

To obtain a corner angle, set the compasses at any distance, and mark off equal lengths E d, E d along E B and E C. Draw perpendiculars to the last two lines through the points d, d, so getting the points f and e. Now take f as centre and $f\,d$ as radius, and swing on to the line E b, thus obtaining point g. (If the point e be also taken as centre, and $e\,d$ as radius, this will likewise give point g; hence it will be seen that the latter point can be obtained by the intersection of the arcs, without the use of line E b.) The angle $f\,g\,e$ will represent the rake of the corner angle-iron, so that a template can be made to this, to which the angle-iron can be opened.

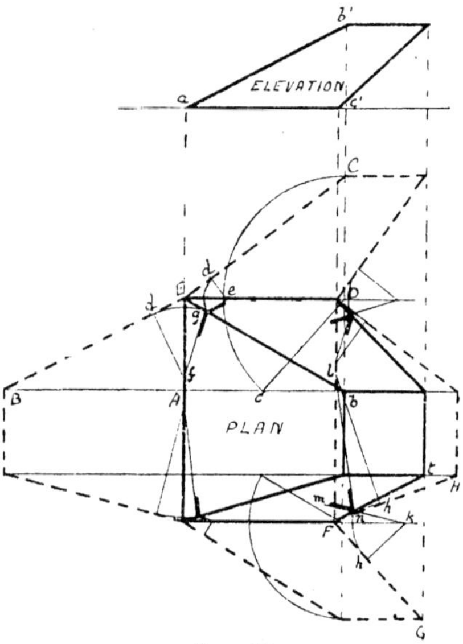

FIG. 55.

The application of the above method to finding the acute angle between the front and side plates is, perhaps, not so easily followed as for the back and side plates; it will therefore be an advantage to go over the construction for one of the front angles. Again referring to Fig. 55: Mark F h along both F H and F G to any convenient length. Draw $h\,k$ square to F G, and $h\,l$ square to F H, giving the points k and l. Take k as centre, and $k\,h$ as radius, and swing down the arc on to the line F t, thus obtaining the point n. Join n to l and k to n, producing the latter line, say, to m. Then angle $l\,n\,m$ will give the rake of the angle-iron for this corner. The construction lines for obtaining the angles on the other pair of corners are also shown; but, after what has already been said there should be no need for further explanation.

Irregular-shaped Bonnet.

In Fig. 56 the plan and elevation of an unequal tapering bonnet or uptake is shown. The bottom and top are both square, but as they are not parallel the bonnet cannot be considered as part of a pyramid.

The plate shapes for back and front can be laid out as in Figs. 54 and 55. The side plates "B" and "D," however, on account of the twist, will require to be set out by the method of triangulation.

Let us take plate "B" first. Draw along B A equal to *b a*. Mark *c a* along the base line from c', and then join a' to c''. Set the compasses to the length of a' c'', and, taking A as centre (on the pattern "B"), draw an arc (shown passing through C). Now open the compasses to the length of

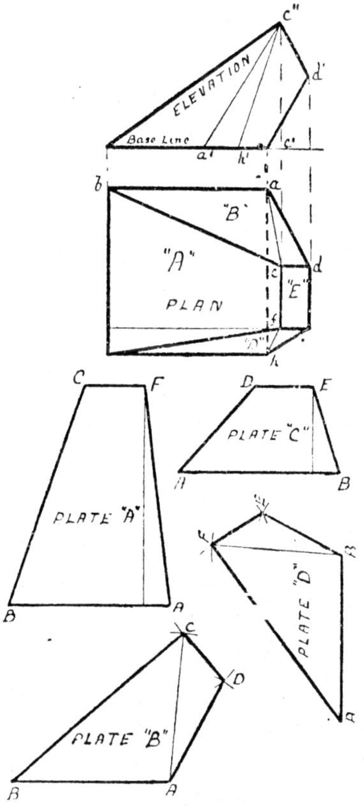

FIG. 56.

B C on the pattern "A," and, using this as a radius from the centre B, cut the first-drawn arc, so obtaining the point C. Fix the compasses to the length $c'' d'$, and, taking C as centre, draw an arc (shown passing through D); then, setting the compasses to the length A D on the pattern "C," use this as a radius from centre A to fix the point D. Thus the pattern is complete.

For the pattern "D," the line A B = $a\ b$. A F will be the same length as the corresponding line on pattern "A," and B E the same as the similarly lettered line on pattern "C." To obtain the length of B F, mark $f\ h$ along the base line from c', and join h' to c''; then $h' c''$ will be the true length of B F. Having the lengths of all the lines, the pattern can be laid out in the same way as for plate "B."

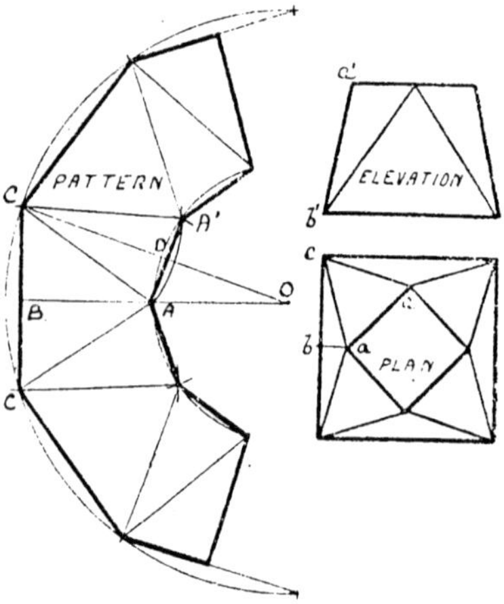

FIG. 57.

To bring the plates "B" and "D" correctly into position, it will be seen that there must be a slight bend about the lines A C and B F. In cases like the above, however, where the twist is slight, there will be no need to kink the plates before bolting together; the screwing up should be sufficient to pull the plates into position.

For articles of any description whose surfaces are twisted similar to above, the method of triangulation can always be applied in the laying out of the plate shapes.

Twisted Square Base.

We will bring this chapter to a conclusion by giving an example of an ornamental tapered base made up of flat surfaces.

A plan and elevation of the base is shown in Fig. 57. On examination it will be seen that the square top is twisted diagonally to the bottom.

The triangle A C C on the pattern is struck out by making A B equal to $a' b'$, and B C equal to $b c$. The point C is then used as a centre, and the arc passing through A A′ drawn. The compasses are next fixed to the distance $a a$, and the point A′ determined by cutting the arc from the point A as centre. The whole pattern could in this way be built up by adding triangle to triangle; but, as the figure is symmetrical, it is better to disect A A′ in D, and draw the lines C D and B A, produced to meet in O, using this latter point as a centre to draw the arc, as shown, upon which the rest of the pattern can be constructed. Five lengths are stepped along the outer arc, the last two being halved, this way insuring the seam lines being in their correct position.

A somewhat peculiar case of the above kind of base is when the top square is the same size as the bottom one, the pattern then coming out as a rectangle, and being built up with triangles, as in the pattern given above.

CHAPTER X.

PAN CORNERS.

THE sheet metal worker is so often called upon to make all sorts of pans that a consideration of the different kinds of corners that can be formed will not be here out of place. The unprofessional workman, too, occasionally wants to make a pan to hold some odds and ends that are lying about the bench or shop, and he should find no trouble in forming the simpler kinds of corners as explained in this chapter.

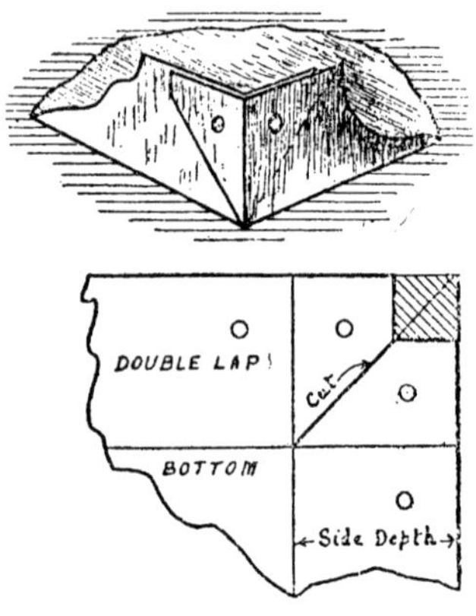

FIG. 58.

The strongest kind of pan corner that can be formed is probably that known as a "double lap," a sketch of which is shown, and also the pattern, in Fig. 58. The pattern for one corner only is given, as the setting out for each corner will be exactly the same. On account of the double lap it will be seen that there are two thicknesses of sheet metal at the corner, one of the laps being turned inside the pan and the other outside. The only marking out that is necessary is to add the depth of the side on to the size of the bottom, the corner or diagonal line being cut along as indicated. It is a good plan to cut the points off the flaps as shown by the shaded part. If made of light tinplate, the flaps can be soldered down after the sides are bent up, and if of strong sheet iron, riveted as seen in the sketch. Without the iron is very strong, such as 16 or 14 gauge, there is no need to put holes in the plate before bending, as the rivets can be drawn right through with the upset, as explained in Chapter XXXV.

In Fig. 59 the same method of jointing the corners as just explained is followed; but in this an edge is folded over along the top of the pan and used for gripping the two flaps, besides strengthening the edge of the sides. If the edge is also left on the flaps, one can be turned under and the other over the side edges, as seen in the sketch. There is no need to rivet this corner, or even to solder it, without the pan is required to hold a liquid.

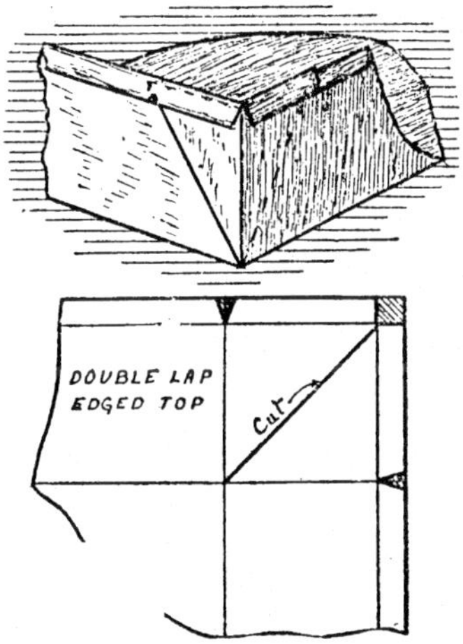

FIG. 59.

Fig. 60 shows a pan corner that is formed by bending over a single lap and riveting. Allowance is made on the pattern for an edge to fold over all round the top of the pan. The corner of the plate will be cut away, as seen by the shaded part on the pattern. If the top of the pan is to be wired, it will be as well to notch the lap slightly larger, as seen by the dotted line. Holes for rivets, if re-

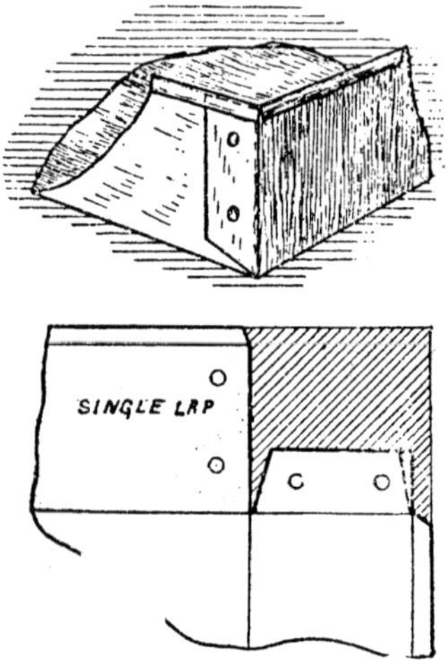

FIG. 60.

quired, will be punched in the plate, as shown on the the pattern.

A pan with a knocked-up corner is illustrated by Fig. 61. In cutting the corner of the pattern, care should be taken that a single edge is allowed on one side, and a double edge on the other. If the pan is to be wired along the top edges then notice must be taken that the laps are properly notched before bending. If the knock-up is required to be on the inside of the pan instead of the outside, then the edges for the knock-up should be folded over in the reverse direction, so that the double edge will come on the inside of the pan.

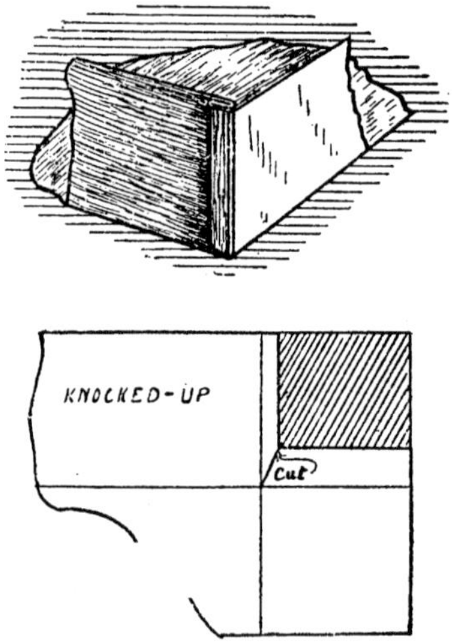

FIG. 61.

The pan corner sketched in Fig. 62 shows the method of doubling up the sheet metal to form a solid, or what is sometimes called a "pig's-ear" corner.

If it is required to form a pan with the sides square to the bottom, without wire or edge around the top, then there will be no need to cut the pattern at all, the corner being formed by bending along the dotted lines, as shown on the pattern.

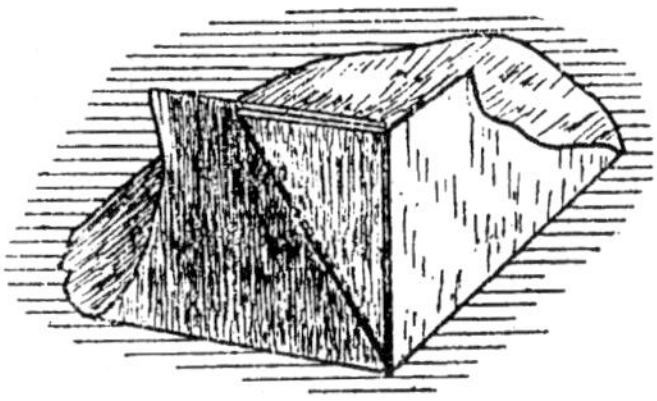

All the above methods of forming a pan corner are applicable to pans having sloping or tapered sides, the various allowances for jointing being put on after the net pattern is marked out, as explained below.

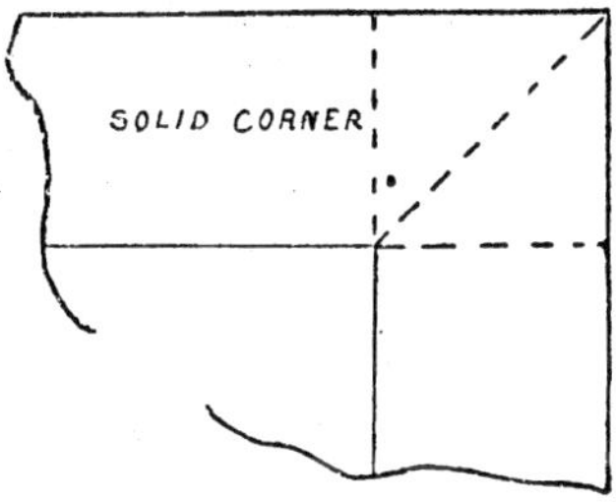

FIG. 62.

Tapered Pan with Solid Corners.

This kind of pan (Fig. 63) is of the baking-tin order; but the method of forming the corners can be adopted in all cases where it is necessary to have a pan that will be liquid-tight at temperatures above the melting point of solder.

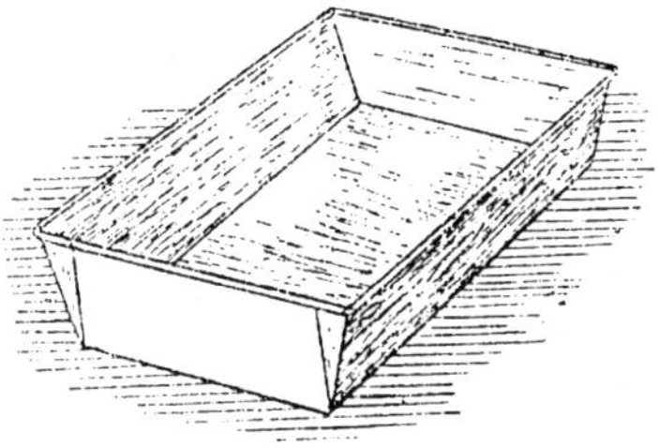

FIG. 63.

A pan may have an equal overhang all round, or its ends may overhang the bottom more or less than the sides do. We will set out a pattern for each case, taking a pan with equal taper for sides and ends first.

Suppose a pan is 12 in. by 9 in. at the top, $9\frac{1}{2}$ in. by

$6\frac{1}{2}$ in. at the bottom, and 2 in. deep. The distance that the top projects over the bottom will be found by deducting the length of the bottom from the length of the top and dividing by two. Thus—

$$\text{Overhang} = \frac{12 - 9\frac{1}{2}}{2} = 1\tfrac{1}{4}\text{ in.}$$

To get the length down the side all we need do is to set out a right-angled triangle (Fig. 64) with height 2 in. and base $1\frac{1}{4}$ in.; the third side, or hypotenuse, will then give

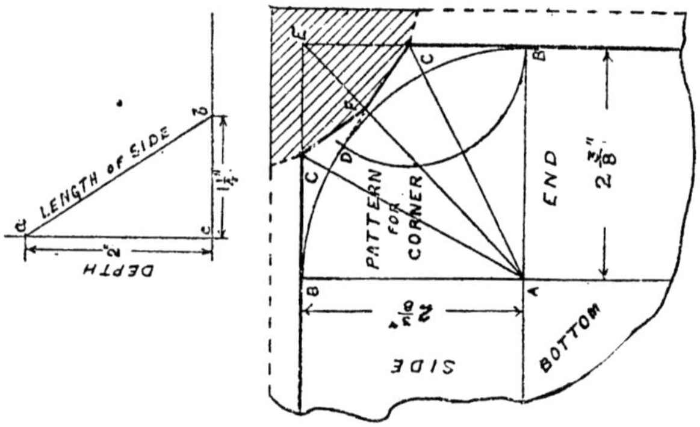

FIG. 64.

us the length down the side of pan. Or, without setting out the triangle, the required length can be calculated thus—

$$\text{Side length} = \sqrt{2^2 + (1\tfrac{1}{4})^2} = 2\tfrac{3}{8}\text{ (nearly).}$$

The size of the bottom is first marked out, and the side length added by marking A B on the pattern (Fig. 64) equal to *a b* on the triangle. The overhang is then set along the sides—that is, B C is marked off equal in length to *b c*. The points C C are joined up to A, and what we might call the net pattern is now complete; for if the piece

C A C be cut out and the sides of the pan bent up the two lines A C, A C will coincide; hence, whatever method of forming the corner is adopted the allowance for jointing must be additional to the net pattern. In this case we want to keep the corner solid by doubling up the sheet to form a flap, which will be folded over on to the end of pan. For the flap to turn over on the end and come flush with the top edge of pan, it is manifest that the angle of the flap must be equal to the angle of the end, and whatever construction is followed to obtain the cut of the corner is with the object of arriving at this result. Two methods can be used, and we will show both—one in this case and the other in connection with a pan of unequal overhang.

Again referring to Fig. 64, bisect the angle C A C, which, in this case of equal overhang, is simply done by drawing the diagonal line A E. With centre A and radius A B, describe the arc of circle marked B D; then, if a line be drawn from C to touch the arc, the point F on A E will be determined, and thus the shape of the top of flap. To accurately draw the line C F, it is not a bad plan to take centre C and radius C B, and thus mark the point D on the arc; then join D to C, and so obtain F.

The allowance for wiring must be added on as shown, and if the sheet is fairly strong, it will be as well to cut the top of the flaps a little lower, and thus avoid the wiring being lumpy where it runs over the flaps.

To obtain the shape of the part to be cut away at the other three corners, without the trouble of marking each out separately, a good plan to follow is to cut out the shaded part as shown, and use this as a template to mark off the other corners.

Pan with Unequal Tapering Sides.

Suppose it is required to make a pan whose dimensions are 19½ in. by 13½ in. at top, 18 in. by 10 in. at bottom, and

2 in. deep. Then the overhang of the sides will be—

$$\frac{13\frac{1}{2} - 10}{2} = 1\tfrac{3}{4} \text{ in.}$$

And that for the ends—

$$\frac{19\frac{1}{2} - 18}{2} = \tfrac{3}{4} \text{ in.}$$

The lengths to add on to the bottom for the sides and ends can be calculated as in the previous case, or obtained

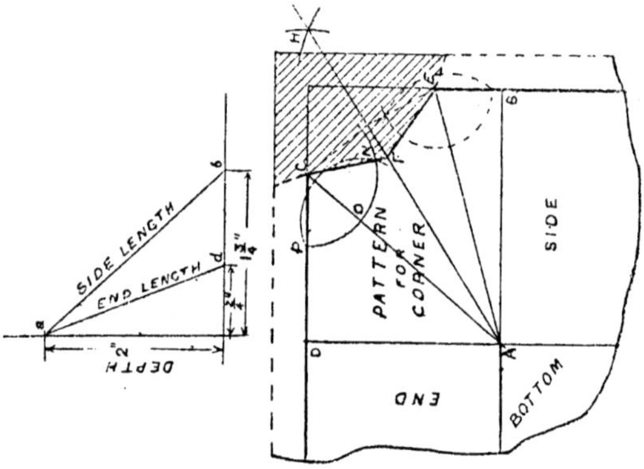

FIG. 65.

by setting out as in Fig. 65. Two lines are drawn square to each other, and the depth of the pan marked up, and the two overhangs along, the lengths of the side and end being obtained from the slant lines. On the pattern it will be seen that these lengths are set out by making A B = *a b* and A D = *a d*. The overhang of the side must now be put on to the *end* and the overhang of the end added to the *side*. That is D C must be made $1\frac{3}{4}$ in. long and B E $\frac{3}{4}$ in. Now, if the setting out is done correctly so far, the lines A E and A C should be equal in length; hence this always gives a check as to the correctness of the work. In bending up, it should be remembered that the lines A C and A E coincide to form the corner, so that for the top of the pan to be level these must be the same length. Now to mark out the lines for the part to be cut away. Bisect the angle C A E by describing two arcs of circles with equal radius from the centres E and C, intersecting in H, thus obtaining the line H A. Decide now whether the flap has to be folded on the end or side, for whichever it has to be turned on will fix the angle of the top line of the flap. In this case the flap is arranged to be folded over on the end. With centre C, and any convenient radius, describe the arc P O N, then cut off O N equal to O P by drawing an arc with O as centre and O P as radius. Join C to N, and produce the line until it cuts A H in F. Join E to F, and thus the part to be cut away is determined. If the flap is to be folded on to the side of pan, then a similar construction will have to be gone through, commencing with point E. This is shown in dotted lines.

It will be seen that in the case of a pan of unequal taper the shape of the corner cut on the pattern depends upon whether the flap is to be turned over on to the side or the end, and if cut to suit one will not fit on the other. The shaded part on the pattern can be used for a template to mark the other corners, as in the previous case.

Double-Flap Solid Corner.

A pan whose sides are square or tapered may have its corner formed by a double flap, as shown in Fig. 66. This is no stronger than the single flap (Fig. 62), but gives a little better appearance to the pan, and is more conveniently made by machinery.

The setting out of the pattern is very similar to Fig. 64. The overhang D C (Fig. 66) is first measured down and the angle A O C divided into four equal angles by dividing the arc K M into four equal parts and drawing the lines O N,

O B and O F. The length O B is next cut off equal to O C. The compasses are then fixed at O, stretched out to D, and the arc D E drawn; the point E being determined by cut-

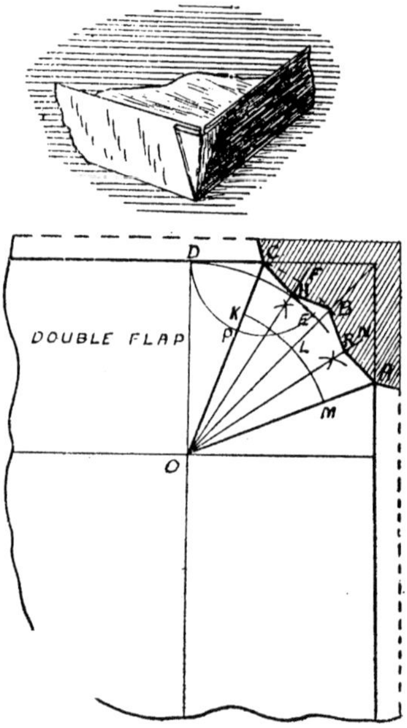

FIG. 66.

ting off C E (as shown by the arc D P E) equal to C D. A straight line is drawn from C to E, and where this intersects the line O F will give the point H. To finish, the line O R is made equal in length to O H.

For a pan with unequal tapering sides the construction would be a little different, but from what has been said in connection with Fig. 65 there should be no difficulty over this.

Working Up a Pan.

After the four corners of the sheet are cut, the bisecting line of the corner should be placed on the hatchet stake, as shown in Fig. 67, and the sheet bent down on each side. Then the sides and ends should be turned down on a square head or pan stake, as seen in Fig. 68, care being taken that the bottom is kept at its proper size, and that its edges are straight. On the same stake the corner flaps can be closed together (Fig. 69); the greatest care being exercised that

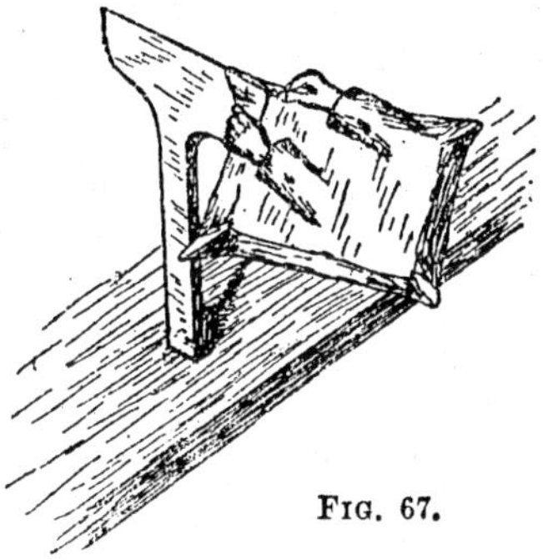

FIG. 67.

FIG. 68.

FIG. 69.

the flaps double up along their centre lines. They should now be slightly bent over on the hatchet stake. The hammering down of the flaps will be done as seen in Fig. 70, and as this is the crucial test of the quality of the sheet metal, and of the operator's skill, some judgment must be exercised in the hammering, or the flap will fracture near the root. To assist in avoiding the breaking of the metal, it should be seen that the flap is fairly well closed together near the root, before proceeding to turn it over. In light sheet metal the mallet must be used carefully, as there is the danger also of the corner of the hatchet or pan stake cutting through the metal.

FIG. 70.

For wiring, the edge of the sheet can be bent over the hatchet stake and the wire slipped in and tucked by the use of the mallet and hammer on the pan stake.

Pan with Moulded Sides.

The making up of a pan with moulded sides, as shown in Fig. 71, is not a difficult matter if the pattern for the cut corner is marked out as accurately as possible. This can be done as seen in Fig. 72.

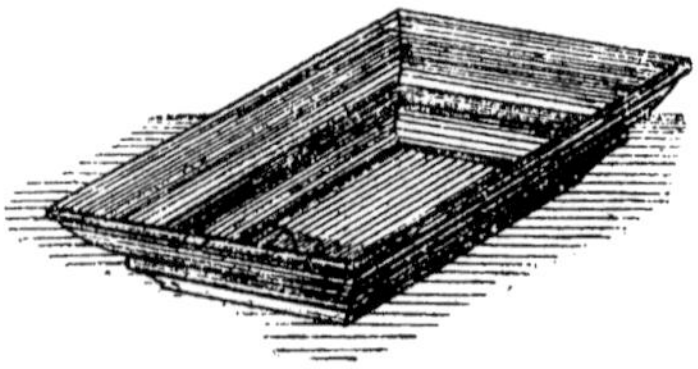

FIG. 71.

For a square or rectangular-shaped pan (Fig. 71), there is a special method which we will show first.

and then afterwards explain a general method that will apply to all cases for pans of the regular polygon shape, such as hexagonal, octagonal, etc.

In all cases the first thing to do is to set out the shape of the moulding (Fig. 72), and divide the curved parts up into a number of equal divisions.

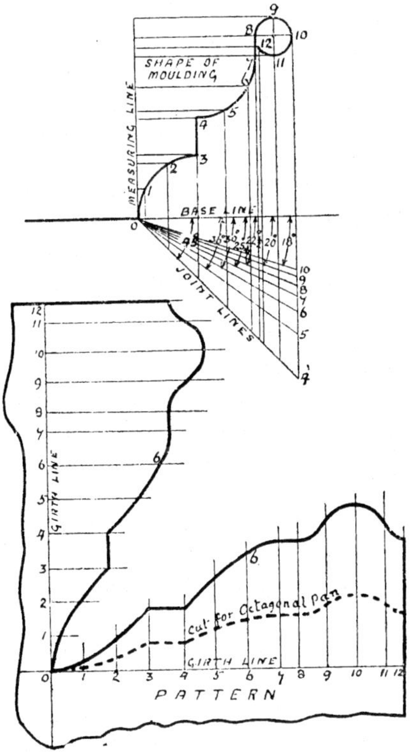

FIG. 72.

The pattern for the corner of a square or rectangular pan can be marked out by first drawing two lines at right angles, and setting along each of these the girth of the moulding by taking the lengths of the numbered parts on the moulding section. Lines square to the girth lines are then drawn from each numbered point, and the length of these cut off equal to that of the line drawn through the same numbered point on the moulding section up to the measuring line. Thus, to give one example, the line marked 66 on the pattern will be the same length as the line 66 on the moulding shape. When all the required distances are marked along the pattern lines, the points are carefully joined up, and thus the corner-cut obtained. It should be observed that any part of the moulding section which is straight will also have straight lines corresponding to it on the corner-cut of pattern.

The general method will apply to all cases, no matter how many sides the pan has or what is the shape of the moulding. It consists in drawing a base line (Fig. 72), and setting off a joint line at an angle equal to 360° divided by twice the number that the pan has sides. Thus, if the pan has four sides, as in the above case, the joint line will make an angle of—

$$\frac{360}{4 \times 2} = 45^\circ$$

with the base line. To cut off the pattern lines to their required lengths, they will be made equal to the lengths of the correspondingly-numbered lines running between the base and joint lines.

All the joint lines for pans having from four to ten sides are shown in Fig. 72, and also the shape of cut for the end of one side of an octagonal pan. This being eight-sided, the angle of joint line will be—

$$\frac{360}{8 \times 2} = \frac{45}{2} = 22\tfrac{1}{2}^\circ$$

The lines are measured between base and joint lines, and their lengths set up from the girth line on pattern, the thick dotted line thus representing the cut for one side of an octagonal pan. In setting out the pattern for a complete pan of this description, the best plan to follow is to first mark out the shape of the bottom, draw lines square to the end of each bottom line, set along the girth, and then proceed to obtain the shape of corner-cuts as explained above.

In shaping the sides of a moulded pan to the required form, it is necessary to be as accurate as possible, if the edges of the moulding are to fit together properly at the corners.

CHAPTER XI.

TRUNKS, BOXES, FENDERS, ETC.

THE setting out of the patterns for a trunk or box is usually not a very difficult matter; for whilst their shapes and sizes are of almost infinite variety, there is very little of a complicated nature in their make-up. The chief thing to which attention should be paid in the construction of trunks is accuracy of workmanship, so that the various parts shall fit together properly.

The bulk of trunk work is now wholly or partly machine-made, and as the sheet iron used is very light, the product turned out is not of a character to stand much knocking about. Hence, so that the sheet metal worker may know how to lay out the patterns for a good strong, hand-made trunk, we have selected just one representative example, which should be sufficient for all practical purposes.

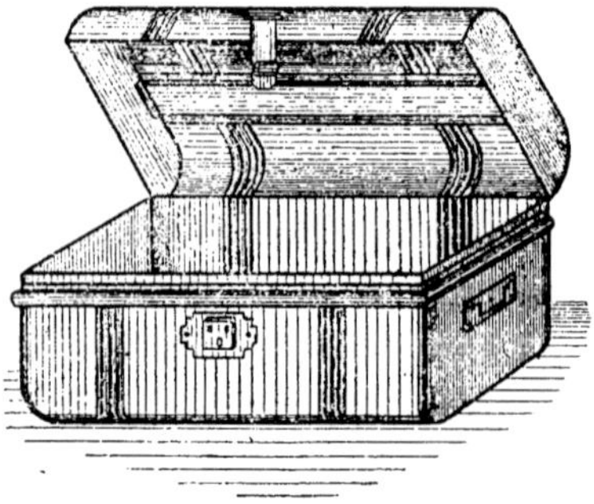

FIG. 73.

A sketch of the trunk or box is shown in Fig. 73, from which some general idea will be obtained as to its shape and make.

Before proceeding to mark out the patterns for the trunk parts, a template for the moulding at the corner should be made. In Fig. 74 the shape of the moulding which runs down the top edge of the trunk body is shown drawn out full size, also the template for corner-cut of same. To mark out the latter, the girth line, 0 0, on the pattern is first laid out by making it equal in length to the semicircle on the section—that is, twice the length of the arc 0 3, or six times the length of one of the small arcs. Through each division point lines square to the girth line are drawn, and these cut off equal in length to the corresponding line on the section. That is 1 1′, 2 2′, etc., on the pattern are the same length respectively as 1 1′, 2 2′, etc., on the section. The lines lettered *o* B are next drawn square to the girth line, and measured off equal to *a b* from the section. The points B B are then joined, and

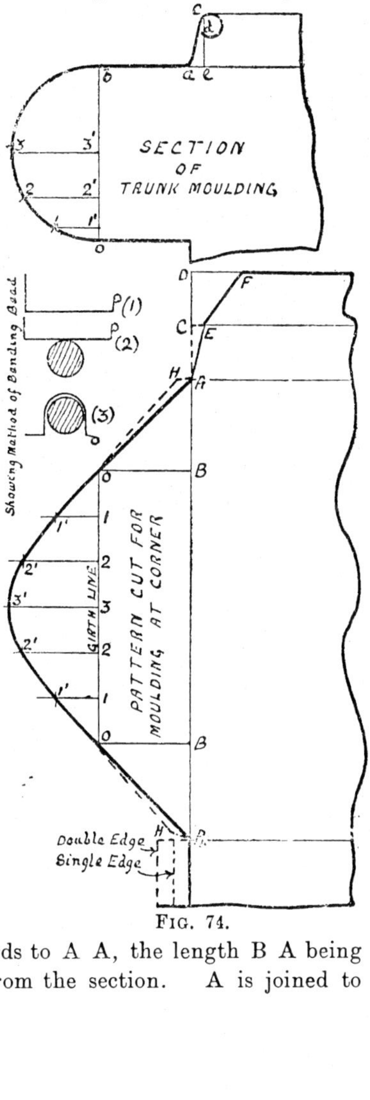

FIG. 74.

the line produced outwards to A A, the length B A being made the same as *a b* from the section. A is joined to

0, and the curve passing through the points 1′, 2′, etc., drawn. A C is now marked off equal to *a c*, and C E equal to *a e*. The allowance for wiring will be C D, this being equal to the length of the arc *c d* measured around the wire on the section. The notch E F can be fairly accurately obtained by making D F equal to D C. The small lap, A C E, shown with a dotted line, should be left on the pattern, as this can be bent around the corner, and will stiffen it somewhat. Also, the allowance marked H *o* should be left on, as this will cause the moulding to overlap a little on the flat parts, and strengthen the corner considerably.

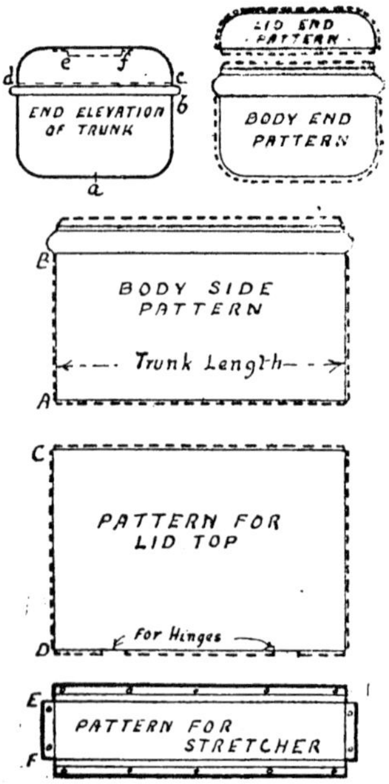

FIG. 75.

The moulding template as marked out above can now be used to scribe out that portion of the patterns for the body, and end as shown in Fig. 75.

The body is usually made up by running a groove along the middle of the bottom, hence the length of A B on the pattern for body will be equal to the length of line measured from *a* around to *b*, as shown in the elevation.

The length C D for the lid-top pattern will be equal to the girth *c d* measured around the lid in the elevation. The body-end and the lid-end patterns will be the same shape as the end elevation plus the allowance for moulding, wiring, etc. Trunk parts are joined together by knocking-up either on the end or side. In light machine-made boxes the knock-up is usually turned over on to the end; but in the stronger hand-made work the knock-up is generally folded over on to the side. The patterns have been marked out for the latter method, hence there is a single edge on the sides of body and lid-top patterns, and a double edge allowed around on the end patterns.

If the trunk is made of black iron, the surface of the sheet about the moulding cut is carefully cleaned and tinned on one side, so that when the corners are mitred the moulding at that part may be filled up solid with solder.

Three reduced sketches are shown on Fig. 74, explaining the method of forming a bead or moulding by hand. The sheet is first bent square, as shown in (1); it is then placed on a round bar (2), and bent over as seen in (3). In a press or moulding machine it can, of course, be put on the sheet in less than one-quarter the time by hand.

A simpler and cheaper kind of corner can be formed by cutting the sides of the body patterns straight and riveting a stout knee on the corners.

The flat parts of the box surface are usually swaged, to stiffen and strengthen them, and also to add a little ornamentation. All swaging must, of course, be done after the body and lid are shaped, but before the ends are fixed on.

In most trunk-lids a stretcher is placed, as shown in Fig. 73, and by the dotted lines *e f* in the end elevation (Fig. 75). The pattern for this can be marked out as seen in Fig. 75. The distances for the width of the stretcher pattern are taken from the end elevation. Thus, E F equals *e f*, and so for the widths of the flanges.

To additionally strengthen the lid, it is a good plan to fix two or three hoop-iron stiffeners across the inside of the lid, passing under the stretcher, bending them to the lid shape and riveting. Also, flat iron stiffening bars are usually fixed lengthways on the bottom of the box, bent square on to the ends, and riveted.

Experience seems to show, however, that there is no better surface protection for a trunk than wood battens, fixed lengthways on the outside of lid and bottom and bolted firmly, using large plate washers on the inside of box for the bolt-nuts to bear upon.

The hinges are made out of strips of sheet iron, doubled over a piece of wire the same gauge as that used for wiring around the lid. Notches are cut out of the lid pattern, as shown in Fig. 75, thus leaving the wire bare when the lid is made up. The hinges are slipped over the wire, soldered to the moulding, and then bent down and riveted to the body.

Moulded Lid or Cover.

A lamp-top, or base, lid or cover for a variety of articles, is sometimes made in the shape shown in Fig. 76. And as this kind of object brings in an important principle, in determining the form of the moulding on two of the cover sides, we will explain the setting out of the moulding section and the pattern for the complete cover.

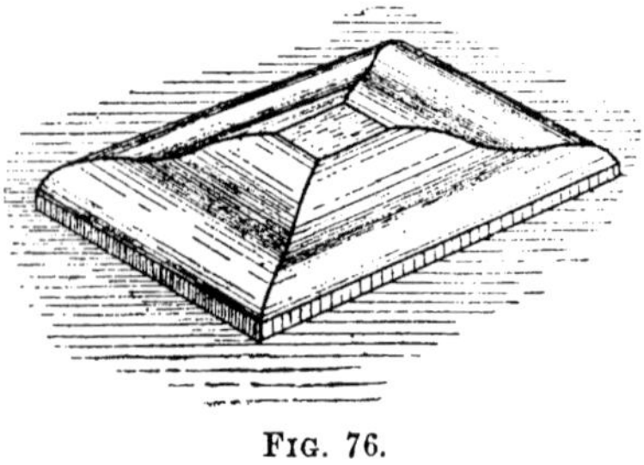

FIG. 76.

A plan of the cover is given in Fig. 77, on which the moulding section for the ends and the projected section for the sides is also shown.

If the width of the end-moulding (0 e) had been the same as the width of the side-moulding ($0'$ e'), then the same shape of section could have been used for both, and the pattern marked out as explained in a previous chapter. But in all cases where the widths are not the same, the shape of moulding for either end or side (whichever is fixed) must first be set out, and the other projected from it. In this way a proper mitring of the corners can be effected.

In Fig. 77 the moulding section on the plan is first marked out as required, this being drawn upon the middle line 0 *e*, or in any other convenient position. The curve of the section is then divided into five equal parts, the division points being numbered 1, 2, 3, etc. Through these points perpendiculars to 0 *e* are drawn, and produced to meet the joint line E O. Then, from each of the points of intersection on this line projectors are run along square to, and through, the middle line 0′ *e*′. To obtain the shape of the projected section, the heights are cut off equal to the length of lines on the moulding section. That is, *a*′ 1′, *b*′ 2′, *c*′ 3′, etc., are respectively set up equal to *a* 1, *b* 2, *c* 3, etc. The curve drawn through the points 0′, 1′, 2′, etc. (called the " projected section ") will give the shape

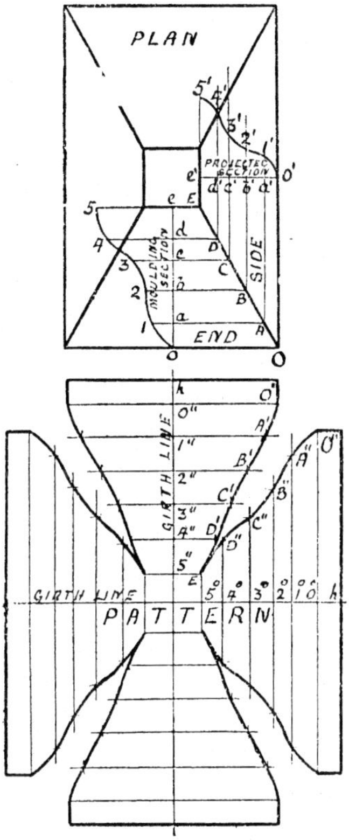

Fig. 77.

of moulding for the side that will exactly mitre on to the moulding as set out for the end.

The pattern for the cover is laid out by drawing two lines square to each other, and along these stepping the respective girths of the two mouldings. Thus, the lengths $5''$ to $4''$, $4''$ to $3''$, etc., will be made the same as the lengths 5 to 4, 4 to 3, etc., on the moulding, whilst the distances 5^0 to 4^0, 4^0 to 3^0, etc., will be the same as $5'$ to $4'$, $4'$ to $3'$, etc., on the projected section. The lengths $0''$ *h* and 0^0 *h* will, of course, equal the breadth of the rim around the cover.

The small square in the middle of the pattern will be the same size as that in the centre of the plan. The lengths of the construction lines for the pattern will be measured from the middle lines 0 *e*, $0'$ e', on the plan, up to the joint line E O. Thus, lines $0''$ O', $1''$ A', $2''$ B', etc., respectively equal 0 O, *a* A, *b* B, etc., and lines 0^0 O'', 1^0 A'', 2^0 B'', etc., are equal, respectively, to $0'$ O, a' A, b' B, etc.

Even curves are drawn passing through each point, and the sheet can then be cut out to the shape as shown.

In making up into the cover shape, care must be taken that the two mouldings are bent to their respective forms, and then there will be no difficulty in jointing. The principle involved in the above example is an important one and worth taking notice of, as it applies to all classes of moulding, beading, etc., where two different sizes have to be jointed together.

Sheet-Metal Kerb Fender.

The work involved in the making of a sheet-metal kerb fender is of such an elementary character that the ordinary workman or amateur craftsman should find very little trouble in making one up to his own design and liking.

A simple form of kerb is shown in Fig. 78, which will give some idea of the shape into which the sheet metal is to be bent.

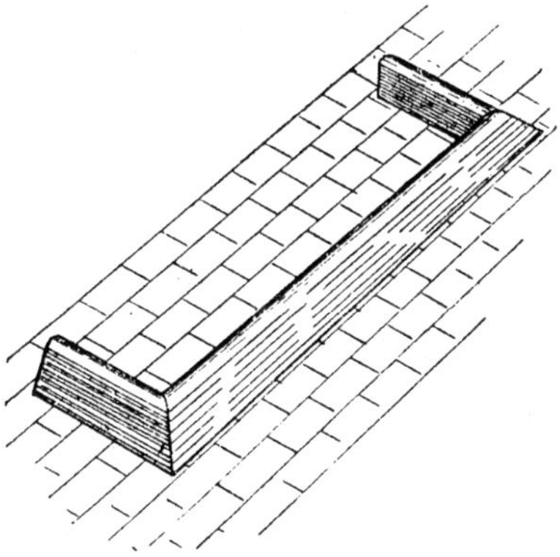

FIG. 78.

The kerb may be made out of hammered or polished sheet copper or brass, or even be made out of plain sheet iron, and afterwards blacked or japanned. The setting out of the patterns for a very simple shape is shown in Fig. 79.

The form of the section, it will be seen, is drawn on the plan of the fender, this being afterwards divided up into four parts, 0 to 1, 1 to 2, 2 to 3, and 3 to 4. The girth line 0′, 1′, 2′, 3′, 4′, on the pattern is first laid out, the lengths of the different parts being taken from the corresponding lines on the section. Lines are then drawn square to the girth line through each division-point, and these cut off the same length as the corresponding lines drawn through the division-points on the section parallel to the front of fender, and running up to the joint lines C B, A 0. On

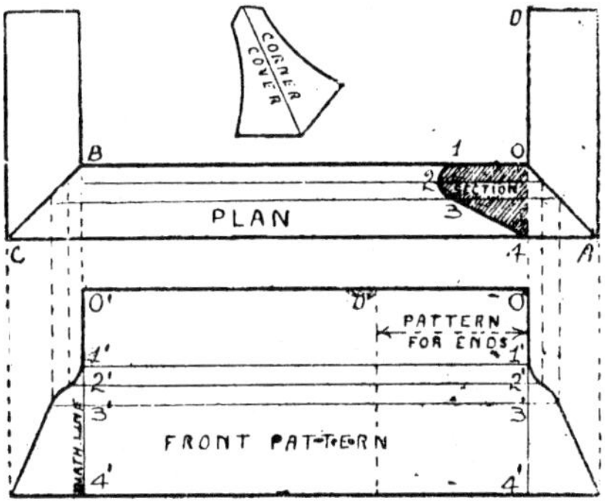

FIG. 79.

Fig. 79 the lengths are shown cut off by the dotted lines projected from the plan on to the pattern.

The cut on the pattern for the end will, of course, be the same as that for the front, hence the one pattern will do for the two parts. The length for the end will be obtained by making 0′ D′ equal to 0 D.

Without the kerb is made of strong sheet metal, it will be necessary to have a wooden core, of the section shape, to which the edges of the sheet metal can be nailed.

A corner cover is sometimes used, and if this is required, some such shape as that shown in Fig. 79 can be adopted, it being bent up over the joint and fastened along its edges with nails having ornamental heads.

If a corner cover is not used, the mitring will have to be done very carefully if the fender is to look neat; but in the event of a cover-plate being attached, there will be no need to bother with accurate joining at the corners.

If the fender is made of copper or brass, then any degree of ornamentation in the way of repoussé work can be put upon it, depending upon the skill of the craftsman and the time at disposal in making.

CHAPTER XII.

CONICAL ARTICLES OF SHORT TAPER.

It is probable that of all the articles that are manufactured out of sheet or plate metals, the larger proportion are conical or circular equal-tapering in shape. It is, therefore, essential that a careful study should be made of the various methods that can be used to obtain the patterns for this class of article.

The simplest form of a conical-shaped object is that of a cap, as shown in Fig. 80. This is, of course, a complete cone, and in this shape is applied in the formation of ventilator and stove-pipe caps, pan-lids, pointers, strainers, candle-extinguishers, etc.

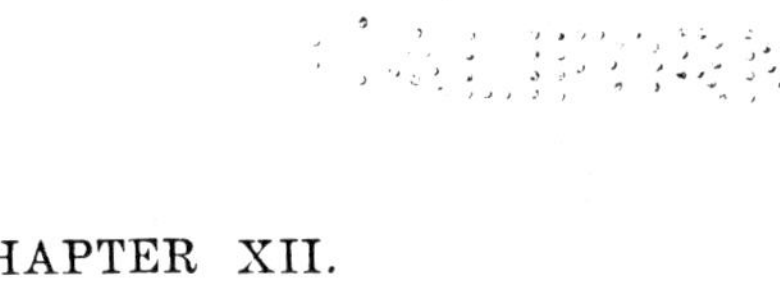

FIG. 80.

The pattern for a conical surface is perhaps one of the easiest patterns that can be developed. Imagine a cone to roll on a flat surface as in Fig. 81, and that as it rolls along, the base of cone marks the outline as shown. Now if the joint line of the cone first of all lies on the line A C, and the cone is then rolled around until

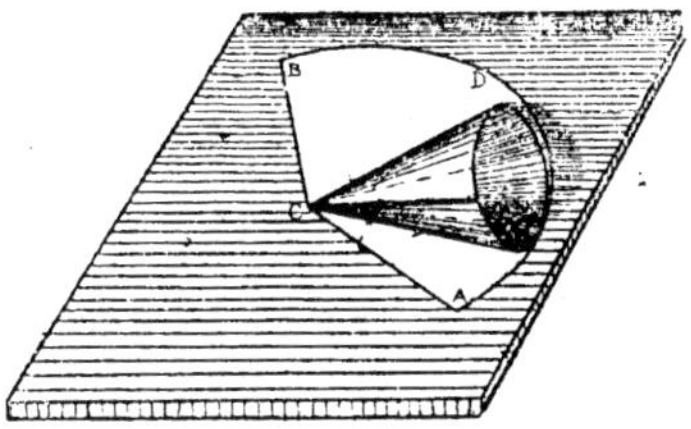

FIG. 81.

the joint comes on the flat surface again, say on the line B C. then it is evident that the whole of the curved surface of the cone will have been in contact with the flat surface, and the sector of circle so marked out will be the development of the cone surface. The radius of this sector of a circle will, of course, be equal to the slant height of the cone, and the length of the arc A D B will be of the same length as the circumference of the base of cone. Thus, in Fig. 82, suppose the diameter of base of cap is 3 in., and slant height 2 in., then it is evident that the radius for describing the pattern will be 2 in. The length of the arc can be marked off in two or three different ways, as will now be shown. On the base line of half-elevation of cone construct a quarter-circle, and divide it into three equal parts, and carefully measure the lengths of one of these parts, and set it along the pattern curve twelve times. Join the points so found to the centre, and allow laps for grooving, riveting, or soldering as shown. Care must be taken that the lap lines are parallel to the end lines of net pattern.

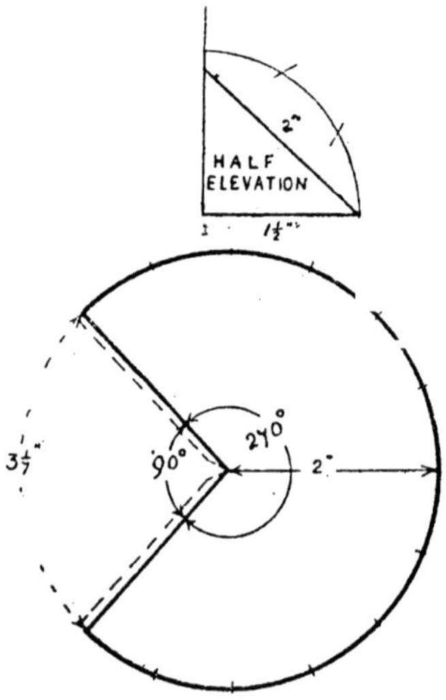

FIG. 82.

The length of curve on pattern can be quite easily calculated from the following rule:—

"Multiply diameter of cone base by $3\frac{1}{7}$."

Thus, in the above example, the length of curve will be:

$3 \times 3\frac{1}{7} = 9\frac{3}{7}$ in. This length is best set along the curve by a steel tape measure, piece of thin wire, or strip of sheet metal.

Sometimes it is convenient to calculate the length of arc

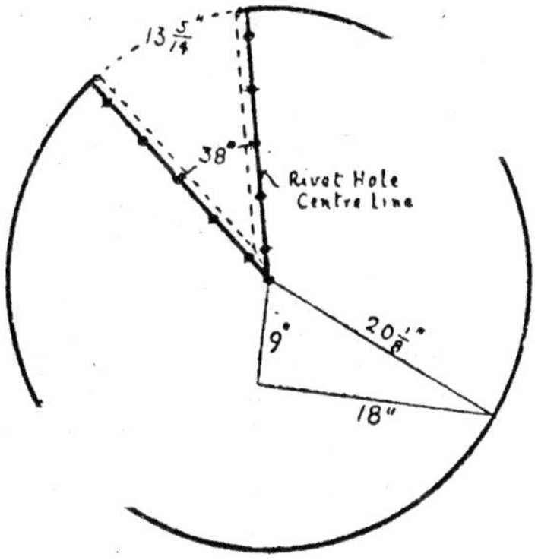

FIG. 83.

on the piece cut out, and this can be done by either of the following rules :—

(1) Deduct the diameter of base of cone from twice the slant height, and multiply the remainder by $3\frac{1}{7}$.

(2) "Multiply the difference of the diameters of pattern and base circles by $3\frac{1}{7}$."

By the use of either of the above rules it will be seen that the length of arc of the sector to be cut out in Fig. 82 will be—

$$(4 - 3) \times 3\frac{1}{7} = 1 \times 3\frac{1}{7} = 3\frac{1}{7} \text{ in.}$$

The cut circle (Fig. 83) shows the pattern for a conical cap the vertical height of which is 9 in. and diameter 3 ft. The slant height of cone or the radius of pattern can be calculated by bringing in the property of the right-angle triangle, previously mentioned, thus—

$$\text{Slant height} = \sqrt{9^2 + (18)^2} = 20\tfrac{1}{8} \text{ in.}$$

Having marked out the pattern circle to this radius, the length of arc to set along circumference to cut piece out can be calculated thus—

$$(40\tfrac{1}{4} - 36) \times 3\tfrac{1}{7} = 4\tfrac{1}{4} \times 3\tfrac{1}{7} = 13\tfrac{5}{14} \text{ in.}$$

The end lines on the pattern for a conical cap may also be set out by the use of degrees. The following rule will give the angle that the end lines make with each other:—"Multiply 360 by the radius of the base, and divide by the slant height." Thus in Fig. 82 it will be seen that the angle—

$$= \frac{360 \times \text{radius of base}}{\text{slant height}}$$

$$= \frac{360 \times 1\frac{1}{2}}{2} = 270^\circ.$$

Sometimes it is more convenient to mark the angle on the piece that is to be cut out, and the degrees for this can be calculated by aid of the following rule:—"Deduct the radius of base from slant height, and multiply the remainder by 360 and divide by the slant height." Thus in Fig. 82 this angle will be—

$$360 \times \frac{\text{slant height} - \text{radius of base}}{\text{slant height}}$$

$$= 360 \times \frac{(2 - 1\frac{1}{2})}{2} = \frac{360}{4} = 90^\circ$$

In Fig. 83 the number of degrees in the piece to be cut out—

$$= 360 \times \frac{20\frac{1}{8} - 18}{20\frac{1}{8}} = 38^\circ$$

The above examples are given to show the application of

this particular method. The rules can be applied in all cases, and will usually give more accurate results than by measuring along the circumference of pattern circle.

Workshop Protractor.

In workshop practice it is a good plan to have a protractor or bevel with which to set out angles. A useful protractor can readily be made out of sheet brass or aluminium, as in Fig. 84. The semicircle being about a foot in diameter. The protractor, as shown, is divided into 10° divisions; but if it be made 12 in. in diameter there should be no difficulty in subdividing into divisions of 1°. The distance apart of the division lines on the circumference would in this case for 1° be about 1-10 in. apart.

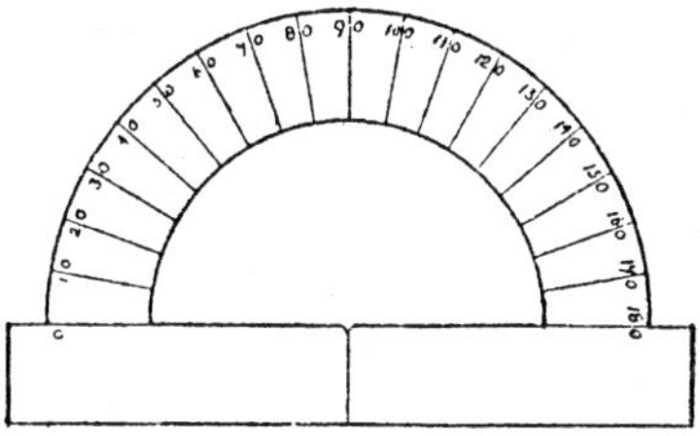

FIG. 84.

Special Conical Shapes.

Before passing from conical caps there are one or two peculiar shapes that give interesting results that are worth while specially considering. Thus, in the case of a cone in which the slant height is equal to the diameter of base, the pattern develops out to an exact semicircle. When the diameter of base is equal to half the slant height, the pattern will be just a quarter of a circle. In Fig. 82, where the diameter of base is one and a half times the slant height, it will be seen that the pattern comes out to three-quarters of a circle. Several other interesting cases can no doubt be discovered by the reader by the careful choice of dimensions for the conical cap.

The Pointer.

An old-time but useful article is the pointer or ale-warmer, shown in Fig. 85. It is an exceedingly handy form of vessel for sticking into a fire, and rapidly heating any kind of thin liquid. It can be readily made out of either tinplate or copper. Where first cost is no great consideration copper will be the better metal, on account of its longer life and superior heat-conducting properties.

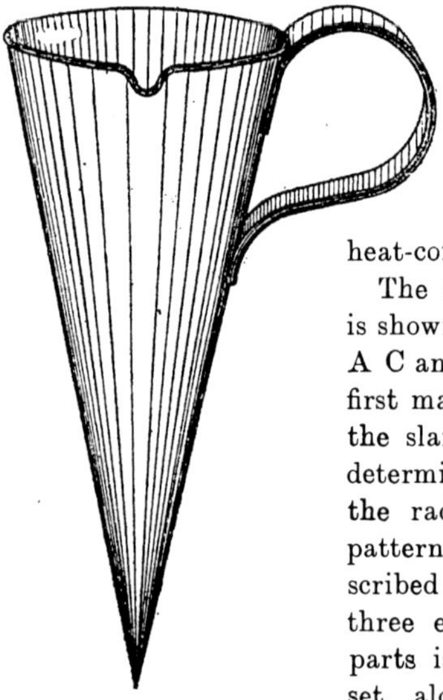

FIG. 85.

The setting-out of the pattern is shown by Fig. 86. The depth A C and half the width A B are first marked out at right angles, the slant height C B thus being determined. This is now used for the radius in marking out the pattern. A quarter-circle is described on A B, and divided into three equal parts. One of the parts is carefully measured, and set along the pattern curve twelve times, the points thus obtained being joined up to the centre C. An allowance must be put on top for wiring and on sides for grooving, as shown by dotted lines. The pattern should be notched for wiring and the laps carefully cut away at C to allow the groove being turned over at the point. If the pointer is made of copper the inside should be properly tinned, notice being taken that the point has a little tin left in it to obviate leakage. If made of tinplate it will be necessary to solder down the inside of seam, and also at the point. To those who have facilities for brazing, it is worth while remembering that a brazed joint will be much the best job if the ale-warmer is made out of copper.

Whilst it is usual to fix a handle, as shown in Fig. 85, it is certainly a distinct advantage to have a handle of the saucepan kind. This can be made by bending a strip of copper or iron into the form of a tapered tube, and flattening the end to fit on to the side of vessel. Iron will be the best metal on account of being a better non-conductor of heat, and also stronger than copper; but, if appearance is the first consideration, then, of course, copper should be used. The handle could also be made by bending a piece of flat or half-round iron or copper tube. Whatever kind of handle is chosen it should be riveted on to the body of the pointer.

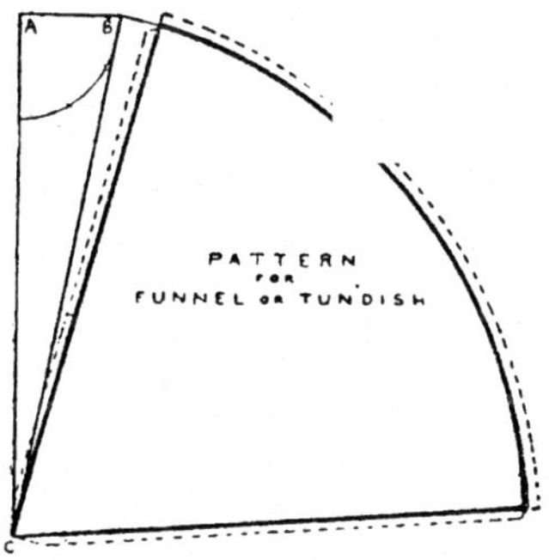

FIG. 86.

Articles Formed from Cone Frustums.

Most articles that are circular, equal-tapering in shape, do not take the form of a complete cone, but come out as a frustum of a cone—that is, the shape that is obtained when the top of a cone is cut off parallel to the base. The bodies of the bulk of tapering articles such as buckets, funnels, coffee-pots, wash-ups, and a host of others are of this character, and their patterns are obtained by considering the surface to be that of a frustum of a cone, or a truncated cone, as it is sometimes called. The shape referred to is that shown by Fig. 87.

To obtain a pattern for this class of vessel is, generally, a very simple matter. The method adopted is to draw a half-elevation to the given dimensions, then produce the slant height until it meets the centre line, and thus obtain the slant height of the cone of which it forms a portion. Having done this, the pattern for the complete cone is set out, and the part cut away that belongs to the top portion of the cone. Thus in Fig. 88, the half-diameter of wide end A B is set along, and the vertical depth A E drawn down square from it. The half-diameter of narrow end E D is marked from E square to A E, or parallel to A B. The slant line B D is produced until it meets the centre line of cone in C. This will, of course, give the top or apex of complete cone. To mark out the pattern the compasses are fixed on C as centre and C B as radius, and the outer curve drawn as shown, the inner curve being drawn with C D as radius. To mark the length along the outer curve a quarter-circle is set on A B and divided into three equal parts, one of these parts being marked along six times (the body of article being made in two pieces). The first and last points are joined up to centre C, and thus the net pattern is determined. It will be noticed

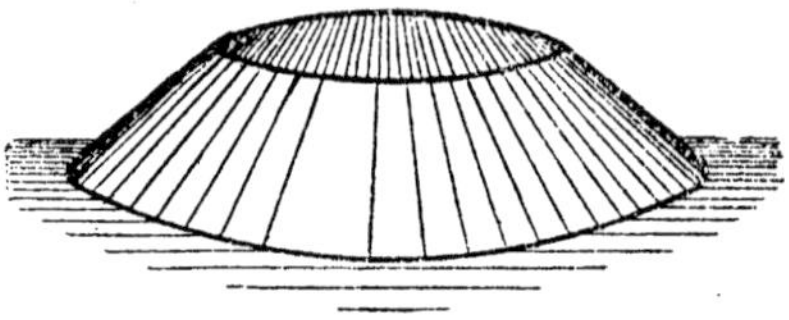

FIG. 87.

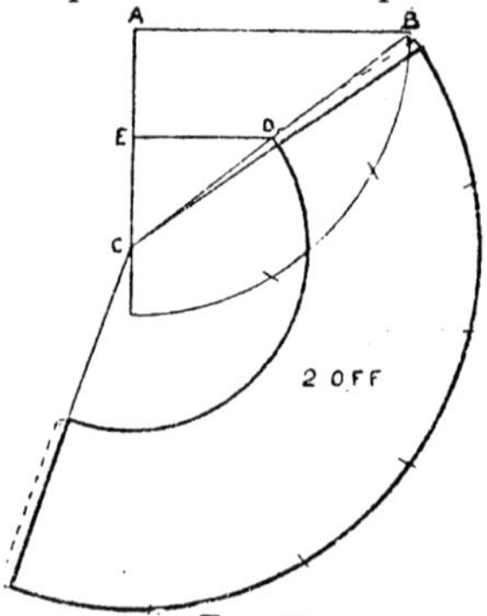

FIG. 88.

that no length is measured along the inner curve, this being cut off proportionately by the end lines. In some cases it is most convenient to mark the length along the inner curve (which, in the above example, would be equal to half the circumference of small end of vessel), join the points so found to the centre, and produce the lines out to cut the outside curve. If measured out accurately the resulting pattern should be the same in both cases; but generally, in practice, it is the best plan to mark the length along the outer curve.

The number of pieces in which the body of any article is made will depend upon its size and shape, the considerations being the economical cutting up of the sheet or plate, and the work on bending or conveniently shaping the plates into their proper form. On the other hand, some thought must be bestowed on the number of joints, or else the extra work on the joints will more than balance the saving in material. No particular rules can be given, as each job must be decided upon its own merits.

The laps for grooving or riveting should be added on to the net pattern, the lap lines being drawn parallel to the end lines. Allowances for wiring, knocking-up, or other form of joint, as required, can be added on, as mentioned in other chapters.

The position of a joint in an article is of some importance. In circular objects the joint is usually the weakest part: hence it is nearly always arranged so that it shall be covered by an ear, lug, or handle, to give it additional strength. In some articles it is arranged for the joint to be at the back, so that, for the sake of appearance, when the article is standing in its position the joint will be hidden.

To Obtain Pattern Lines by Calculation.

It is useful to know that the lengths of lines required in setting out the pattern for any conical article may be obtained by calculation, without previously having drawn an elevation of the object. An example of this method is shown by the calculations for the pattern in Fig. 89. Suppose the article to have the following dimensions: Top 9 in. diameter, bottom 2 ft. diameter, and depth 1 ft. 6 in.; then the height of complete cone, of which the vessel is a part, can be calculated as follows:—

$$\text{Height} = CD = \frac{\text{radius of base} \times \text{depth}}{\text{radius of base} - \text{radius of top}}$$

$$= \frac{12 \times 18}{7 \cdot 5} = 28 \cdot 8 \text{ in.}$$

And slant height of cone—that is, radius of outer curve of pattern can be found by the use of the property of the right-angle triangle, thus—

$$CB = \sqrt{(28 \cdot 8)^2 + (12)^2} = 31 \cdot 2 \text{ in.}$$

The radius for the inner curve of pattern can be found as above, or from the following rule:—

$$CA = \frac{\text{radius of top} \times \text{slant height of complete cone}}{\text{radius of base}}$$

$$= \frac{4 \cdot 5 \times 31 \cdot 2}{12} = 11 \cdot 7 \text{ in.}$$

The length round outer curve of pattern will, of course, be—

$$24 \times 3 \cdot 1416 = 75 \cdot 4 \text{ in.,}$$

so that it will be seen that the complete pattern of the article, whether it is made up in one or more pieces, can be set out from calculated dimensions. This method is especially useful in large work, or where a high degree of accuracy is required.

Capacity of a Conical Vessel.

Seeing that we have the above calculations before us, it will be as well to go over what is usually considered to be a somewhat difficult task—that is, to find the capacity of a circular tapering vessel. The simplest plan to adopt is to calculate the volume of the complete cone, and then to deduct from it the volume of the small cone, which we can imagine is cut off the top.

The rule for finding the volume of a cone is:—" Multiply

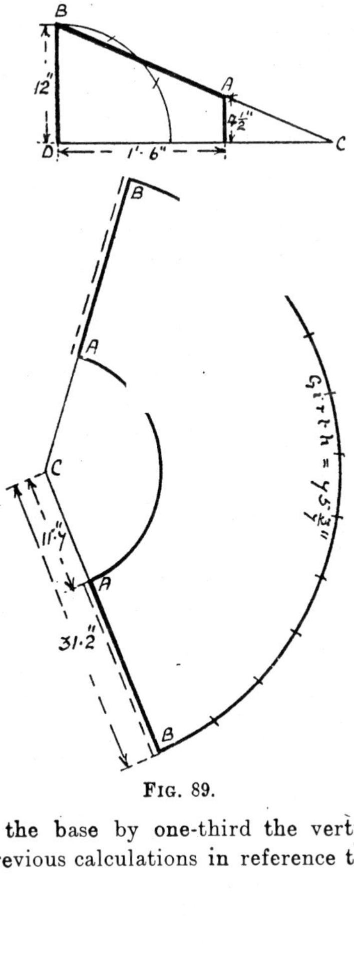

FIG. 89.

the area of the base by one-third the vertical height." From the previous calculations in reference to Fig. 89, we have both the height of complete cone and of the small cones cut away. The volume of the vessel then is—

$$(12)^2 \pi \times \frac{28 \cdot 8}{3} - (4 \cdot 5)^2 \pi \times \frac{10 \cdot 8}{3}$$

$$= \frac{144 \pi \times 28 \cdot 8}{3} - \frac{20 \cdot 25 \pi \times 10 \cdot 8}{3}$$

$$= 1{,}309 \pi = 1{,}309 \times 3 \cdot 1416 = 4,\ 114 \text{ c.in.}$$

The number of cubic inches in an imperial gallon is 277·274. This is a most awkward number to use; but as it has been fixed by Act of Parliament (in 1826) as the volume of a gallon, we have to make the best of it. The capacity of the above vessel will therefore be—

$$\frac{4{,}114}{277 \cdot 274} = 14 \cdot 84 \text{ gal.}$$

The capacity may also be calculated, but not quite so accurately, by remembering that a cubic foot contains as near as possible $6\frac{1}{4}$ gal. Thus—

$$\frac{4{,}114}{1{,}728} \times 6\tfrac{1}{4} = 14\tfrac{7}{8} \text{ gal. (nearly).}$$

As a gallon of fresh water is usually taken to weigh 10 lb., the weight of water in the vessel will be—

$$14 \cdot 84 \times 10 = 148 \cdot 4 = 148\tfrac{1}{2} \text{ lb. (nearly).}$$

Funnel Patterns.

The setting out of the patterns for a funnel is illustrated by Fig. 90. The half-elevation is drawn in the usual manner, and the slant heights of the cones forming the different portions thus obtained. The lettered lines on the various patterns correspond to the lines with the same letters that are taken from the elevation to form the radii in drawing out the curves. A thimble, as D E, is often soldered on to the spout and body to give additional strength; it looks much better if hollowed a little, as shown by the dotted line. A small flange is allowed for on the spout for soldering inside the body of funnel; there is no need for this, though, if a thimble is fixed over the spout

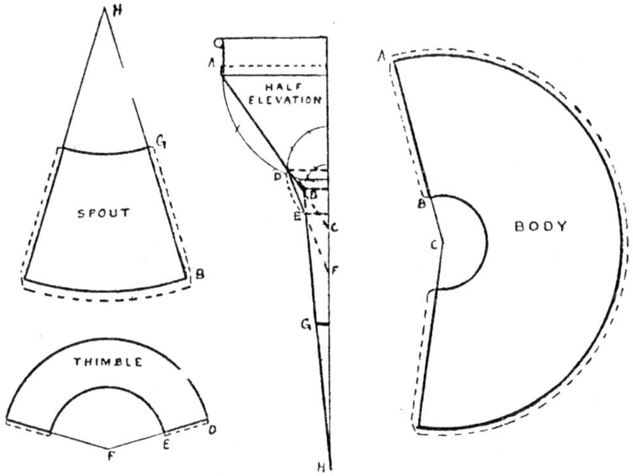

FIG. 90.

and body joint. A flute is often creased in the spout to allow the air to pass out of the vessel whilst being filled. No pattern is shown for the rim, as this is but a straight strip.

Half-Round Tapered Article.

There are a great number of articles whose shapes are built up by portions of a cone surface and that of some other solid or plane figure. One such is sketched in Fig. 91. This is a semicircular-ended conical vessel with a flat back, made to fit against a wall as shown. The pattern for it is set out as explained by Fig. 92. The half-elevation A B C D is first drawn to the required dimensions; this figure also giving the shape of half the flat back. The cone part of the vessel is developed in the ordinary way, using O as the centre. The remaining half of the back can be added on to the other end of the pattern in a variety of ways. This is as good as any:—With centre O and radius O A draw the dotted curve A *a*, and with centre *b* and radius equal to A B draw a small arc cutting the dotted curve in *a*; join this point to O. With centre O and radius O D draw the inner dotted curve and thus determine *d* as shown. Join *d* to *c* and *a* to *b*, and thus the net pattern is completed. Add laps as required for grooving, knocking-up, and wiring; notch as in figure, and the working pattern is complete.

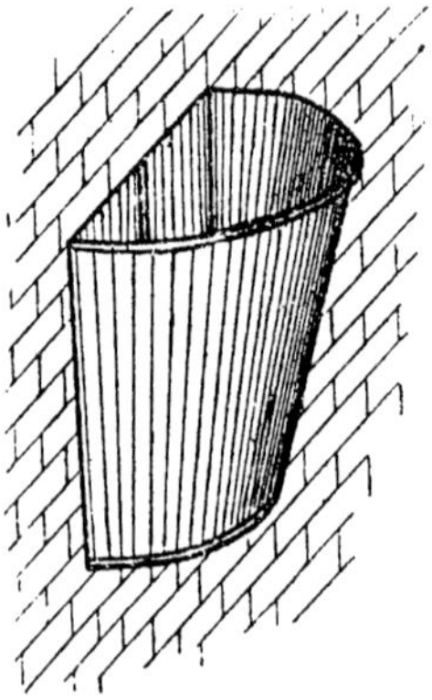

FIG. 91.

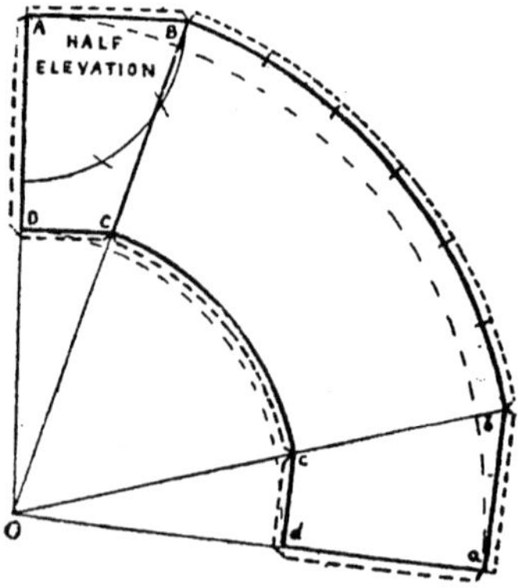

FIG. 92.

Wiring.

In wiring an edge, care should be taken to draw the one end of wire along from the end of sheet, and to let the other end in, so that the joint in the wire will not coincide with the sheet-joint. This will be best understood by reference to Fig. 93.

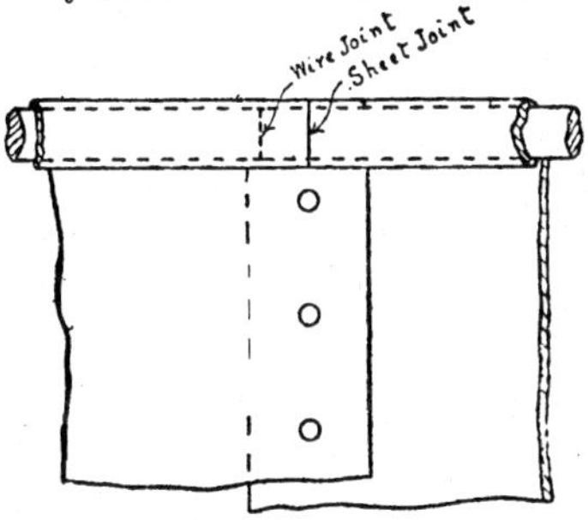

FIG. 93.

Conical Plate Work.

For plate work of a conical character, the centre line of the section should be taken as the slant height of the conical part required for developing the pattern. This is illustrated by line A B in Fig. 94. If the edges of the plates

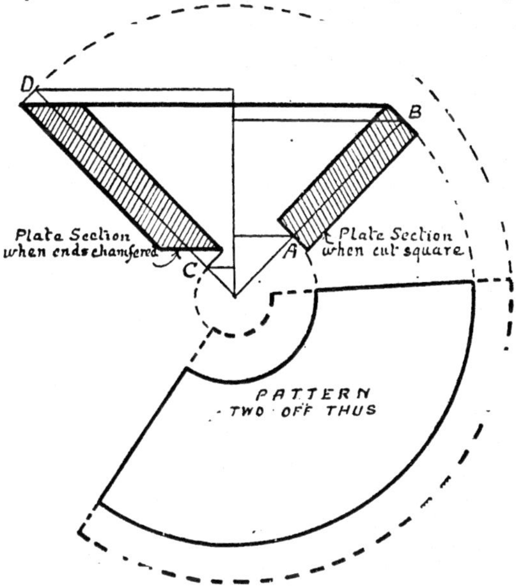

FIG. 94.

are to be chamfered, as shown by the section of plate on the right-hand half of figure, then the line C D will have to be used for the width of pattern; the plates being sheared along the two dotted curves. This is, however, a somewhat exceptional case, and has a very limited application.

If the lengths of lines required for the pattern of plate are calculated, as in Fig. 89, then the whole of the calculations must have reference to the centre line of the plate section. This centre line really represents the section of an imaginary cone which passes through the centre of the plate.

As explained in Chapter XXXII, the greatest care must be taken to properly allow for the thickness of metal.

CHAPTER XIII.

CONICAL VESSELS OF LONG TAPER.

To develop the pattern for a circular article which has very little taper, by the ordinary method, is somewhat inconvenient in practice on account of the long radius required. A way that it can be done, and a plan that is often adopted, is by the

Method of Triangulation.

This is nothing more or less than the method used by surveyors in measuring-up the exact shape and area of land. The sheet and plate metal worker should be most familiar with the application of this system to the scores of cases that crop up in his own particular line. The use of this method of triangulation is a plan that can be followed in obtaining the shapes of patterns for any and every kind of job where it is possible to obtain the development of a pattern. Also, in some cases, where the pattern is not strictly developable, it will give us considerable aid in obtaining an approximation. It is not by any means in all cases the shortest way of getting out a pattern; but this defect is more than compensated by its universal application. Essentially, the method consists in dividing up any surface, for which a pattern is required, into a series of triangles and then obtaining the true lengths of the sides of each triangle and plotting or setting-out their true shape in the flat. If the three sides of a triangle be given, then one shape of triangle only can be marked out from these. Thus, suppose three links, A B, B C, and

C D (Fig. 95) are hinged together to form a triangle, and one link, say A C, is held fast, then we shall find it impossible to alter the shape of the triangle by pushing it either one way or the other. Thus, it will be impossible to move the two sides so as to cause B to come into the position B^1: hence the triangle will remain of constant shape.

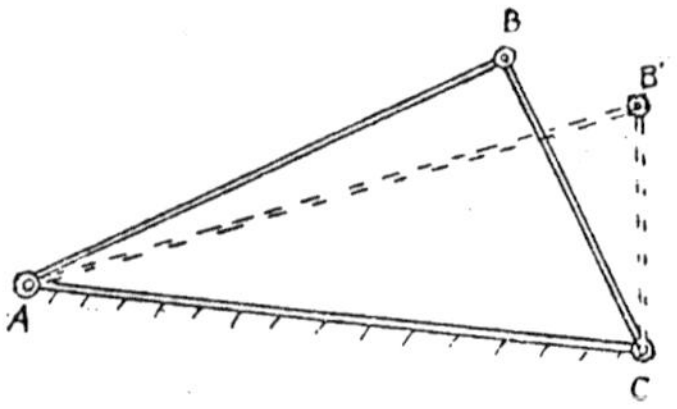

FIG. 95.

To illustrate the above, suppose we wish to reproduce the area (*a*) in Fig. 96, either full size or to scale. Divide the figure up into triangles as in (*b*). Now carefully measure the sides of the triangle A, and reconstruct it as in (*c*), then obtain the lengths of the two remaining sides of the triangle B, and thus construct this triangle on A. Continue the process by adding triangles C, D, E, etc. These triangles will give a series of points, all of which will lie on the curved outline. Join the points together with an even curve, and the figure (*a*) will be reproduced in (*c*).

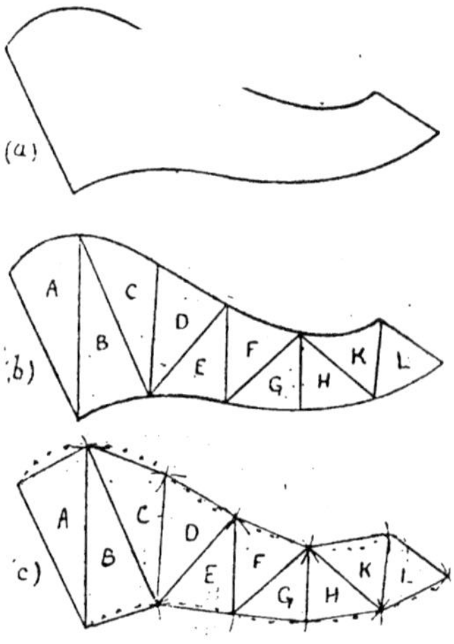

FIG. 96.

In applying the above kind of work to the development of patterns, the operation is not quite so simple, as before the pattern can be laid out, some construction work is necessary to obtain the true lengths of the sides of the triangles.

A practical application of the method will now be shown in connection with the setting out of a pattern for a conical vessel of long taper. A sketch of the article is given (Fig. 97) showing lines drawn on the surface to represent its division into triangles. In this case the method is quite easily applied, as only two different shaped triangles are required—viz., *a o* 1 and *a o b*. In practice the plan followed for setting out the pattern is shown in Fig. 98. A half-elevation and a quarter-plan are drawn as indicated, the latter consisting of two quarter-circles respectively representing a quarter of the top circle and a quarter of the bottom. These quarter-circles are divided into three equal parts, and numbered and lettered as shown. Join *a* to 1, and using *a* as centre swing this length on to the base line. The length A 1 will then

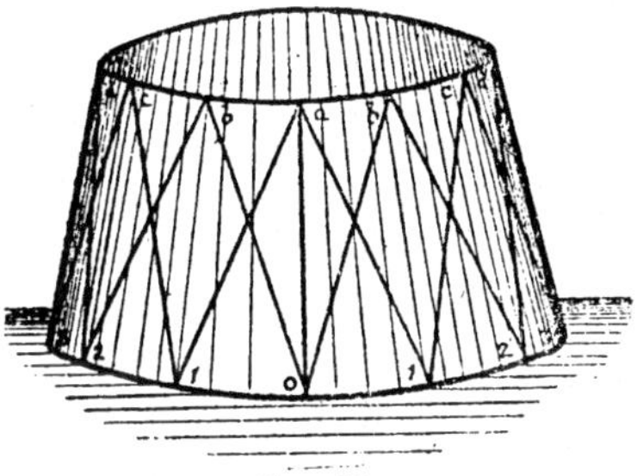

FIG. 97.

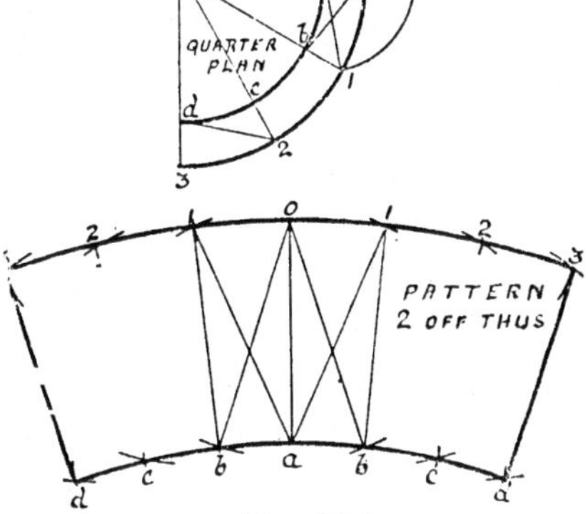

FIG. 98.

give the true length of the diagonal line required in setting out the pattern. To mark out the pattern, draw a centre line, *a o*, and make it equal in length to the line A *o* in the elevation. Set the compasses to a radius equal in length to the curve *o* 1 in plan, and from centre *o* on pattern mark arcs as shown, and with radius equal to curve *a b* do the same from centre *a* on pattern. Now stretch out compasses to length A 1 in elevation, and from centres *o* and *a* cut the arcs in 1 1 and *b b*. In the same manner determine points 2, 3, and *c d*. Join the points up with an even curve and the net pattern is complete. Allowances for jointing, etc., can be added as required. It should be remembered that in practice there is no need to draw any lines on the pattern, those shown being put there to illustrate the method. It is most important that lengths of lines should be found with some degree of accuracy, or else the resulting pattern will not be of much use. This system of "triangulation" will be further explained in connection with more difficult patterns in subsequent chapters.

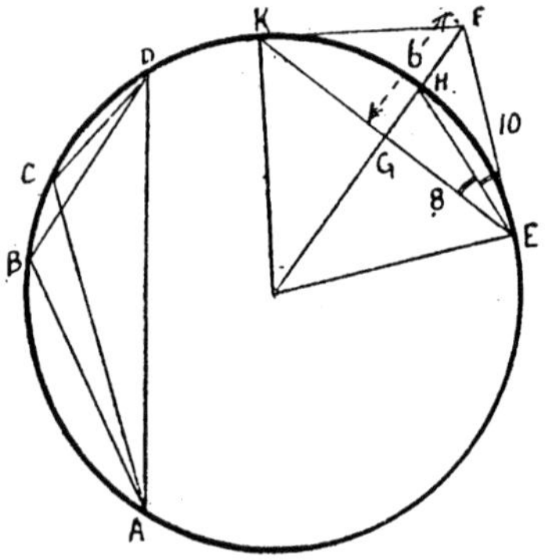

FIG. 99.

Segment of Circle Method.

There are one or two methods that might be of much use in marking-out patterns for circular equal-tapering articles, but which, unfortunately, are little known in practice. They depend upon a few important properties of the circle, which we will mention before proceeding to show their application in the setting-out of patterns.

Referring to Fig. 99, in the segment A D C B it will be found that if points such as B and C be joined to the extremities of the chord A B the angles A B D and A C D will be equal; that is, "angles in the same segment of a circle are equal," and conversely if pairs of lines be drawn all containing the same angle, as in Fig. 103, then the points of intersection of the lines will lie on the arc of a circle. The way this can be used in practice is explained by Fig. 100. If two strips of timber or two lengths of hoop-iron be screwed or riveted together to form a bevel, and one arm of the bevel be allowed to slide along the nail at A and the other along the nail at B, then the scriber or pencil which is held against the joint will mark out an arc of a circle A C B as shown.

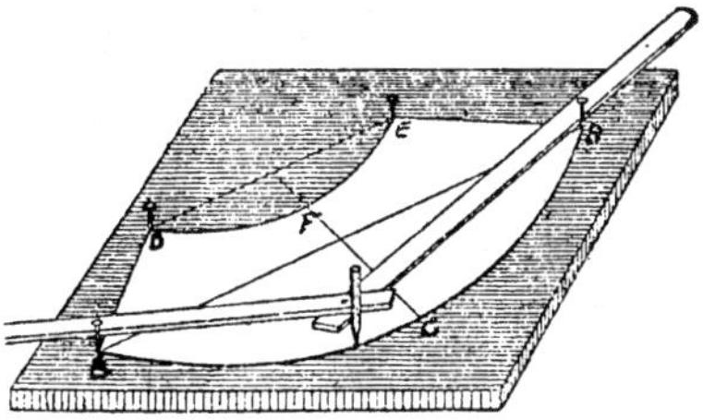

FIG. 100.

Referring again to Fig. 99. If a chord, E K, be drawn, and a tangent, E F, then the line E H, which is drawn to the middle point of the arc E H K, will bisect the angle between the chord E K and the tangent E F. It is also useful to remember that the line G F is divided by the line E H into two parts in the ratio of E F to E G. Thus, suppose E F = 10 and E G = 8. Then—

$$FH = \frac{GF \times EF}{EF + EG} = \frac{6 \times 10}{18} = 3\tfrac{1}{3}$$

And—

$$GH = \frac{GF \times EG}{EF + EG} = \frac{6 \times 8}{18} = 2\tfrac{2}{3}$$

From this it will be seen that if we have the lines K E and G F or K E and E F, a third point, H, can always be determined which lies on the arc of a circle. This particular property we shall find of much use in setting out patterns for circular articles of long taper, and we shall be able to obtain the pattern without the use of the pattern circle centre.

Fig. 101 indicates a pattern so set out. A centre line is drawn as shown, and the girth of the large end set along the line A B, half being measured along each side of the centre line. Then with centres A and B and radius equal to the slant height of the pipe or vessel, arcs of circles are drawn. Two lines are now set down parallel to the centre line, and at a distance apart equal to the circumference of the small end of the tapered pipe or article. Where these lines cut the arcs, as previously drawn, will give the points E and D. The line B K is drawn square to B D, and the line B C drawn to bisect the angle A B K. Thus we now have three points, A, C, and B, which will come on the curve of the pattern. Set the bevel to the angle A C B, and fix pins or nails at A and B, and slide along, thus marking the pattern curve as previously explained by reference to Fig. 100. The curve at the small end can be described in the same manner, the bevel being kept set at the same angle or rake as used at the large end. There is really no need to determine the point F, as shown by the construction lines, the bevel giving the correct height of curve. It must not be forgotten that the

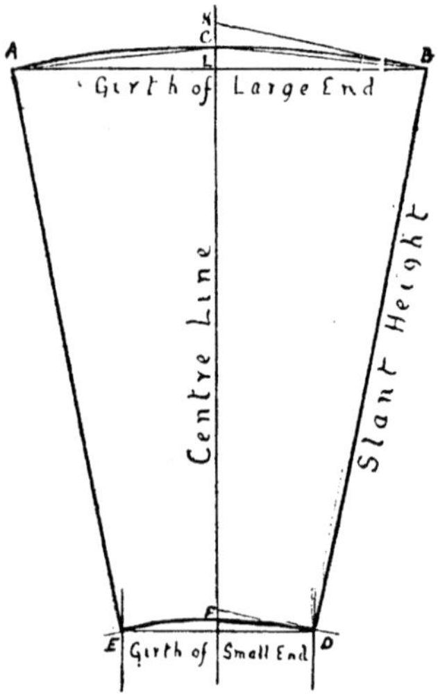

FIG. 101.

length of the bevel arm should not be less than the line A B. When the article has little taper, the point C will come as near as possible at the middle point of L K; there will, therefore, be no need to bisect the angle A B K. The pattern will come slightly wider than it ought to be on account of the curve being a little longer than the girth lines. But when the difference between the end diameters is small, and the article long, it will be quite near enough for ordinary work in practice.

The exact size and shape of pattern, however, can be marked out by this method, and the reader who can follow a few simple calculations will readily understand the construction involved. The essential thing to determine is the exact length of the chord A B (Fig. 102) which is required to give the correct length of the arc, A C B. To obtain this, let us take, for the sake of clearness, an actual example. Suppose we require to set out a plate for a pipe 4 ft. diameter at one end, 3 ft. diameter at the other, and the slant length 5 ft. The slant height of the complete cone, of which this is a frustum, can be obtained from the following rule:—" Multiply the diameter of the large end by the slant height, and divide by the difference of the diameters." Thus, the slant height of cone equals—

$$\frac{\text{large diameter} \times \text{slant height}}{\text{difference of diameters}} = \frac{4 \times 5}{4 - 3} = 20 \text{ ft.}$$

The angle made by the two outside lines of the pattern with each other can be found by the following rule:—" Multiply the diameter of large end of the pipe by 360, and divide by twice the slant height of cone." Thus, the angle between end lines of pattern equals—

$$\frac{\text{large diameter} \times 360}{\text{twice slant height of cone}} = \frac{4 \times 360}{2 \times 20} = 36°.$$

The angle between the chord A B (Fig. 102) and the end line, B D, of pattern can be calculated from this rule:—

"Deduct half the angle between end lines of pattern from 90°." Thus, angle between chord and end line equals—

$$90 - \frac{36}{2} = 72°$$

Having obtained the above particulars, let us set out the pattern, first going over a construction which will give us the correct length of A B. Draw a line across, and mark a point B anywhere upon it. Now set off a line B D at 72°, and make it 5 ft. in length. Drop a perpendicular on A B from the point D which will cut off a length H B. The half-length of the chord A B will be equal to—

$$\text{L B} = \frac{\text{H B} \times \text{slant height of cone}}{\text{slant height of pipe}}$$

$$= \frac{\text{H B} \times 20}{5} = 4 \times \text{H B},$$

which means to obtain the point L three additional lengths of H B must be marked along from H. The line L F is drawn square to A B, and points A and E determined. To scribe out the curves, we shall first have to obtain the angle at which to set the bevel. The rule for this is: "Deduct half the angle between the end lines of pattern from 180°."

$$\text{Bevel angle} = 180 - \frac{36}{2} = 162°.$$

Set the bevel at this angle; fix pins at A and B, and slide along as in Fig. 100. To mark the curve at the bottom of the pattern, fix pins at E and D and slide bevel along, keeping it at the same angle as used for describing the curve at the top.

Those readers who can understand the use of mathematical tables will be able to calculate the lengths of A B and E D in a simple manner, and thus set out the pattern quite easily. Referring to a table of chords, the chord of 36° is given as 0·618. The length of A B will therefore be—

$$0·618 \times 20 = 12·36 \text{ ft.,}$$

and—

$$E\,D = 0·618 \times 15 = 9·27 \text{ ft.}$$

For the sake of comparison we will calculate the length of the arc A C B, which will, of course be—

$$4 \times 3·1416 = 12·57 \text{ ft.}$$

Thus the difference in the length of the chord A B, and the arc A C B, will be as near as possible 2·5 in. Hence, where accurate work of the above description is required, it will be necessary to follow the last-named method. Fig. 102 (*a*) shows the various angles set out; but in practice the only lines required are those on the pattern, as indicated in (*b*).

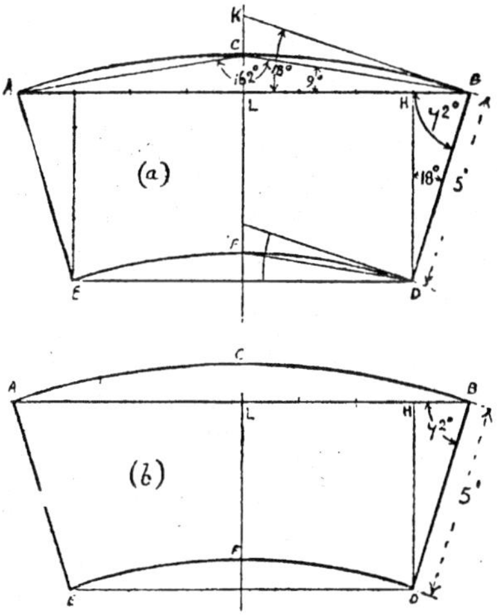

FIG. 102

If it is required to build an article up in several pieces, as in large plate work, the pattern can be subdivided when set out, or a pattern for the required segment can be marked out by either of the methods explained.

When the angle in the segment of the circle is determined, it is sometimes convenient to obtain a few points that would lie on the curve, and then join them up by bending a lath of timber along the points and scribing along. Points that would lie on the curve can be obtained in the way illustrated by Fig. 103. Thus, suppose the angle in the segment is 100°, then, as the three angles in a triangle are together equal to 180°, the sum of the base angles must be—

$$180 - 100 = 80^\circ.$$

Thus, if a line making 10° with A C is set at one end, and one at 70° with C A at the other, then the point of intersection of the two lines will give a point on the curve. In the same way, further points can be obtained by marking

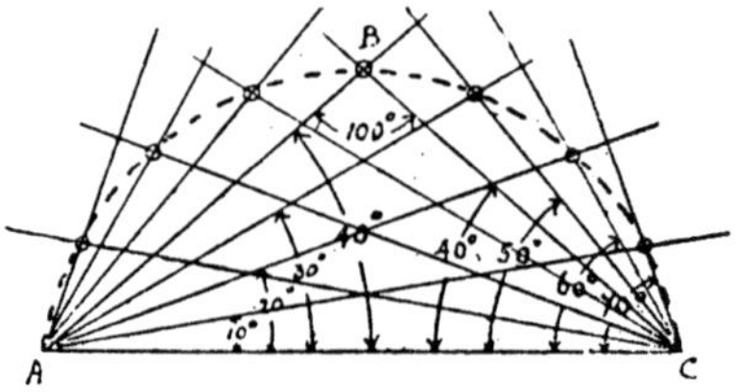

FIG. 103.

angles of 20° and 60°, 30° and 50°, and so on, as explained by the diagram.

Possibly this chapter is a little more difficult to follow than the preceding ones, on account of the calculations introduced; but the reader who is interested should make an effort to understand all that has been stated. It is exceedingly important to the sheet and plate metal worker, especially to the latter.

CHAPTER XIV.

PART CONE SURFACES.

There are a great number of articles and parts of articles whose patterns can be developed as some portion of a cone surface. To explain the method that is followed in setting out patterns for this class of work, it will perhaps be as well to first go over the development of the surface of a

Cone Cut Obliquely.

This is shown in Fig. 104, the base of the complete cone being marked 0 6, and the apex C. The part of the cone

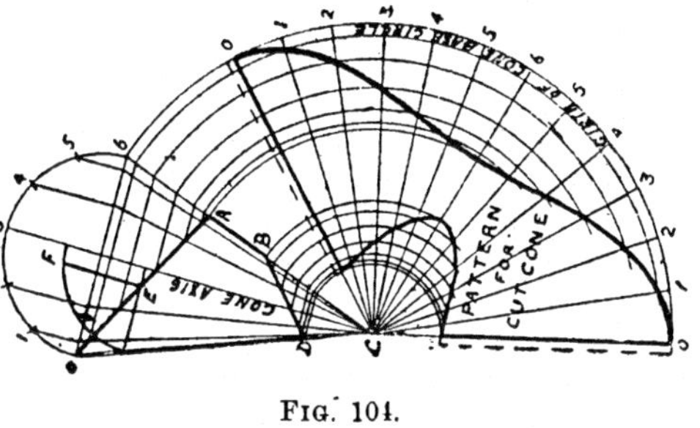

Fig. 104.

for which a pattern is required is shown by the thick outline 0 A B D. It is best to imagine the base circle of the cone divided into twelve equal parts, and lines drawn to the apex from each of the division points. The cone being cut away at both ends will limit the surface lines to a definite length, and these we can obtain the true length of, and thus determine the shape of pattern that would exactly envelop the cut cone. This will be carried out by first describing a semicircle on the cone base, and dividing it into six equal parts; running lines down perpendicular to the base from each division point, and then joining up to the apex of cone. From the points where the radial lines intersect 0 A and D B, perpendiculars to the axis are drawn to the outside line of the cone; thus the true lengths of all the lines are projected on to the outside of cone, and can be measured off as required.

Careful observation should be taken of this method for obtaining the true lengths of lines, as it is applicable to all cases in which the surface of the article comes out as any portion of a cone.

In setting out the pattern, the development of the complete cone is first drawn (Fig. 104), the girth of the cone-base circle either being calculated or marked along by using one of the parts of the semicircle twelve times. Whilst there will be only twelve lines on the surface of the cone, it will be seen that there are thirteen on the pattern, the two outside lines coinciding to form the seam when the sheet or plate is bent into shape. After the radial lines are drawn, their required lengths can be marked off by taking the distances already projected on to the side of cone. Thus, to follow one point only, the line which is brought from 4 on the semicircle down to the cut on cone and then run to the outside will, when swung on to the pattern, cut off points on line 4. So with all the other lines. When the points are joined with a free curve, the net pattern is complete.

For purposes of shaping, it should be noted that the slant ends of the conical pipe are elliptical in form, their respective long diameters being 0 A and D B. To obtain the short diameter of the large ellipse 0 A will be bisected in E, and a line drawn through the point square to the centre line of the cone. On this line a quarter-circle is described, then the line E F which is drawn parallel to the axis of cone will give half the short diameter of the ellipse. This method for obtaining the width of the cone at any part should be taken particular notice of, as it very often comes in useful in getting out shapes of holes in the flat.

Circular Spout Pattern.

The pattern set out in Fig. 104 has a large application in spouts for all kinds of articles, such as tea and coffee-pots, tin kettles, water-cans, etc. One simple application is shown in Fig. 105, the setting out of the pattern being practically the same as in the former case. The line C 4 on the spout is made the same length as C 0. The line 0 4 is taken as the base of the cone, a semicircle being described upon it and divided into four equal parts. Perpendiculars are drawn from each point on to 0 4, and then the radial lines from the apex of the spout cone to meet the body of the article. From each point where the radial lines meet the body, lines square to the centre line are drawn to the outside lines of the spout, these giving the lengths required for the pattern, which is marked out as shown.

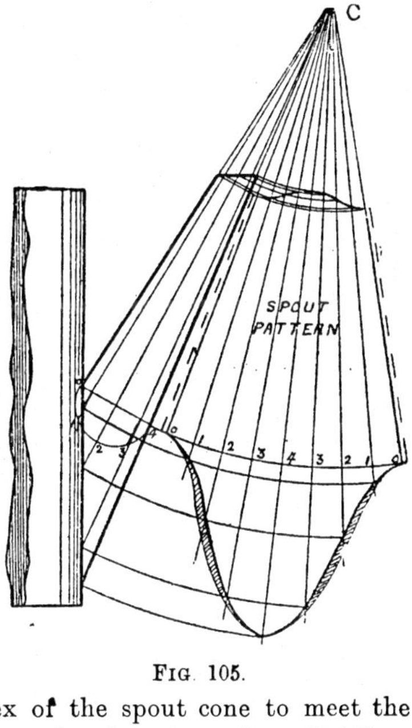

FIG. 105.

When the spout is rounded up, the end should come elliptical and lie level against a flat surface. If the body of the article is circular, this shaped end of spout will, of course, not be correct, openings being left on each side. If the body be of large diameter and the spout comparatively small, then the defect will be inappreciable; but if otherwise, then another method must be adopted which will give a spout fitting around the part of body about hole, and thus leave no opening. This further method, which is somewhat difficult, will be shown in connection with other work later on (see page 141).

The spout pattern, as struck out in Fig. 105, can, after a little experience in setting out, be quickly altered to suit a round body by the addition of parts something like those shown by the shaded strips.

It will be noticed that the seam is arranged to come on the top side of spout; if required on the underside—the centre line of pattern would then become the outside line of same.

Round Hopper on Pipe.

Another interesting application of the cut cone is that of a hopper, as shown in Fig. 106. This is really the junction of a cone and a circular pipe, the axis of the cone being perpendicular to, and also meeting, the centre line of the pipe.

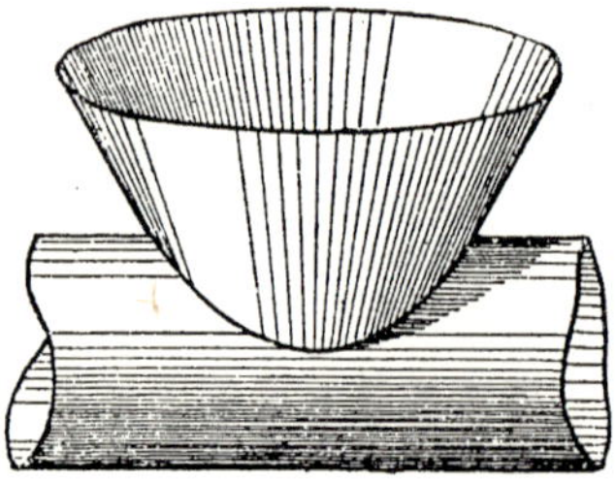

FIG. 106.

The striking out of the pattern for the hopper and the obtaining of the shape of hole in the pipe-plate are illustrated by Fig. 107. A half-elevation of the pipe and cone is drawn, and on the base of cone a quarter-circle is described. This is divided into three equal parts, and lines run up from each of the points, and then joined to the apex of the cone. Where these lines intersect with the bottom part of the pipe, lines square to the cone axis are run to the outside line. To mark out the pattern, the full cone is first set out in the usual way, and the radial lines drawn as shown. All the lengths to form the cut are taken, as before mentioned, from the outside line of the cone in elevation. Thus, lines C 0, C1, etc., on the pattern will be the same length as the lines bearing similar distinguishing marks in the elevation. As each quarter of the pattern is exactly the same, some of the lengths, it will be seen, are used four times over. In large work a part pattern like the shaded portion is all that is necessary to set out, as this for marking out the full plate can be folded four times, or the plates cut in sections to form the hopper.

Laps have been added for wiring, grooving, and stretching; but the allowances for any particular job will, of course, depend upon the size of hopper, and whether made of sheet or plate metal.

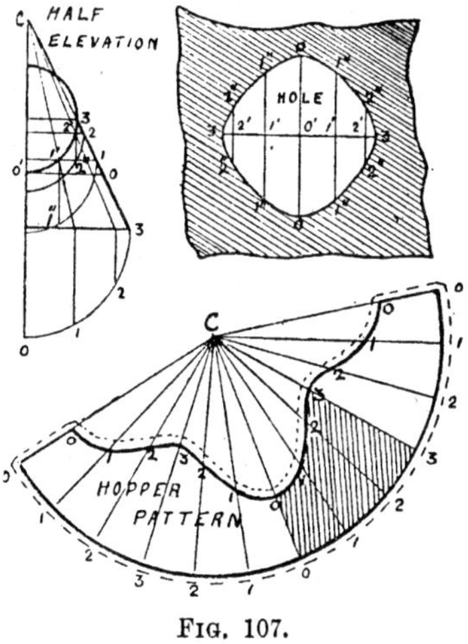

FIG. 107.

Before proceeding to set out the shape of hole in the flat, the widths required for the different parts must be first determined on the elevation. Thus the width of hole at the middle of pipe will be equal to the diameter of the cone at that part; that is, twice the line 0′ 0 which is drawn across the half-cone. The width at 1′ will be found by drawing a line across through the point as shown, and on it describing a quarter-circle, then drawing a line down parallel to the axis of cone, this line giving half the width of hole at that point. A similar construction will give the width at point 2′.

Now in marking out the hole (Fig. 107), the length 33 is made up by the three arcs with the same numbers on the pipe in elevation. Cross lines are drawn, and the widths, as previously obtained, set above and below the line 33. There should be no difficulty in following the setting-out of the hole, as the lines on it are figured in identically the same manner as those from which they are obtained in the elevation.

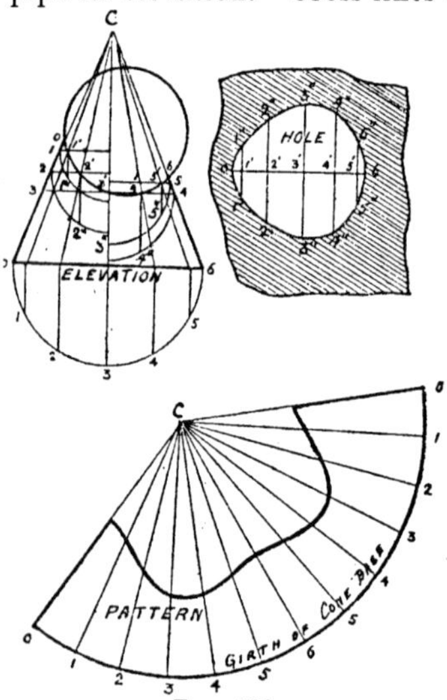

FIG. 108.

Holes are rarely set out in the flat in the shop, the cumbersome method of chiselling them out after the plate is bent very often being followed. But with a little practice holes can quickly be struck on the plates, and this will often save a lot of trouble after shaping. When the pattern is for a stock article the extra time taken in marking out the hole properly is always well repaid.

Offside Circular Hopper.

After mastering the setting-out in connection with the last hopper the reader should find no trouble with this, the methods being identical. It will be noticed (Fig. 108) that some of the lengths are marked down one side of the cone, and some down the other. The two halves of the pattern being the same, lines will be used twice over, as seen by the numbers. On account of the hopper being on one side of the pipe, the hole in the plate will be egg-shaped, as drawn.

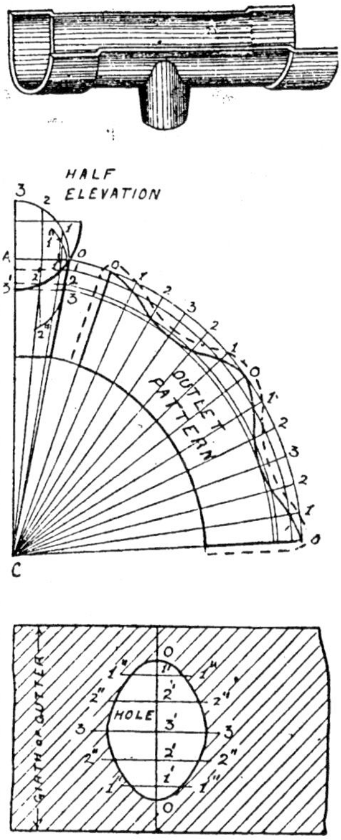

FIG. 109.

Half-Round Gutter Nozzle.

Another application of the geometry of the cone and cylinder is in the making of a pattern for an outlet or drop fitting on to a half-round gutter, as shown in the sketch at top of Fig. 109. In this case A 0, on the half-elevation, will be the base, and C A the centre line of the cone. The quarter-circle is described on the line A 0, and divided into three equal parts. Taking C 0 as the radius, the base curve of cone is swept out, and the twelve parts to make up the girth set along. The points on the gutter curve are now projected on to the outside line of cone, and then carried round to cut off the radial lines on the pattern. The points so determined are joined up with a free curve, and the net pattern completed. A lap is allowed along the top for flanging, and also strips left on the ends for grooving or riveting. In the case of galvanised sheet-iron gutters the flange thrown off will be about ¼ in., the hole being cut in the gutter, the outlet squeezed in tightly and well soldered around the flange. In gutters of heavier material, such as 16 and 14 S.W.G., the outlet is attached to the outside of gutter by fixing a ring of the same material to form a flange of about 1½ in. wide over the outlet and riveting or bolting to the gutter through, say, six holes. By carefully taking the thickness of metal into account, as explained in Chapter XXXII., the holes for gutter and ring can be set out and punched in the flat. The shape of hole in the gutter-plate is also shown marked out in Fig. 109. The lengths and widths are determined as before, a quarter-circle being drawn below the line for the half-width at 2^1, and one above the line to obtain the half-width of hole at 1^1.

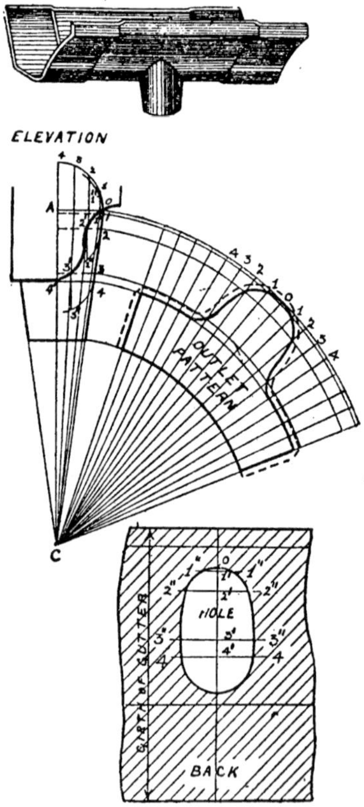

Fig. 110.

O.G. Gutter Nozzle.

To develop the shape of pattern for an outlet for this class of gutter requires perhaps a little more intricate work than in that for a half-round gutter. The method followed, however, is really the same as in that case. As before, the quarter-base circle is constructed on A 0 (Fig. 110), and, on account of the more irregular gutter curve, is divided into four instead of three parts as formerly. Sixteen spaces will therefore be required to make up the girth of complete cone. The lengths necessary to cut off the radial lines on pattern are projected as before on to the outside line of cone, and then swung on to the pattern. Thus, to find the point on the line C 2 on the pattern, the radius C 2 on the side of the cone is used. So, in the same way, the other points can be determined. The back half of outlet, it will be seen, is flat on the top; hence this part on the pattern for the four spaces at each end will be a plain curve, with C 4 on the elevation as radius. The method of jointing and attaching to gutter will be the same as with the half-round gutter.

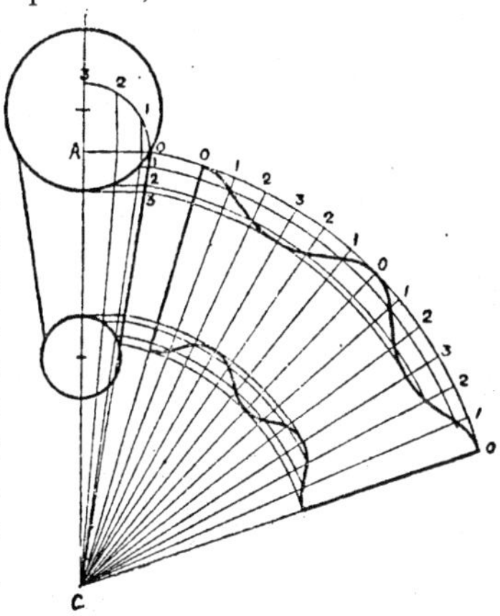

FIG. 111.

The shape of hole can be found as in the previous cases; but it should be noted that the part of hole which comes on the flat portion of bottom of gutter is a semicircle.

Tapered Connecting Pipe.

Fig. 111 shows the method of striking out a pattern for a circular-tapered connecting pipe which joins together two lines of parallel piping. The lengths of lines on the pattern are run around from the elevation as in the former cases, the same plan of lettering and numbering being followed. No laps or allowances of any kind are put on the pattern, as these can be arranged to suit the particular job in hand. The holes in the pipes are not shown set out, as these can be obtained by the method explained in connection with Fig. 107.

Galloway Water-Tube.

A conical Galloway water-tube for a boiler furnace being formed by part of a cone surface comes in very conveniently at this stage of our work. A section of flue and cross-tube is shown in Fig. 112.

To set out the pattern there is need only to draw a half-elevation of the flue-tube and conical pipe, both sides being the same. In arranging the size of the Galloway tube, care should be taken that the diameter across the flange at the small end should not be greater than the diameter of hole in flue-tube at large end, as the end of conical pipe must pass through this hole. It will be noticed that the flange at the small end fits on the inside, and the flange at the large end on the outside of the flue-tube.

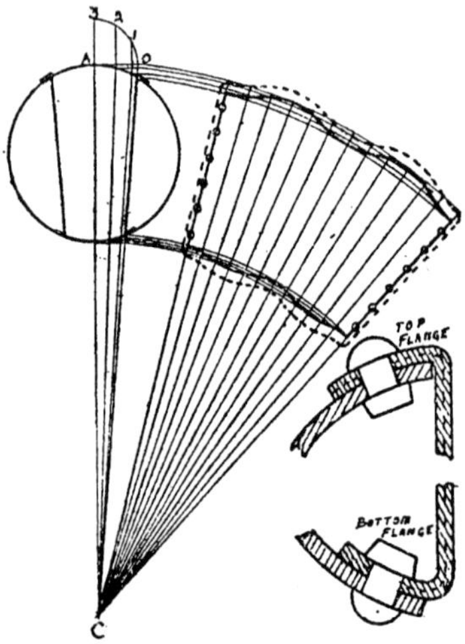

FIG. 112.

When going over plater's work, we explained that the thickness of metal must always be taken into account by using the centre lines on the plate section. In this case, therefore, to get a correct pattern, the section of water-tube with flanges must be accurately drawn, and the pattern developed from the middle lines of the plate section. The tube flanges are shown in Fig. 112, the centre dotted line on the shaded plate section representing the line from which the pattern would be set out.

The pattern is marked out as before explained, the line A 0 being taken as the base of cone. Laps will be allowed on the sides for riveting, the end lines of the net pattern forming the centre lines of the rivet-holes. Allowances are also put on the top and bottom of pattern for flanging, the width allowed being slightly greater than the length of dotted lines on flange section, to cover for draw. Some thought should be bestowed on the thinning of the plate corners, so that a good job may be made where the two thicknesses of plate come on the flange.

The holes in the flue-tube can be cut out in the flat, and if the plates are punched the rivet-holes for the water-tube flanges also put in.

In all conical work it should be particularly remembered that the lengths of lines required to fix points on the cut of pattern are taken from the *cutside* line of cone. There are many more difficult examples of cut cone work in sheet and plate metal, and a few of these wili be given in further chapters.

CHAPTER XV.

ARTICLES FORMED BY CONES CUT OBLIQUELY.

Conical Pipe Fitting on Slanting Round Pipe.

In addition to the example given in Chapter XIV. of a conical hopper fitting squarely on to a cylindrical pipe, we have yet to deal with the more difficult case of a conical pipe fitting obliquely on to a round pipe.

Two applications of this would be in the case of the hopper on the slanting pipe, as in Fig. 113, and also the foot of the coal-scoop as seen in Fig. 116.

FIG. 113.

The developing of the pattern and the setting out of the hole are illustrated by Fig. 114. Before the pattern can be marked out, a side elevation of the cone and pipe must first be drawn, and on this the elevation of the joint line, or some points on the joint line, shown. To do this, mark down the outline of the cone and pipe on the side elevation, construct a semicircle on the cone base and divide it into six equal parts, after which, run lines square to the base and on to the cone apex as shown. Now to determine points on these lines which shall be on the joint line, it will be necessary to draw the half end-elevation.

From each point on the cone base, and also from the apex, run dotted lines along parallel to the centre line of pipe. Produce the end line of pipe both up and down, and using this as a base line to measure from, cut off the dotted lines equal in length to the lines with the same number on

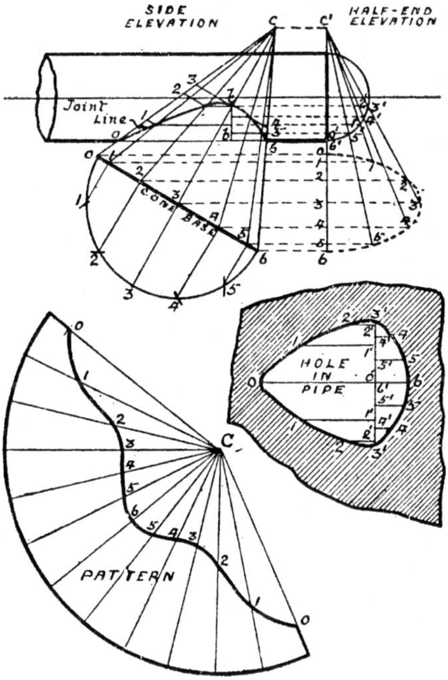

FIG. 114.

the cone-base semicircle. Thus the dotted lines on the half end-elevation numbered 1 1, 2 2, 3 3, etc., will be the same length as the lines 1 1, 2 2, 3 3, etc., on the cone-base semicircle. If the new-found points on the end elevation be joined up, a half-ellipse will be formed as shown. (There is really no need to do this in practice, as the fixing of the points is all that is required.) Join the points to C′ the apex of the cone, as seen in the figure, and from the points where the lines cross the semicircle on end of pipe draw lines along parallel to the centre line of pipe. Where these lines intersect the respective lines having the same number on the cone in side elevation will give points on the joint line. Thus, take point 4 on the half-ellipse, follow the line up towards the apex of cone, and we come to point 4′ on the semicircle. Now go along the line drawn through this point parallel to the centre line of pipe, until it intersects the line drawn through point 4 on the cone-base; this will give one point on the joint line. In the same manner the position of every other point can be followed out. In Fig. 114 the points are joined with a free curve and an elevation of the joint line thus determined. There is, however, no need to draw in this curve, the fixing of a few points being quite sufficient to enable us to obtain the lengths of lines necessary for the striking out of the pattern. Through each point on the joint line draw lines square to the axis of cone, and thus project the true lengths of lines on to the outside line of cone, as previously explained.

The pattern is set out by first marking down the development of the complete cone, dividing up into twelve parts and setting the lengths along from the sides of the cone in the side elevation. Thus lines C 0, C 1, C 2, etc., on the pattern will be the same length as the lines C 0, C 1, C 2, etc., measured from the apex down the sides of the cone on the side elevation.

For a hopper it will be necessary to lay out the shape of the hole in the pipe. The width of the hole can be obtained from the semicircle on the end of pipe, which should be set down by drawing a straight line and marking along it the lengths of arcs 0′ 1′, 1′ 2′, etc., as seen on the hole in Fig. 114. Through each of these points lines square to 3′ 3′ should be drawn. To obtain the distances to set along these we must again refer to the side elevation. Draw the line *b b* perpendicular to the centre line of pipe, and use it as a base line from which to measure in obtaining the lengths for the different parts of the hole. The centre line 0 6 of the hole will be made up by marking 0′ 0 equal to *b* 0 and 0′ 6 equal *b* 6. The line 1 1′ on the hole will be of the same length as the line measured from point 1 on the joint line up to *b b*, and so the lengths of all the other construction lines for the hole will be obtained by measuring to the right or left of *b b* up to the points on the joint line. The points thus determined are carefully connected with an even curve, and the shape of the hole thus obtained.

The reader with little knowledge of geometry will think the above a somewhat complicated case; but with care in following the correspondingly numbered lines, anyone who can use a rule and a pair of compasses ought to be able to set out pattern and hole from the description given. Anyhow, the problem is well worth studying, for in all work where circular and tapered pipes have to be joined together the same principle is involved.

Spout for Cylindrical Vessel.

In the last chapter we dealt with the striking out of a pattern for a round spout fitting on to a flat surface, and pointed out a rough way in which the pattern could be altered to suit a circular body. We will now go over a method which will give an accurate shape of pattern for a round spout of any size fitting on to a circular body of any dimension.

An elevation of spout and body is shown in Fig. 115, the spout being drawn relatively large to the body, to more clearly show the method of construction. Although the appearance of the articles in Figs. 113 and 115 are not very similar, yet the same principle in marking out the pattern underlies each case. Both come under the heading of a "cone fitting on to a cylinder."

To obtain the shape of the complete cone of which the spout is a part, the side lines are extended, C 0 being made equal to C 4. On the cone-base, a semicircle is described and divided into four equal parts; lines run up from each

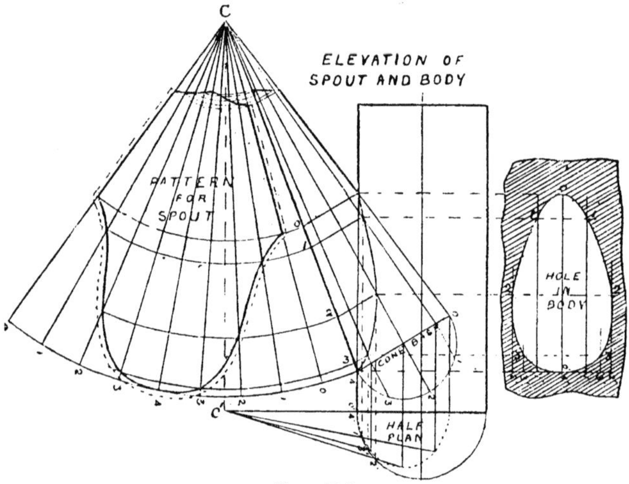

FIG. 115.

point square to the base and then joined to C, the apex of the cone. The half-plan is now drawn, C′ being the plan of the cone apex and the dotted half-ellipse representing the plan of half cone-base, the same as the half-elevation in Fig. 114. The points on the half-ellipse are joined to C′, and from the points where the joining lines cross the semicircle projectors are run up to intersect the corresponding lines in the elevation of cone. In this way the joint line of spout and body is determined. Lines from the points on the joint line, and also from the lip of spout, are drawn square to the centre line of cone, and thus all the true lengths required in marking out the pattern projected on to the bottom side of spout. The pattern for the complete cone is first set out by marking along for the cone-base girth eight lengths, each equal to the length of one of the arcs on the cone-base semicircle. The required lengths to obtain points on the cut at top and bottom of the pattern are swung around from those on the bottom side of spout.

The shape of the hole in this case (Fig. 115) is shown projected on the right-hand side of figure. The lengths 0′ to 1′, 1′ to 2′, etc., for the widths of the hole at the different parts are taken from the lengths of arc with the same distinguishing numbers on the semicircle in the half-plan.

Any allowances for seam or throw-off must be put on as shown by the thick dotted line around the pattern. Whilst it may appear that for the correct marking out of the pattern for so simple an object as a spout the work is somewhat complicated, it should be borne in mind that for stock articles it certainly pays to have patterns as accurate as possible.

Coal Scoop.

It will perhaps not be out of place at this stage of our progress to show the setting-out for all the parts in a complete article; and after having gone over the two previous cases, we shall find no difficulty in applying the same principles to the coal scoop as shown in Fig. 116.

FIG. 116.

The setting-out of all the details that go to the making-up of a small sheet-iron scoop are seen in Fig. 117. And for the benefit of amateurs and others who wish to make up such an article, all the dimensions are given. In the left-hand top

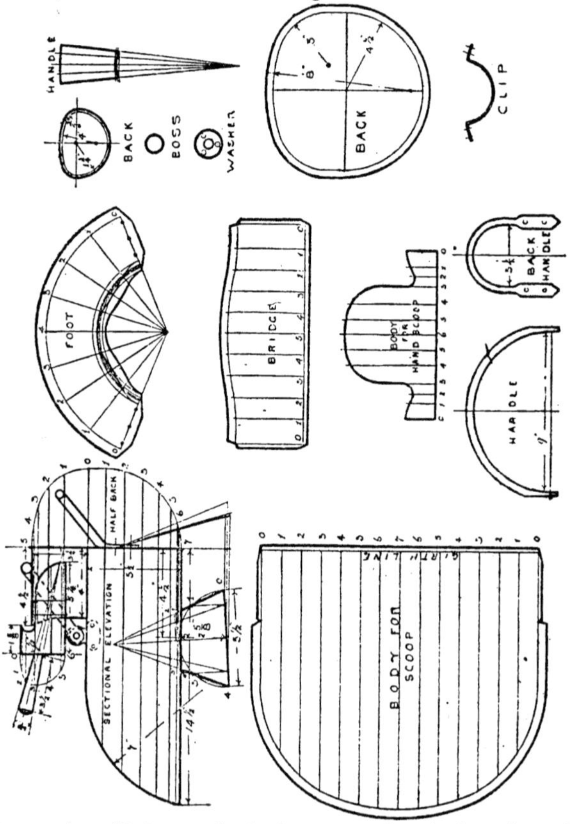

FIG. 117.

figure it will be noticed that a sectional elevation of the scoop is shown, which also includes a hand scoop.

Now for the patterns. The girth line of the body is made up by adding together the lengths of arcs as divided out and similiarly numbered on the shape of half-back shown in the elevation of the scoop. The lengths of lines to form the shape of pattern for mouth of scoop are measured from the back up to the mouth of scoop in elevation, and set along the corresponding lines on the pattern. In identically the same manner the pattern for the bridge is marked out. Allowances for wiring along the fronts of bridge and body must be made, and also for grooving at sides, and a single edge for paning down at back. It should be noticed (Fig. 117) how the body of scoop is notched where the wiring and groove come together.

The body and bridge of hand scoop are combined in one

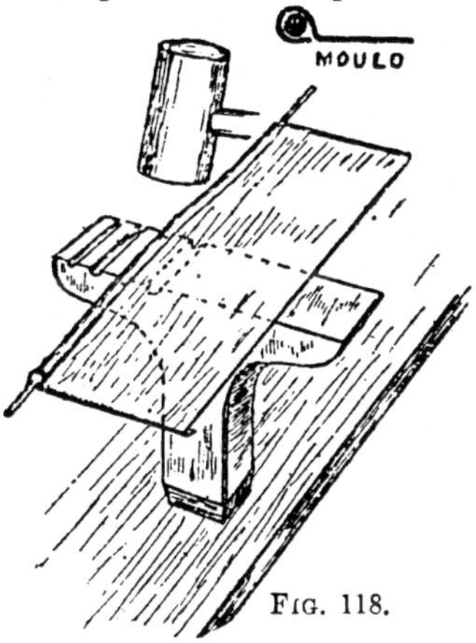

FIG. 118.

pattern, which is set out in the same way as that followed for the body of scoop.

The pattern for the foot of scoop is laid out, as explained in connection with the pattern shown in Fig. 114.

The edges of body, bridge, and foot of scoop are wired, and the back is edged over and paned down. The foot is flanged outward, and riveted to the body by four rivets, two on each side. The bridge is wired along the front edge, and then sunk into a suitable groove on the creasing iron, as seen in Fig. 118; the raw edge of the metal then coming under the bridge, as seen in the sketch at the top of the same figure. With neat wiring, however, which has the edge of the sheet properly tucked in, there is no need to reverse by creasing, as good wiring always looks bolder than creased work. It should be stated, though, that in forming a small mould to reverse wiring by creasing is, perhaps, a little quicker than to carefully shape a wired edge. In wiring the bridge a sufficient length of wire must be left overhanging each end to bend square, and come under the ears, so that the wire may be jointed at these places.

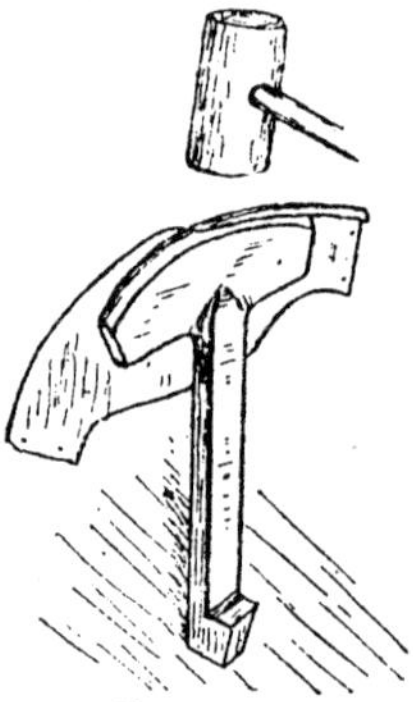

FIG. 119.

The bridge and body should be shaped by rolling or bending and then grooved together. The edge around body is afterwards edged over on a curved top hatchet stake (Fig. 119) for wiring, and the wire inserted on a side or rounding stake (Fig. 120) and properly tucked in. On the back of body an edge about $\frac{1}{8}$ in. wide is thrown off on the hatchet stake, as seen in Fig. 121.

FIG. 120.

A sketch of an

edging stake, explaining the operation of edging around the back, is shown in Fig. 122. Stakes of this description are usually made of wrought iron, the working edge being steel-faced. The edge of the stake should not be too sharp for sheet iron, or else there will be danger of the edge cracking when the back is paned down. The width of edge turned over should be about $\frac{1}{8}$ in. Care should be taken that the back is so edged as just to slip on the body of scoop.

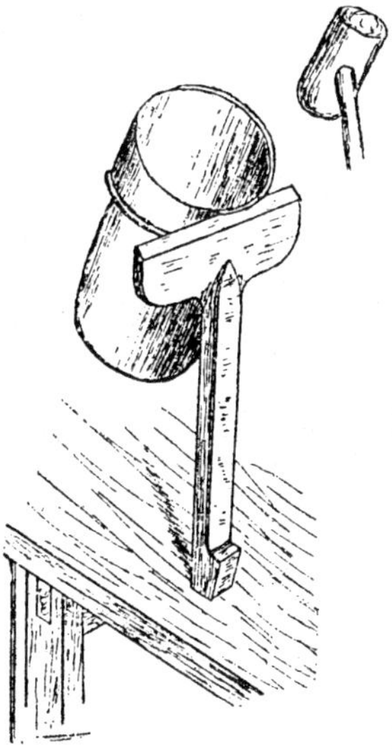

FIG. 121.

The foot is edged over for wiring, as seen in Fig. 119, and the wire run in, as shown by Fig. 120. The whole of the edging and tucking, as mentioned above, can be done in a jenny or burring machine if the operator possesses one.

The foot can be passed through the rolls, or bent on a bar to bring it into shape, and after being riveted the flanged stretched off, as explained in previous chapters.

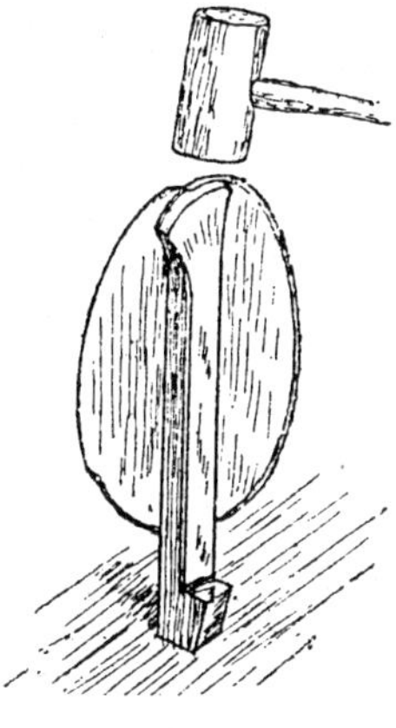

FIG. 122.

Whilst, for the sake of the professional workman, the various tools have been described as above, it is as well to point out that the amateur who is desirous of doing a little work in this way can carry out the whole of the operations on a single iron bar.

In practice, as a rule, the handles are forged out of fluted iron, but the amateur can readily form them out of, say, ½ in. brass tube. The tubes can be bent to the required shape by first loading them with lead, and after bending melting it out.

It will be a good plan to make the hand-scoop out of sheet brass, soldering the body together under the bridge, edging the back, slipping on and soldering around. Before polishing, all the superfluous solder should be carefully scraped away. To fasten the handle, the washer is slipped on the end, a small edge thrown off, and then soldered around and riveted on to the body.

The clip to carry the hand scoop is made of a strip of brass with the edges folded over, bent to the proper shape and then riveted on to the bridge of the scoop, as seen in Fig. 117.

The scoop can be japanned, gold lined, or the surface protected and decorated in any other way to suit the individual taste.

It is hardly necessary to point out that there are scores of different shapes and sizes in coal scoops of the above character; but the reader should find no difficulty in adapting the methods of setting out and working up, as explained, to a great number of the designs.

CHAPTER XVI.

HIP AND SPONGE BATHS.

Egg-Shaped Oval.

ARTICLES are occasionally required to be made up in the form of an egg-shaped oval; hence a knowledge of how to describe that figure will be useful to a sheet and plate metal worker. The construction, which should not be difficult to follow, is shown in Fig. 123. The long diameter is first marked down and the point O obtained by making A O equal to half the width of the oval. The short diameter, C D, is drawn at right angles to A B passing through the point O. A semicircle is then described upon C D. Join B to D and make E F equal to O C by running round the dotted curve and line, as on the figure. Cut off B P equal to D E, and so

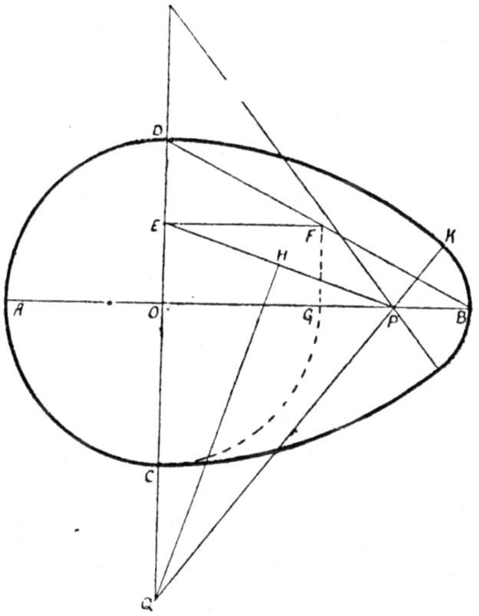

FIG. 123.

fix the point P for the centre of the end curve. Join P to E and bisect the line P E in H. Draw H Q square to E P, thus obtaining the centre, Q, for the side curve. Join Q to P and produce to K. The line Q D will now be used as the radius for the side curve and P B for the end curve, both of these curves meeting in K. The object of putting in the line Q K is, of course, to determine the meeting-point of the side and end curves. If the construction is carefully carried out there should be no difficulty about the curves meeting at the point K.

Oxford Hip Bath.

This is a good example of an article which follows the egg-shaped oval form. A sketch of the bath is shown in Fig. 124, and on careful consideration of this and Fig. 125, the reader should find no trouble in understanding the composition of the bath surface.

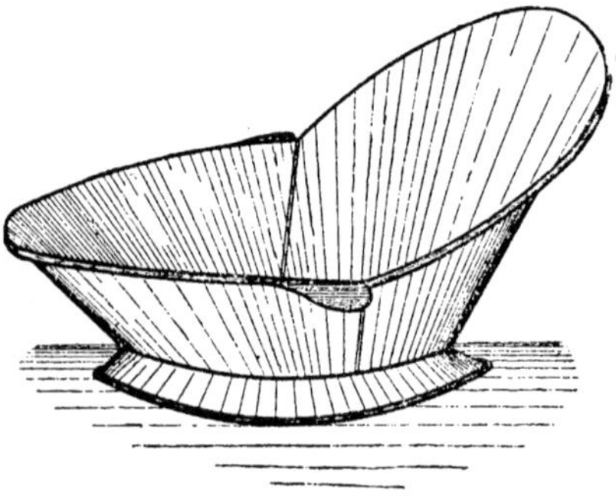

FIG. 124.

The body of the bath is usually made up in three parts, the back and two side-pieces; the joints being respectively at the two sides and down the middle of the end.

On examining the plan and elevation in Fig. 125 it will be seen that the back of the bath is formed of part of a right cone, whilst the sides and end are built up from portions of two oblique cones. To obtain the construction lines for the back pattern first produce the lines 3 *a* and *t k*, to meet in the point *c*. This will give the apex of the cone of which *t* 3 may be considered to be the half-base. On *t* 3 describe a quarter-circle and divide it into three equal parts, dropping perpendiculars from each division

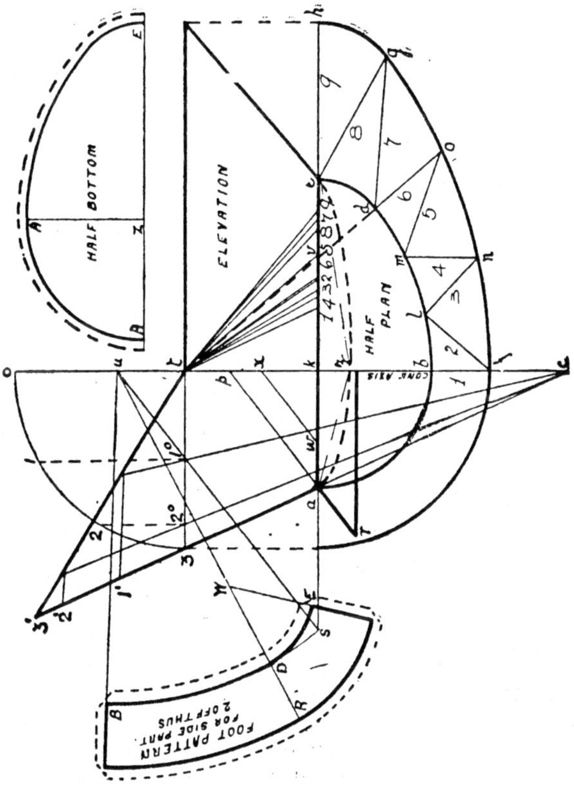

FIG. 125.

point on to the line *t* 3. Join *c* to each of the points 1° and 2° and produce to meet the top line of the back *t* 3′. The pattern for the half-cone is now developed by using *c* 3

11

as a radius, and setting along a girth line (Fig. 126) equal to twice the length of the quarter-circle in the elevation. Radial lines are then drawn, passing through C and each numbered point on the girth line; these being cut off equal to the corresponding lengths taken from the eleva-

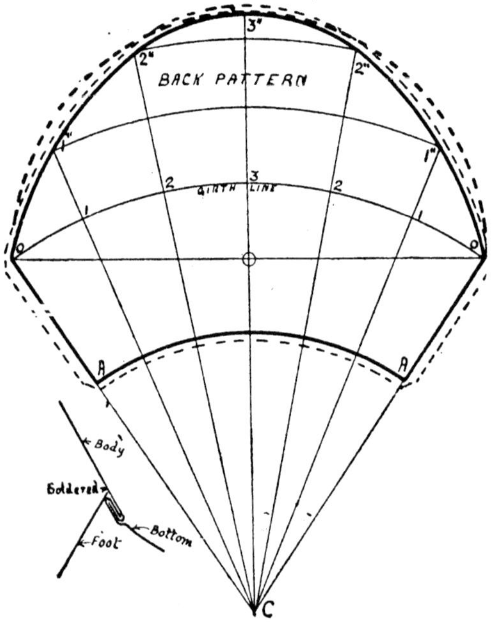

FIG 126.

tion. Thus C 1″, C 2″, and C 3″ are respectively equal to *c* 1′, *c* 2′ *c* 3′ from the elevation. The points are joined up with a curve and thus will give the outline for the top part of back pattern. The radius C A for the bottom part will be taken from *c a* in the elevation. Allowances are put on the pattern for an edge, to which the bead is attached, a lap on the sides for grooves, and a single edge for the knock-up around the bottom. When worked up, the back, as set out, will come level across the top; in practice, however, the shoulders are brought round a little, and to do this the upper portion of the back pattern is very often formed by describing a semicircle on the line 0 0. The thick dotted line shows this semicircle on the pattern in Fig. 126. It will be seen that this latter method is much easier for marking the back pattern out, and gives a bolder look to the bath when made.

If it is required to make the bath so that the top of the back is to come some other shape, then all that is necessary is to draw an elevation of the particular shape, instead of the line $t\ 3'$. The construction lines $c\ 1^0$, etc., would then be run up to meet this curve.

Instead of marking the side pattern (Fig. 123) out by the methods shown in Chapter XVIII. in connection with oblique cone surfaces, it will be simpler, in this example, to strike it out by the general method of triangulation. Turning again to the plan and elevation in Fig. 125, the curves $b\ d$ of the bottom and $f\ g$ of the top are each divided into three equal parts, and the lines numbered 1, 2, 3, etc., drawn, thus dividing the plan of the side and half-end into eight triangles. Imagine these are the plans of triangles, which lie on the surface of the bath, and it will readily be conceived that the pattern for this part can be obtained by adding together the true shapes of the eight triangles. To get the true lengths of the sides of the triangles set each

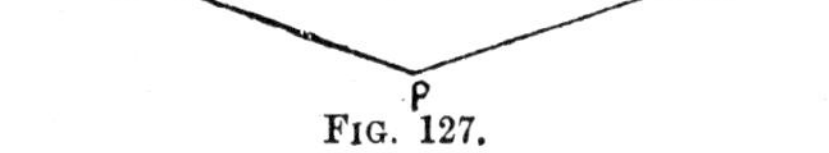

FIG. 127.

of the numbered lines along from *k*, as shown. That is, make *k* 1, *k* 2, *k* 3, etc., equal respectively to the lines numbered 1, 2, 3, etc., on the plan. These points are then joined to *t*, and the lengths of lines for the pattern will be measured from the respective points up to *t*. Now turn to the side pattern (Fig. 128). Line number 1 is set down equal to 1 *t*, line number 2 equal to 2 *t*, and B L equal in length to the curve *b l* on the plan. The length F N equals *f n*, and the line number 3, or L N, equals 3 *t*, and so on for the other six triangles. When the points are connected up and the proper allowance added, as in the back plate, the pattern is complete.

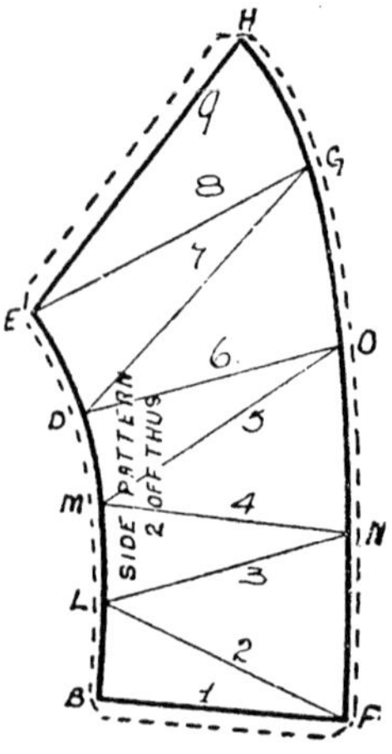

FIG. 128.

The foot being equal-tapering can have its pattern struck out with very little trouble. The radius for the part of foot to go around the back of bath will be taken from *p a*. So that, on the pattern (Fig. 127), P A equals *p a* and P R equals *p r*; the length of the inner curve A A being equal to twice the length of the quarter-circle *a b* in plan. The side pattern of the foot (Fig. 125) is developed by first marking off *k s* equal to *t d* (the point *t* in this case being the centre from which the side curve of egg-shaped oval is described), and then drawing *u s* parallel to *p a*. The curve S B for the pattern is described with radius equal to *u s* and the part D B cut off equal in length to the arc *d b* in the plan. The distance *k w* is now set along the same length as *e v* (the radius for the

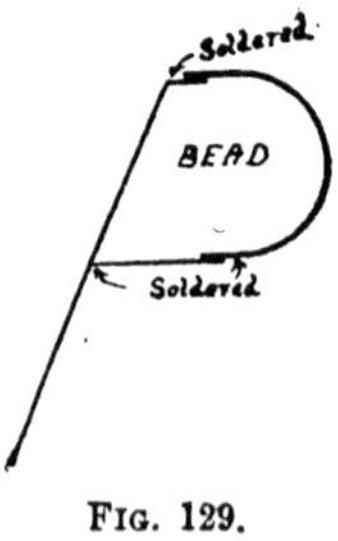

FIG. 129.

small end of oval) and *x w* drawn parallel to *p a*. The centre W is determined by setting D W equal in length to *x w*; the arc D E then being drawn from this centre and cut off the same length as the curve *d e* in the plan. The width of the pattern will, of course, be the same as that for the back part, that is, D R will equal *a r*.

Allowances must be put on to both the foot patterns to cover for grooving at ends, wiring at bottom and for slipping over the knock-up on bottom of bath.

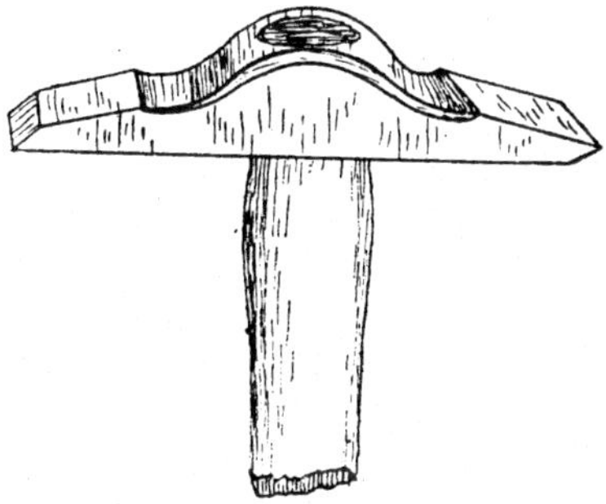

FIG. 130.

A pattern, showing the half-bottom, is drawn at the top of Fig. 125; the lines Z A and Z E being respectively equal in length to the lines *z a* and *z e* on the elevation. Allowance for a double edge is made all round the bottom to cover for knocking up on to the body.

A sketch, showing the arrangement of body, bottom, and foot, is shown at the lower part of Fig. 126. A good deal of care wants exercising in attaching the bottom to the body. After the bottom has been hollowed to the proper shape a flange about ½ in. wide should be set down all round. A single edge is then turned up to fit on the edge around the bottom of bath. After the bottom is slipped on, the paning down can be done with a paning hammer (Fig. 130), or, as is more generally the case, with a sheet metal worker's common hammer, as shown in Fig. 131. The knocking up of the joint can, of course, be done in the usual manner on a bench bar or otherwise.

A sketch explaining the arrangement of the bead is also shown on Fig. 129. The bead is usually made from a strip of sheet metal, being bent and curved in a beading machine. It can, however, be quite easily blocked up to the required shape with a suitable hammer on a lead block.

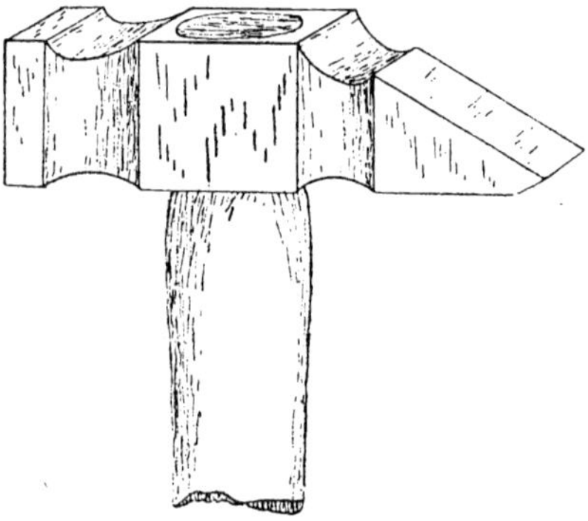

FIG. 131.

It is soldered to the edge of body and filling-in pieces are also soldered to the bead and body, as shown.

Sheet-metal lugs are fastened to the side of bath, as seen in Fig. 124; but these present no difficulty in marking out their shape or making.

Sponge Bath.

The patterns for a sponge bath (Fig. 132) can be laid out by one or other of the several methods already shown in connection with cone-work. The only part that is not conical and that calls for attention here is the lip or spout. The pattern for this is shown marked out in Fig. 133. An elevation of the lip fitting on to the side of the bath is drawn and also a half-plan, showing the shape of the top of lip. The arc *a b*, on Fig. 133, represents a part of the top of bath. The lip curve is divided into, say, six equal parts, and perpendiculars dropped from

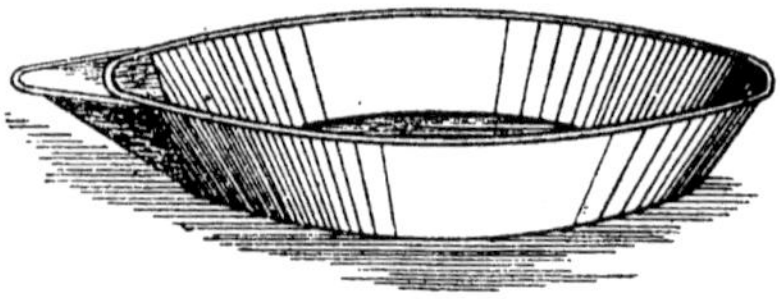

FIG. 132.

each of the division points, 0, 1, 2, etc., on to the line 0′ 6. The joint line is then drawn, passing through the point 0′,

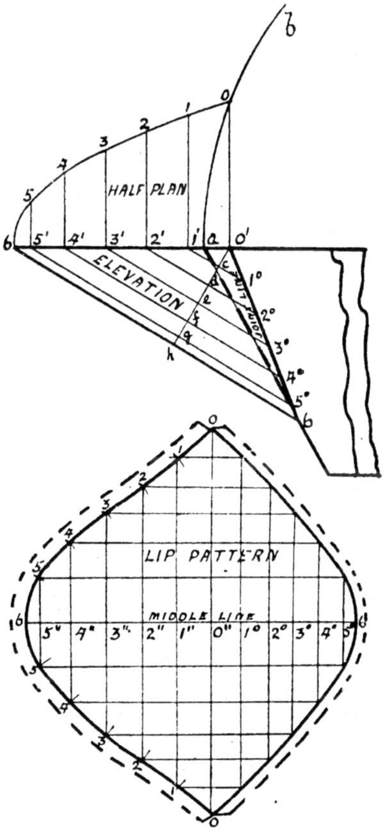

FIG. 133.

as shown. Through the points 1′, 2′, etc., lines are drawn parallel to 6 6. For the pattern the middle line is set down equal in length to 6 6 on the elevation, the intermediate points being obtained by making 0″ 1″ equal to *c* 1′, 0″ 2″ equal to *d* 2′, etc.; and then, on the other side of 0 0 making 0″ 1° equal to *c* 1°, 0″ 2° equal to *d* 2°, and so on for the remaining distances. Through each of the points on the middle lines perpendiculars are drawn, as shown by the lines 0 0, 1 1, etc. Now fix the compasses to the length of one of the arcs on the half-plan of lip, and with this distance, commencing at 6 on the life-hand side of pattern, cut off points 5, 4, 3, etc., up to 0. Through each of the points so found draw lines parallel to 6 6, and where these intersect the perpendiculars already drawn through 1°, 2°, etc., will give points on the curve for the right-hand side of pattern. Join these points with a regular curve, add allowances for wiring and a flange, and the pattern is complete.

CHAPTER XVII.

OVAL ARTICLES OF EQUAL TAPER.

Construction of Equal-Ended Oval.

THERE are many articles made out of sheet and plate metal that are either oval or elliptical in shape. Not that these two figures are identical, although they are often confused with each other. The ellipse is a figure in one quarter of which we may suppose every small part of the curve is of a different radius, the curvature of the end being most acute, and the curve becoming flatter as it approaches the middle point of the side of the ellipse. The oval, however, although somewhat similar in shape to the ellipse, is a figure which is built up entirely of arcs of circles. Equal-ended ovals can be drawn by using several arcs of differing radii that are a very good approximation to an ellipse. An oval, however, to the sheet and plate metal worker has distinct properties of its own that make it particularly suitable for use in those cases to which it can be applied. When an article is required to be elliptical in shape, the oval should not be used, as there are convenient methods for the development of this class of work. (Shown in Chapter XXI.)

The most useful shape of oval is that which is made up of *two* different arcs of circles, the one with small radius forming the ends, and the flatter curve joining on to make the sides. This can be set out entirely by construction, or partly by calculation and construction. Both methods will be shown. First by construction: Draw a line A B

(Fig. 134) equal in length to the long diameter of the oval, and through the middle point O of this diameter draw a line at right angles. Make O C and O D each equal to half the small diameter of the oval. From A mark off A E equal in length to C D. Divide E B into three equal parts. Now set the compasses at a radius equal to *two* of the parts, and with O as centre, mark points Q, Q. Then with O again as centre, and the compasses set to length

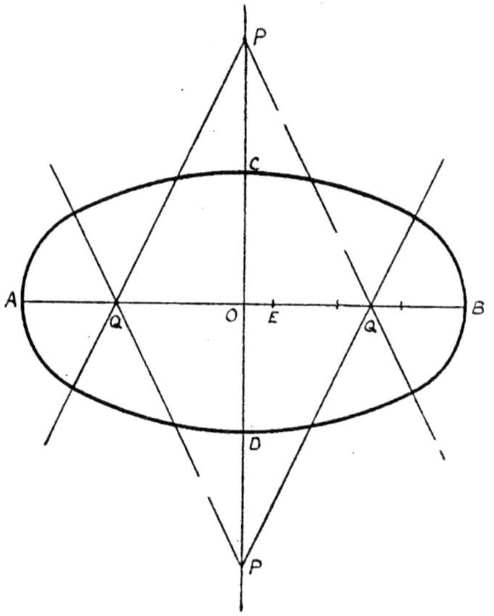

FIG. 134.

Q Q, mark points P, P. It will be seen that O P is equal to *four* of the parts into which E-B has been divided. The points Q and P will be the centres from which the arcs will be described. Join P to Q, and produce the lines through as shown. Now with centre Q and radius Q B describe the end arcs, and with centre P and radius P C describe the side arcs. If carefully and properly drawn, the arcs should meet and run into each other on the lines drawn through P and Q. The object, indeed, for which these lines are drawn is to determine the meeting points of the curves. They also serve another purpose, which we shall see when drawing out the pattern for an oval equal-tapering vessel.

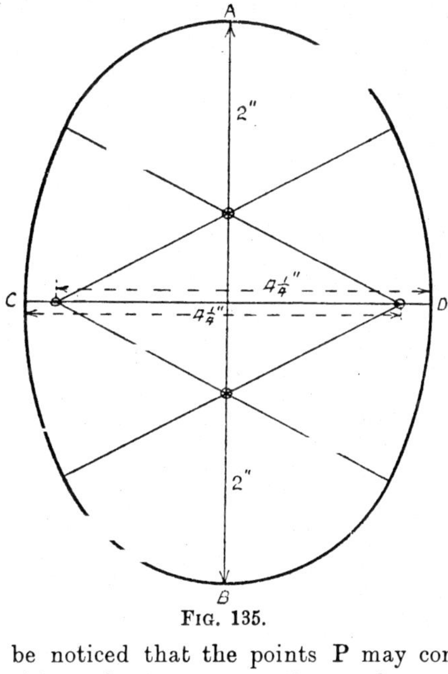

FIG. 135.

It should be noticed that the points P may come either within or without the figure, according as the oval is broad or narrow.

The second method consists in calculating the radii for the arcs and then setting out the figure. The rules for finding the radii are as follows:—

To find radius for sides: "From eight times the long diameter deduct five times the small diameter, and divide the remainder by six." In Fig. 135, the long diameter

A B = 6 in., and the short diameter C D = $4\frac{1}{2}$ in., therefore the radius for the side is—

$$\frac{8\,AB - 5\,CD}{6} = \frac{48 - 22\frac{1}{2}}{6} = 4\tfrac{1}{4}\text{ in.}$$

To find radius for ends: "From four times the short diameter deduct the large diameter, and divide the remainder by six." The radius for the ends, therefore, in Fig. 135 will be—

$$\frac{4\,CD - AB}{6} = \frac{18 - 6}{6} = 2\text{ in.}$$

After having marked the centres, it is generally a good plan to draw in the lines as before, so that the exact point of contact of the curves may be known, as these come in useful later.

The above methods can, of course, only be used in the case of ovals that are the same shape at each end, the egg-shaped oval demanding special treatment.

Pattern for Oval Articles.

Having gone over the construction of ovals, we can now turn our attention to the development of oval equal-tapering articles, or those in which the overhang for the sides is the same as for the ends. Such an article is shown in Fig. 136. It is best to think of the surface of this article as being built up of parts of the surfaces of two different-sized cones, but whose taper is the same. Thus, referring to Fig. 137, the large dotted circles represent the bases of the cones, part of whose

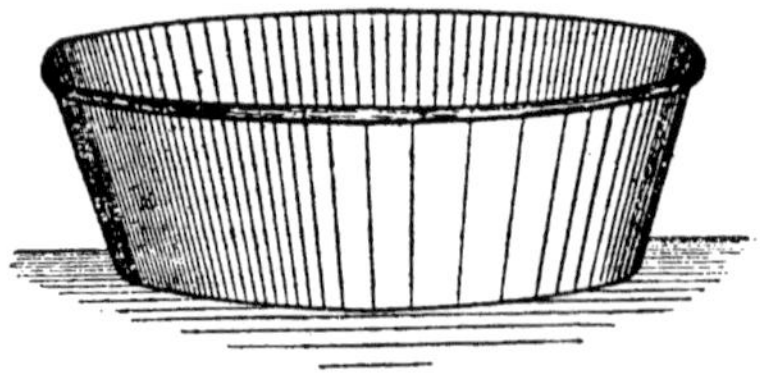

FIG. 136.

surfaces go to form the sides of the oval vessel. The small dotted circles show the bases of the cones from whose surfaces the end portion of the oval article is formed. The plan of the axes of these cones, it will be seen, coincides with the points P and Q.

The fitting together of the cone parts is exhibited in Fig. 138, which is a sketch of a model showing a part of *one* large cone with the *two* parts of the smaller cones fitting on to form the end portion of the oval object. The front side, which would be the part of the other large cone, it may be imagined, is removed to better explain the fitting together of the side and ends. The three upright lines show the axes of the cones. It should be noticed that the small and large cone surfaces join together in a common line (shown by the dotted lines at the back): hence the two

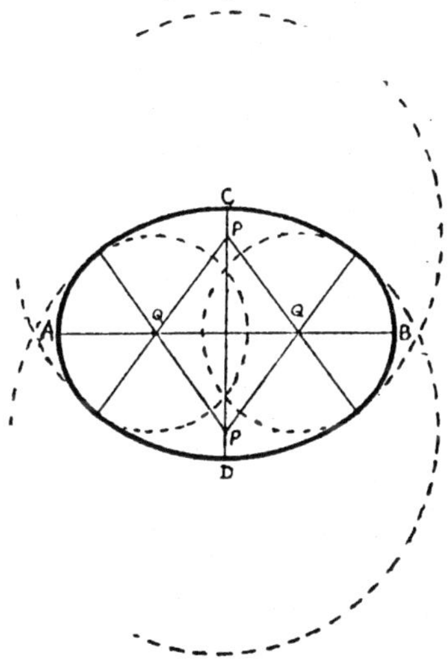

FIG. 137.

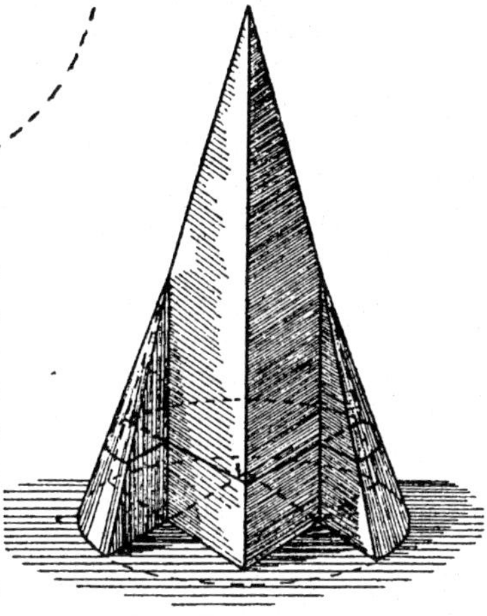

FIG. 138.

curved surfaces fit together without showing lump or hollow. Part of the small oval which forms either top or bottom of the article, as the case may be, is also shown on the model in this figure.

When it is thoroughly understood how the surface of the oval equal-tapering article is built up, the development of the pattern is not at all a difficult matter.

It will, perhaps, be easier to follow if we fix some definite dimensions, and work out the problem completely from these. Thus suppose an oval article is 32 in. by 20 in. at the top, 22 in. by 10 in. at the bottom, and 7 in. perpendicular depth. It should be noted that the dimensions must be such as to give the same overhang all round, and these can be checked by using the following rule:—"The length of bottom deducted from the length of top must be the same as the width of bottom deducted from the width of top." In this case it will be seen that the overhang is—

$$\frac{32 - 22}{2} \text{ or } \frac{20 - 10}{2} = 5 \text{ in.}$$

Calculating the radius for the sides of the large oval by the before-mentioned rules, this will be—

$$\frac{8 \times 32 - 5 \times 20}{6} = 26 \text{ in.}$$

and for the ends—

$$\frac{4 \times 20 - 32}{6} = 8 \text{ in.}$$

As each quarter of the oval is exactly the same, there is no need only to set out just one quarter, and this can be done in the usual way (Fig. 139). The same centres can be used for marking out the quarter of small oval for the bottom, the radii for sides and ends being in each case 5 in. less than those used for the top. For purposes of getting out the body pattern, there is really no need to set out the

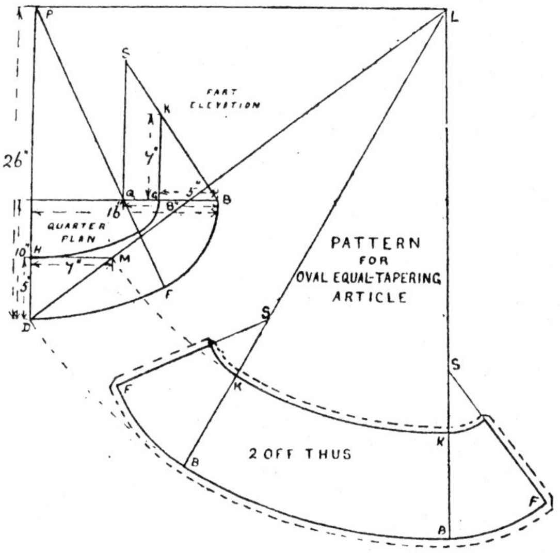

FIG. 139.

quarter-oval for the bottom, its only use being to obtain the size and shape of the bottom plate.

Having marked out the oval, the depth 7 in. should be set up from the point G, and also along from the point H, as shown in Fig. 139. B is joined to K, and produced until it meets a perpendicular through Q in S. Also D should be joined to M and produced until it meets a line which is square to D P in L. By referring again to Fig. 138, it will be seen that the points L and S (Fig. 139) represent the apexes of the large and small cones respectively. Half the large cone being given as a side elevation on D P, and half the small cone being shown as a front elevation on Q B. The line S B then will give us the slant height of the small cone, and thus the radius for the development of its surface; the line L D serving the same purpose for the large cone.

The pattern can now be struck out. With centre L and radius L D draw an arc, and along it mark off a portion, B B, equal in length to twice the arc D F. Join the points B B to L, then with centre L and radius L M draw the bottom curve, K K. The part of the pattern thus set out will give the side portion of the article, or we may imagine it to be the development of the part of large cone. The ends can now be added by opening the compasses to the slant height of small cone B S, as shown in the elevation, and marking it along the lines B L in the pattern, thus obtaining the centres, S S. The curves for the end part of pattern are now set out from these centres, using radii S B and S K. The outside curves, which are shown marked B F, are now cut off equal in length to the curve B F on the plan. This is best accomplished by bending a piece of wire along the curves, as before mentioned. Particular notice must be taken that the points F are joined to the centres S. There is no need to trouble about the length of curve for the bottom of pattern, as this will be cut off to the correct proportion by the radial lines as drawn. This may be tested by measuring the length of the curve, and seeing if it is equal to twice the length of the bottom quarter-oval.

Position of Joints.

The pattern thus drawn out is, of course, for one-half of the oval vessel, two pieces off this being required to form the body of the article. It will be necessary to add laps as required for grooving, wiring, and knocking up.

It will be seen that the joints are at the end of the article, the reason for this being that that part having the sharpest curvature, it will be somewhat stiffer and stronger than the sides, hence the best position for placing the joint. Another reason that assists in determining the position of a wire joint is that, if possible, it should be covered with a lug, ear, or handle. In the present case, if the vessel is to be used as an oval tub or bath, the handles would be riveted over the end grooves, and thus materially assist to strengthen this part. Perhaps a further reason for fixing the grooves at the ends is that the pattern comes out much flatter when the joints are in this position than when on the sides; consequently the material will cut up to greater advantage. The economical cutting up of sheets and plates should always be taken into account when thinking of the position of joints. Also when stock sizes of sheets or plates are being used a little thought bestowed on the pattern will often save a large amount of waste in material.

When it is required to make the pattern for the body of an oval article in one piece, a little consideration of Fig. 139, will show how this can be done.

Before concluding it might be here pointed out that the lengths of both the radii used on the pattern can be calculated by the method shown, in connection with the cone, in Chapter XII.

Bodies of oval articles are usually shaped by bending in the rolls to the curvature of the ends, and then flattening out the side parts.

CHAPTER XVIII.

ARTICLES OF UNEQUAL OVERHANG.

MANY articles may be circular or partly circular in section, also having the property of their surfaces tapering to a point, and yet not be formed of a portion of a right cone (a cone whose axis is perpendicular to its base). Their patterns, however, can be readily developed when the surface is considered as a portion of an

Oblique Cone.

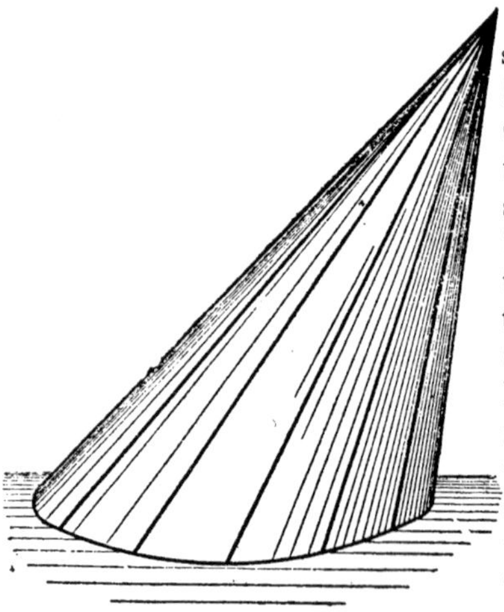

FIG. 140.

A cone of this description is one whose base is circular and whose axis is inclined to the base. Such a solid is shown in Fig. 140. Particular attention should be paid to the development of the surface of this kind of cone, as many articles can have their patterns readily set out for them when it is observed that they are formed from parts of the surfaces of oblique cones.

In marking out the pattern for an oblique cone, the principle involved is to imagine the circumference of the base divided up into a number of equal arcs and the division points joined up to the apex of the cone, thus dividing its surface into a number of triangles. In Fig. 140 these lines are shown, the whole surface of the oblique cone being divided up into twelve triangles.

The setting out of the pattern for a complete cone is

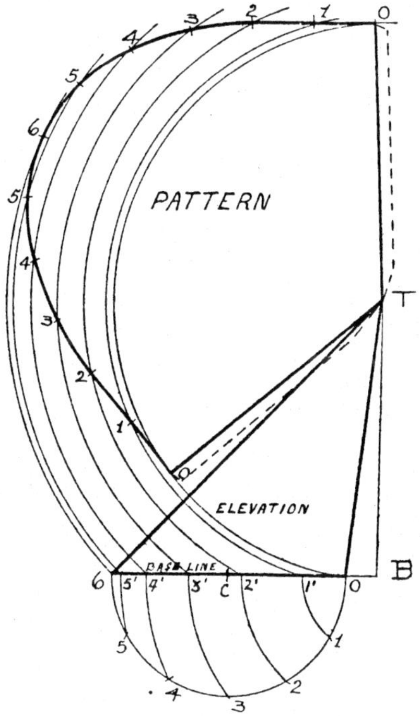

FIG. 141.

shown in Fig. 141. An elevation of the cone is drawn, and on its base a semicircle described and divided into six equal parts. The line T B is drawn perpendicular to the base line produced. Then with B as centre, and B 1, B 2, etc., respectively as radii, the numbered points are swung round on to the base line as shown. Then using T as centre, and T 0, T 1^1, T 2^1, etc., as radii, arcs of circles for the pattern are drawn around. The compasses are now set to a distance equal to one of the six parts into which the base semicircle has been divided, and commencing at some point on the inner arc, say 0, the lengths are stepped from one arc to the other right up to 6, and then back over the arcs again to 0. If the points 0 0 are now joined up to T, and a fair curve drawn through the other points, the net pattern is complete. Allowances for the seam are shown by the dotted lines, the joint of same coming at the underside of the cone.

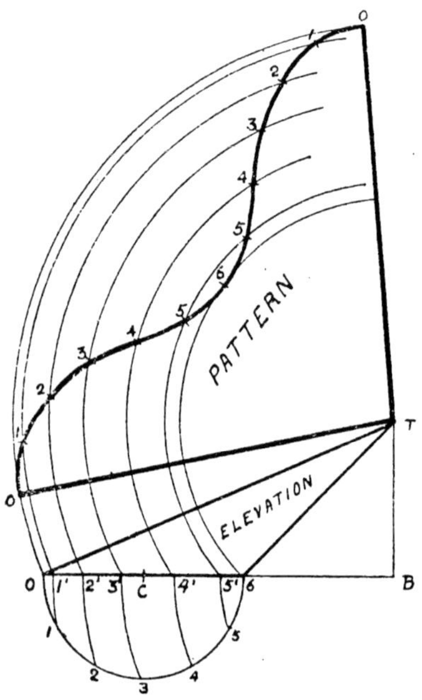

FIG. 142.

The pattern for a circular oblique cone which has a large amount of overhang is shown set out in Fig. 142. The distance that the top overhangs the centre of the bottom will be equal to C B, as in the last example. The height of the top above the base will, of course, equal B T. So that where these sizes are given, together with the diameter of base, the elevation of the cone will be drawn by first putting in the lines T B, B C, and then marking off the base 0 6. It will be observed that in this case the seam is arranged to come down the centre of the back of cone.

Tapered Connecting Pipe.

The frustum of an oblique cone can very conveniently be used to join together two circular pipes of unequal dia-

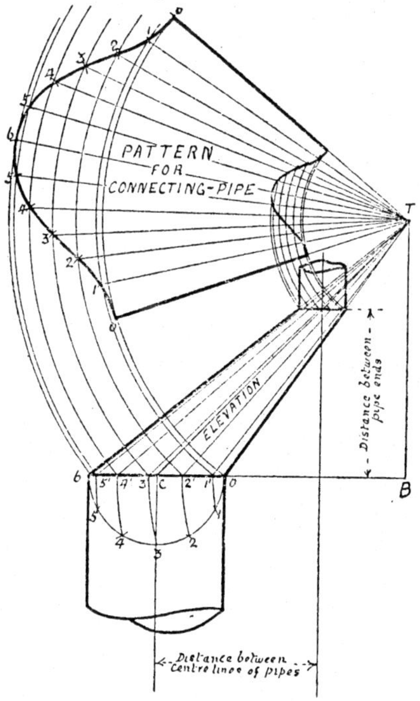

FIG. 143.

meter, whose centre lines are parallel, and whose ends are cut square. A connecting pipe of this description, together with the pattern development, is illustrated in Fig. 143.

The elevation is first drawn by setting in the centre lines, the distance between the pipe ends and the pipe diameters. To complete the oblique cone, of which the tapered pipe is a part, the back and throat lines are produced until they meet at T. To obtain the pattern the surface of the whole cone is developed as in the last two examples, and the portion of the top of cone cut away, as will be explained. After swinging the points 1, 2, 3, etc., on to the base, they are joined up to the top of cone, and where these lines cross the joint line at the top of connecting pipe, will determine the lengths of lines required to mark along to obtain the shape of cut at small end of pattern. In the figure all the lines are shown swung around on to the pattern lines. The points so obtained are joined up with an even curve, and thus the net pattern is finished. No allowances are shown in Fig. 143; but these can be put on to the sides and ends of pattern to suit the method of jointing adopted.

Before passing from this it should be pointed out that whilst the ends of the connecting pipe in the above case are circular, a section of the tapered pipe taken perpendicular to its centre line will be elliptical, and consequently when the overhang is great the pipe will be very flat, as in Fig. 142, and its area restricted. If it is required to have a tapered pipe of circular section, then the method shown in Chapter V. will have to be followed.

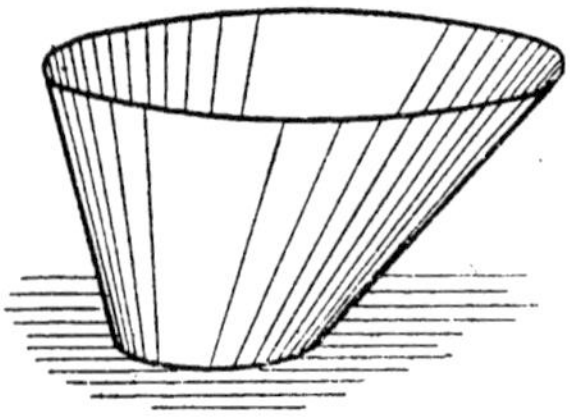

FIG. 144.

Unequal Tapering Circular Article.

Any article whose top and bottom are circular, parallel, and of unequal overhang, such as Fig. 144, can have its pattern developed as a frustum of an oblique cone. The pattern for such an article is shown set out in Fig. 145. Although the shapes seem somewhat different, the same method of construction for obtaining the pattern lines can be followed as in Fig. 143; the lines being denoted in exactly the same manner. No allowances for wiring or

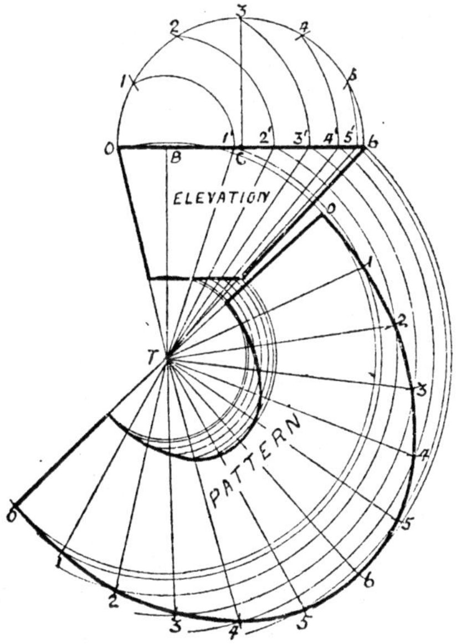

FIG. 145.

jointing have been added to the net pattern; but these can be put on as required.

Tapering Y-Piece.

The oblique-cone method can be used for setting out the pattern for the connecting pipes in a tapering Y-piece, as shown in Fig. 146. Here the problem resolves itself into jointing up two small pipes to one large one, the ends of the straight pipes all being square to their centre lines.

The setting out for the pattern is explained by Fig. 147. In practice there is no need to draw a complete elevation; all that is needed is the shape of a connecting pipe; but in the present case the full elevation is shown to better explain the way the form of a connecting pipe is obtained. To make the setting out plainer, definite dimensions in this case have been taken. The diameter of the main pipe is 17 in. and that of the branch pipes 8 in.; the distance between centre lines of branch and main pipes 10 in. and the distance between their ends 13 in. These dimensions are set out, as shown in Fig. 147, and to obtain the correct position of the joint line of the connecting pipes their back lines are run down to the edges of the main pipe, as seen by the dotted lines, the point of intersection 6″ giving the top end of the joint line. The pattern for the frustum is set out, as explained in connection with Fig. 143, the toe portion being afterwards cut away. To do this we must first get the lengths of lines required. Join B to 5, and where this line cuts line 3 C swing about B on to the base 0 6, and from this point project up and thus obtain point 5″ in the elevation. In the same way find point 4″. Then

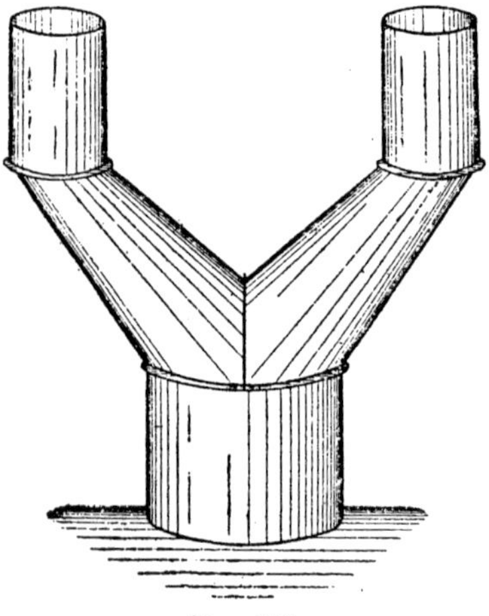

FIG. 146.

with T as centre, swing these points on to their corresponding lines in the pattern. Join the new-found points with an even curve, add allowance for jointing as required, and the pattern is ready for use.

Multiple-way Piece.

If it is desired to join more than two branch pipes on to the main pipe, then the above method will still hold good. The first thing to do is to obtain the plan of a joint line; thus in Fig. 147 the line 3 C on the semicircle, it will be readily seen, is the plan of the joint line when there are two branch pipes. To obtain the position of the plan of one of the joint lines when there are more branch pipes than two, a line will have to be drawn through C, making an angle with 0 C equal to—

$$\frac{180 \text{ degrees}}{\text{number of branch pipes}}$$

Thus, suppose there are three branch pipes, then the angle of the joint line will be—

$$\frac{180}{3} = 60 \text{ degrees.}$$

In this case the plan of the joint will be the line 2 C (Fig. 147). Now for the pattern. Where the line 2 C crosses the lines B 3, B 4, and B 5, swing on to the base line with B as centre. From the base line project up on to the correspondingly numbered lines in the elevation, thus obtaining the points 3^0, 4^0, and 5^0. Now with T as centre swing these points around on to the pattern and draw in the curves. The thick dotted curve thus shows the cut for the toe of pattern when three branches are required to be jointed to one main pipe.

In the same way as above, after fixing the position of the joint-line plan, a pattern for a connecting pipe for any number of branches can be set out. On examining Fig. 146, it will be seen that all the joints are paned down. They may, of course, be knocked up if the material is sufficiently malleable to stand the operation. The allowances for this method of jointing will be a double edge on the end

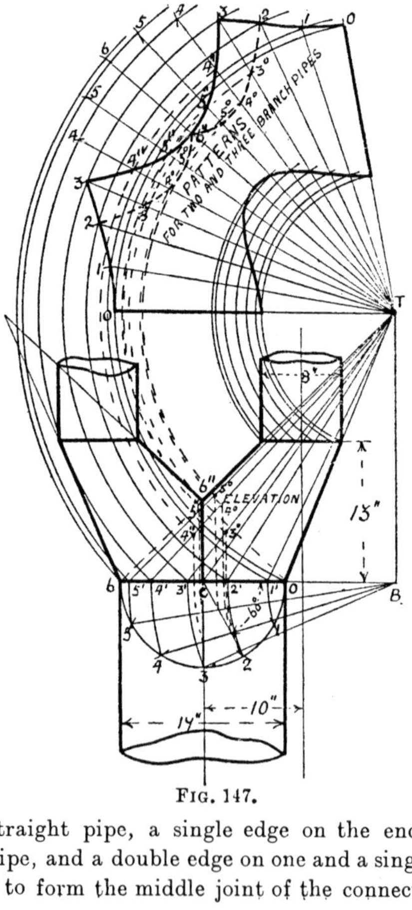

FIG. 147.

of each straight pipe, a single edge on the ends of each tapered pipe, and a double edge on one and a single edge on the other to form the middle joint of the connecting pipes.

CHAPTER XIX.

IRREGULAR TAPERING ARTICLES.

Article with Round Top and Semicircular-ended Oblong Bottom.

THIS is an article (Fig. 148) which belongs to the family of the oblique cone, for its rounded surface at the ends is formed of two half-frustums and its flat sides of two triangles.

Its pattern is shown struck out in Fig. 149. A quarter-plan and a half-elevation are marked out as seen, and the point B obtained by joining 3 to *c*, and producing to meet the base line. The points 1, 2, and 3

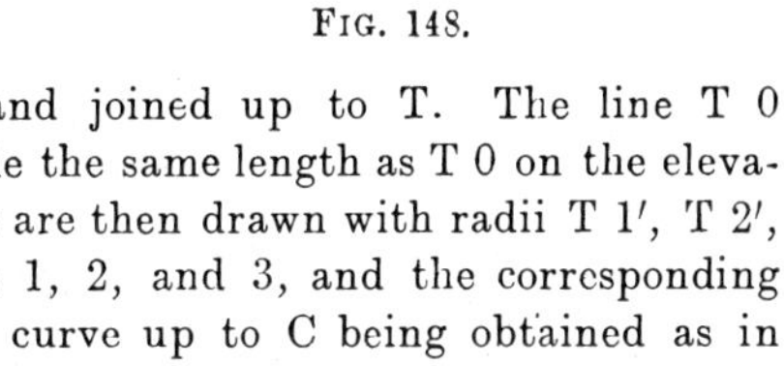

FIG. 148.

are swung around and joined up to T. The line T 0 on the pattern is made the same length as T 0 on the elevation. Arcs of circles are then drawn with radii T 1′, T 2′, and T 3′, the points 1, 2, and 3, and the corresponding points for the inner curve up to C being obtained as in Fig. 143. Now take C on the pattern as centre and C 3 as radius, and mark the arc as shown, cutting off a point upon it by making line 3 3 equal in length to twice the line 3 4 from the plan. Thus we now have the two flat triangles added on to the first portion of the pattern. The two centres for the remaining lines can, of course, easily be fixed when it is remembered that their lengths are exactly the same as the lines used in marking out the first part of the pattern.

Without the object is very small the body of the article will be made up in two pieces, the seams being at the ends

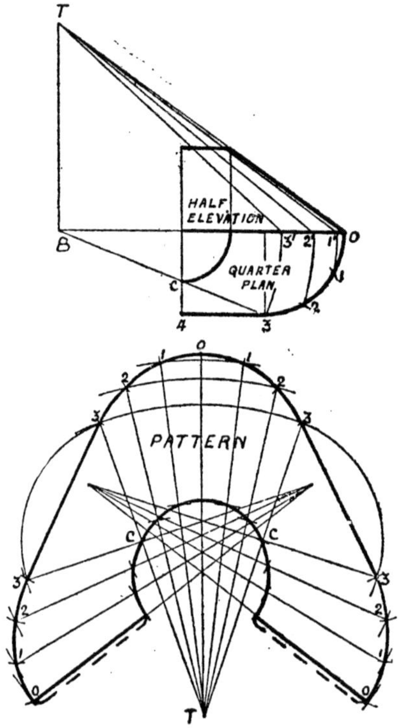

FIG. 149.

and coming along the lines through 0 on the pattern. The sheet and metal plate worker, however, should find no difficulty in making up an article in any number of pieces if he can strike out the pattern for the complete body.

Article with Circular Top and Oblong Bottom with Rounded Corners.

The surfaces of many articles are what might be described as of a compound character—that is, they do not follow the surface shape of any one particular solid, but are built up of parts of surfaces of one or more solids, perhaps combined with one or several plain figures. One simple example of this has been shown in Fig. 148, and we shall now give two more typical cases of this class of pattern-marking.

In all cases it should be borne in mind that the first thing to do is to carefully analyse the surface so as to determine how it is formed. In Fig. 150 an article is shown whose top is circular, and whose bottom has straight sides, with corners formed of quarter-circles. On examination it will be seen that the rounded parts are each a quarter of a frustum of an oblique cone, and that the flat surfaces are triangles. In setting out the pattern (Fig. 151) a quarter-plan is first drawn, as shown, and the point B obtained by joining 1 to a and 4 to c, and producing the lines until they intersect. Through B a line is drawn parallel to a o, and the points 1, 2, 3, and 4 swung on to it, as in the previous cases: these are then joined up to T. It, perhaps, should be here mentioned that the point T in this example is found by produc-

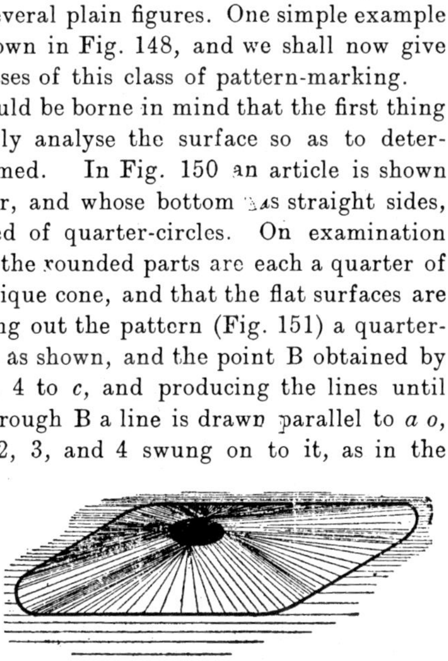

FIG. 150.

ing the back line o^1 a^1 until it intersects the perpendicular drawn up through B. In marking out the pattern the line 0 A is made equal in length to a^1 o^1 on the elevation. A line is drawn square to it through 0, and cut off on each side equal to 0 1 in the plan. The line 1 A is produced, and the point T determined by making 1 T equal in length to

T 1^1 from the elevation. The points 2, 3, and 4 on the pattern, and also those for the curve A C, are now obtained as in the other cases of the oblique cone. C is now taken as centre and C 4 as radius, and the arc drawn as shown, the next point 4 being fixed by making line 4 4 equal to twice the length of *d* 4 in the plan. The other points will be found, as previously explained, in connection with Fig. 149, each quarter of the pattern being the same shape. If the article is made up in two or more pieces, it will be an easy matter to mark out the portion of pattern required.

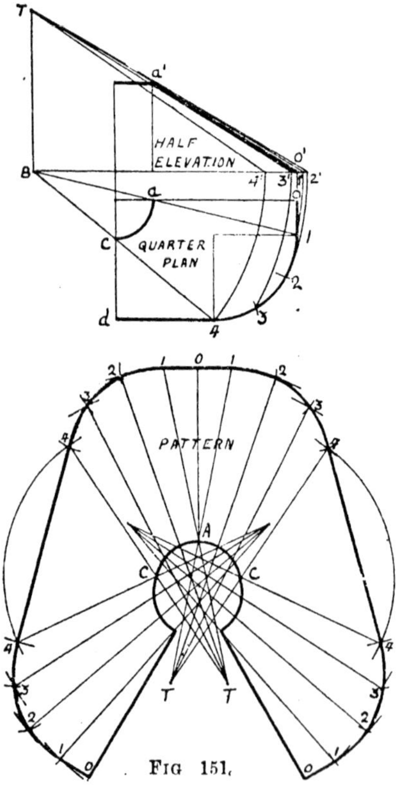

FIG 151.

Hood with Round Top and Flat Back

A pattern for a hood whose top is circular and bottom rectangular with two square and two round corners (Fig. 152) can be obtained on the same principle as in the previous articles. An inspection of the hood surface in the sketch will lead us to see that it is built up by two quarter-frustums of an oblique cone and a flat triangle for the front, a flat triangle and a quarter-frustum of an oblique cone for each side, together with a flat triangle for the back.

To get the length of the pattern lines, a half-plan and a half-elevation are set out (Fig. 153), the point B and the true lengths of lines being obtained as in the last example. The setting out of the pattern up to the lines C 2 will be

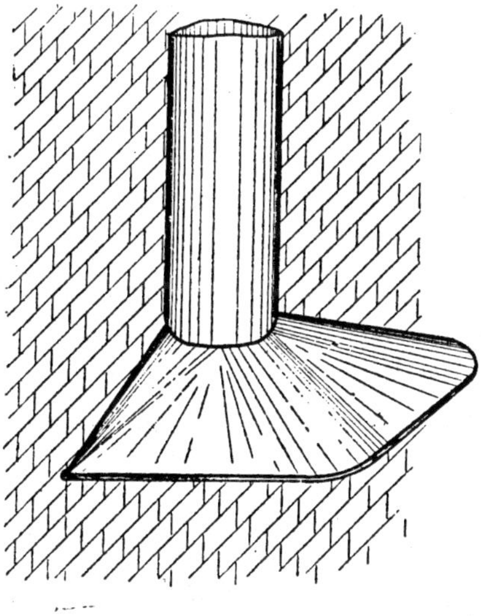

FIG. 152.

exactly as in Fig. 151. Before proceeding any further with the pattern, the lengths of lines required for the remaining portion will have to be determined. To do this, set along distances 1^1 c^1, 1^1 d^1, and 1^1 0^1 in the elevation respectively equal to 1 c, 1 d, and 1 0 in the plan, and join up to a^1. On the pattern, the line 2 1 is equal in length to the line with the same figures in the plan; and the lines C 1, D 1, O^1 1 are respectively equal to lines c^1 a^1, d^1 a^1,

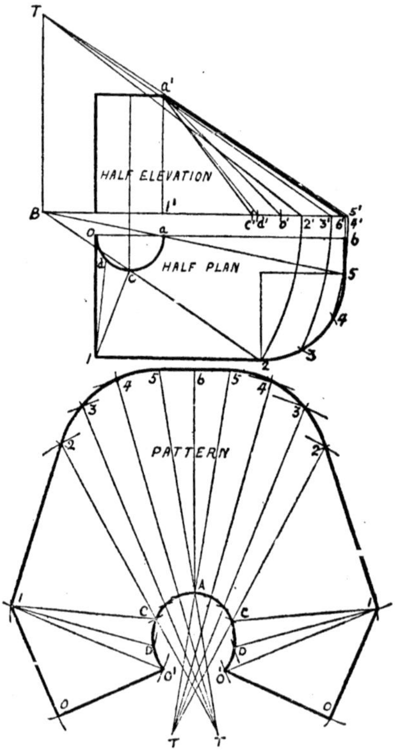

FIG. 153.

and 0^1 a^1 in the elevation. The distance between C and D and D and O^1 on the pattern will, of course, be equal to the lengths of the arcs *c d* and *d* 0 on the plan. To form the last triangles of the pattern, the lines 0^1 0 will be made equal in length to the back line in the elevation, or the line 1^1 a^1 and the line 1 0 equal to the line with the same figures in the plan.

It is perhaps as well to point out that the patterns for all the objects mentioned can be struck out by the method of "triangulation." But whilst this method is general in its application, it is not so convenient for the particular cases as those shown.

Article with Round Top and Oval Bottom.

A vessel of uneven taper having a circular top and an oval bottom can have the pattern for its body set out in a similar manner to that of several of the objects previously dealt with. Examination of Fig. 154 will show that its surface is formed of parts of two different sized and shaped oblique cones. The points T T show the tops of the oblique cones that are used for the end parts of the article, and the point *t* the apex of one of the oblique cones used for obtaining the side parts of the body surface.

A quarter-plan and two half-elevations of the vessel are first set out as shown in Fig. 155, the quarter-oval being marked out as explained in Chapter XVII. The points Q and P show the centres for the end and side curves of the oval, and point 3 where the two curves meet. Having constructed the quarter-oval, divide each of the two curves into, say, three equal parts. Draw the line *c a* parallel to Q 3, then joining 3 to *a* and producing to meet the centre line in B. Run along a line from B square to the centre line, and where this intersects the end line *o c* produced in T, will give the apex of the oblique cone which forms the end part of the article. The apex of the oblique cone, which forms the side part, can be determined by running up a perpendicular to the base line through *b*, and producing line 6 *d* to meet it in *t*. With *b* as centre and *b* 5, *b* 4, *b* 3, respectively, as radii, swing around on to the base line, thus obtaining points 3′, 4′, and 5′. In the same way, taking **B** as centre, swing points 3, 2, and 1 on to the centre line, giving points 3″, 2′, and 1′.

To mark out the pattern, set down the line 6 *t* equal in length to the line 6 *t* in the elevation. Then, using *t* on the pattern as centre and radii *t* 5′, *t* 4′, and *t* 3′, draw

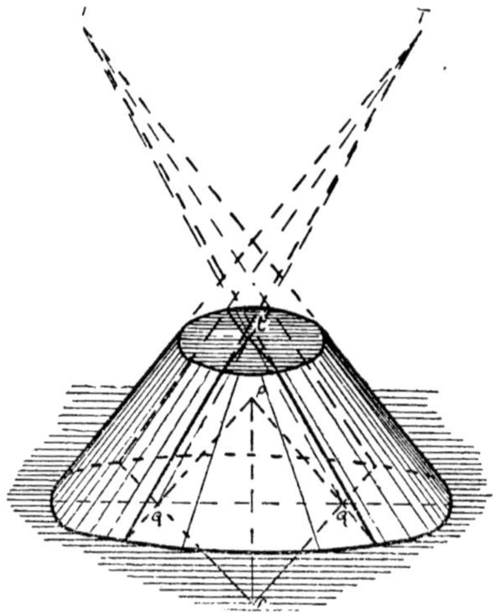

FIG. 154.

arcs of circles as shown. Now set the compasses to the length of one of the three arcs on the side oval, say, 5 6, and commencing at point 6 on the pattern, step from one arc to the other, marking points 5, 4, and 3. Join the points up to *t*, and produce 3 *t* to T, making 3 T equal in length to the line 3″ T in the side elevation. Then, using T as centre and radii T 2′, T 1′, and T 0, from the elevation, describe arcs as seen. Now fix the compasses to the length of one of the arcs on the end of quarter-oval, say, 0 1, and commencing at point 3 on the pattern, step off points 2, 1, and 0, joining these up to T.

To obtain the necessary points for the inside curve of pattern, set the distances along from *t* and T respectively, as measured along the corresponding lines from *t* and T in the elevations down to the top line of the article—that is, the lines *t* D, T A, and T E on the pattern will be respectively equal to lines *t* *d*, T *a*′, and T *e* on the elevations, and so with the other points.

The pattern is set out for one-half of

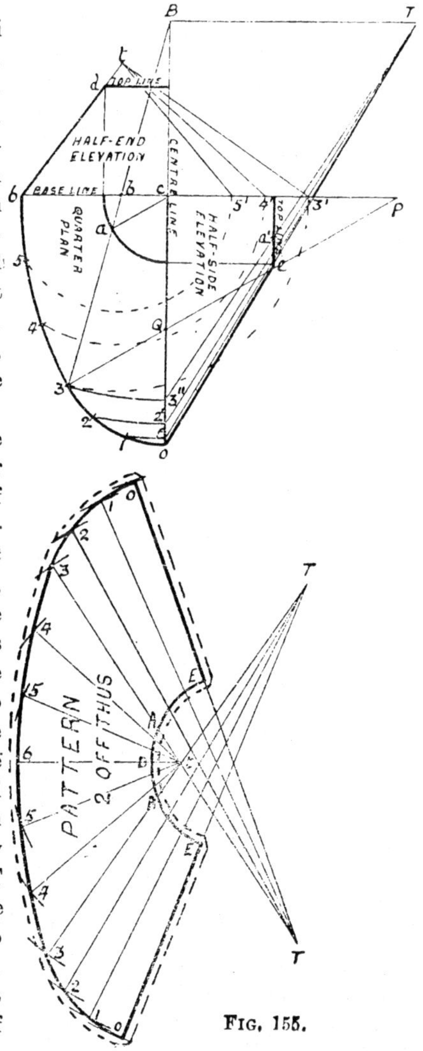

FIG. 155.

the body only, the joints coming down the middle of the ends. Imagining that the article is turned upside down, the allowances are put on for wiring around the top edge, knocking up a bottom, and for a grooved or riveted seam.

Irregular Tapering Article with Oblong Semicircular-ended Bottom and Round Top.

In addition to those dealt with in the last chapter, there are a number of hoods, hoppers, or body parts that are formed in a somewhat different manner. Thus Fig. 156 represents an article whose top is circular and bottom oblong with semicircular ends; but in this case the centre of the top is vertically over the centre of one of the semicircular ends. On examination it will be seen that the left-hand part of surface is formed of half of a frustum of a right cone and the right-hand part of half of a frustum of oblique cone, the side parts being flat triangles. Perhaps the building up of the surface will be better understood on referring to Fig. 157, where the half-pan is shown.

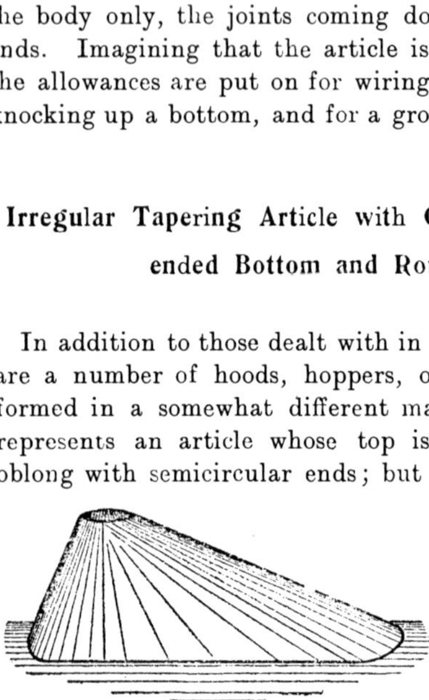

FIG. 156.

To obtain the lengths of the pattern lines, the apex T (Fig. 157) of the oblique cone is obtained by joining 4 to *a* and producing the line to B, then running up a perpendicular to meet the line 7 *d* produced to T. The apex *c* of the right cone is found by producing the slant side to meet the centre line. Taking B as centre, the points 4, 5, and 6 are swung on to the base line, and then joined up to T.

To mark out the pattern, the first line set down is T 7, this being of the same length as the similarly-numbered line in the elevation. Then taking T as centre and radii respectively equal to T 6′, T 5′, and T 4′ in the elevation, the three arcs are drawn. The compasses are now set to the length of one of the arcs in plan, say 4 to 5, and commencing at point 7 on the pattern, the points 6, 5, and 4 are struck off. The points on the inner curve of pattern to form top of article will be found by marking the distances along the lines from T, equal to the lengths of lines measured from T in the elevation down to where they cross the top line. Thus, T D on the pattern is equal to T *d* in the elevation, and so on for the remaining points. The compasses are now set to the length of line 3 4 in the plan, and with 4 on the pattern as centre, arcs are drawn, these being cut by making A 3 equal to the slant length *o e* in the elevation. The line 3 A is now produced, and A C set off equal in length to *e c* in the elevation. The point

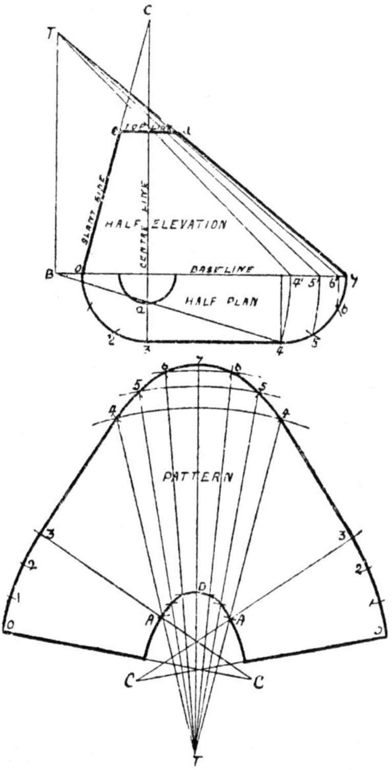

FIG. 157.

C is now used as a centre, and the part 3 to *o* struck out in the usual way for a right cone development.

In laying out patterns for articles of this description, it should be noticed that the straight line and curved parts

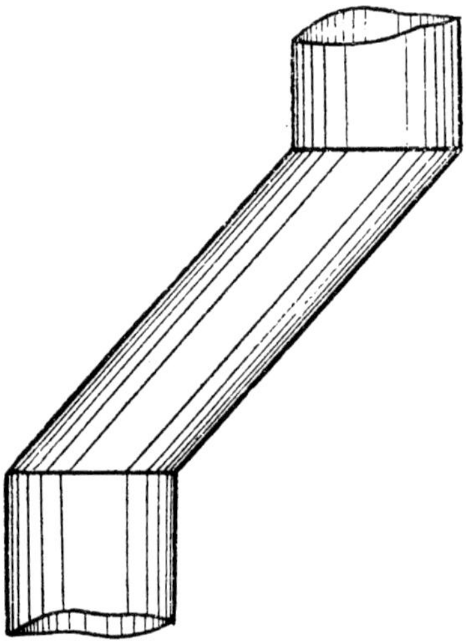

FIG. 158.

run into each other without break or unevenness. It is also as well to check the setting-out by testing if lines 3 A and 3 4 (Fig. 157) are square to each other, as they should be if the pattern be marked out correctly. For a large hood or hopper the body can be made up in as many pieces as will be suitable to the size of sheets or plates used.

CHAPTER XX.

ARTICLES OF OBLIQUE CYLINDRICAL SHAPE.

Oblique Cylinder.

THE shapes of some unequal tapering articles may be made up wholly or partly of the surface, or some portion of the surface, of an oblique cylinder; by which is meant a pipe whose ends are circular and inclined to its centre line. Such a cylinder is shown as an

Oblique Connecting Pipe.

in Fig. 158. It should be remembered, in dealing with this, that although the ends of the oblique pipe are circular, a cross-section of the pipe will be elliptical in shape; hence the more inclined the pipe becomes the flatter it will be, and the smaller its passage area. For this kind of connecting pipe it will be observed (Fig. 158) that the straight pipes are the same size, and also that their ends are cut square.

We shall first set out the pattern for a pipe of this description, and afterwards give a couple of examples illustrating its application to irregular tapering objects.

The elevation of the connecting pipe only is shown in Fig. 159, that being all that is necessary to obtain the development. A semicircle is described on one end of the pipe as in the figure, this being divided up into, say, six equal parts, and perpendiculars drawn to the end line 0 6, as seen. Through each of the last found points lines are run along parallel to the centre line of the pipe. Now to sweep out the pattern. Run lines down through points, 0, 1′, 2′, etc., square to the pipe, and then carefully setting the compasses at a distance equal to the length of one of the six parts of the semicircle, step distances 0 to 1, 1 to 2, etc., from one line to the other on the pattern. The curve on the pattern for the other end of the pipe can be set out in the same way, or lines run down and points from the first curve projected across. The better plan in the workshop is to mark one curve out carefully (or one quarter of it will do, as shown by the shaded part in Fig. 159), cut out in sheet metal, and use this as a template for the other end. The pattern, too, should be set out quite distinct from the elevation, as methods of projection are useless in workshop practice, and are only used in this descriptive way to better explain the connection between the elevation and the pattern. Allowances for jointing are put on as shown by the dotted lines.

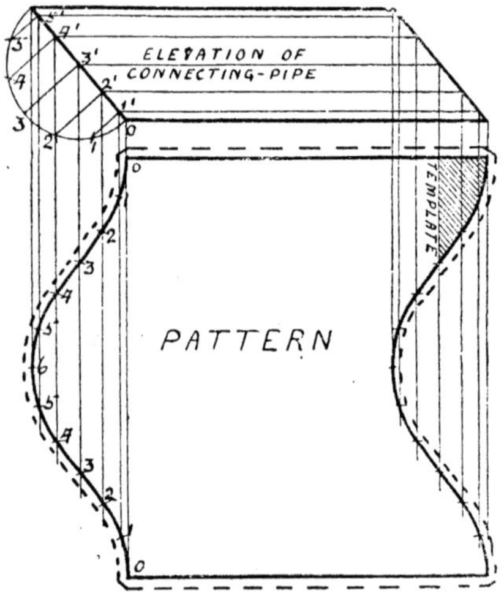

FIG. 159.

Multiple-way pieces having portions of an oblique cylinder for the branch pipe connecting tubes can be set out in a somewhat similar manner to those shown in the last chapter.

Funnel with Central Circular Top and Oblong Semi-circular-ended Bottom.

In this particular example (Fig. 160) it should be noted that the diameter of the top and the width of the bottom are equal; hence its curved surface is formed of two upper halves of an oblique cylinder, together with two upright triangles for its flat sides.

The development of the surface to form the pattern will be followed by referring to Fig. 161. A half-elevation is drawn, and a quarter-circle described on the base, the radius of this, of course, being equal to the half-width of funnel. The quarter-circle is then divided into three equal parts, and lines through each part run square up to the base line. Through the points on the base lines are now drawn parallel to the slant end line. Make *a* 0′ square to *t a*, and then set up distances *a* 3′, *b*′ 2′, etc., equal to the corresponding lines *a* 3, *b* 2, etc., on the quarter-circle. Join 0′ to 3′ with an even curve. This will be a quarter of an ellipse, and will give the half-girth of rounded ends.

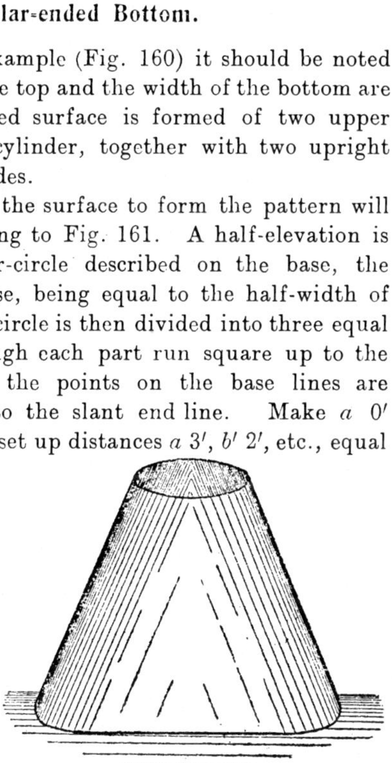

FIG. 160.

To mark out the pattern draw in a centre line as shown (Fig. 162), and a girth line square to it. For the length of the girth line, set along distances equal to the lengths of the separate parts of the girth-curve in elevation, the points being numbered in the same manner. Draw lines square to the girth line through each point, and then mark off distances 2′ 2 equal to *b*′ *b*, 1′ 1 equal to *c*′ *c*, and 0′ 0 equal to 0′ 0 taken from the elevation. The parallel lines on the pattern will now be cut off at the same length, that is to equal *a t* from the elevation. The triangle 3′ T 3′ can be easily constructed, for the line 3′ 3′ will, of course, be equal to twice the length of line *a d*. The setting out of the end parts of pattern will be nothing more or less than a repetition of the construction followed for the middle portion. For the pattern to be accurate it should be noticed that its curves run into the straight lines without lump or hollow. On examination it will be seen that all the curves are exactly the same shape, and in practice the pattern would be marked out by making a small template (like the shaded part of the end), and marking all the curves at top and bottom from this.

The above method of laying out the pattern has been purposely arranged somewhat different to that shown in Fig. 159, but either method can be applied in both cases, the choices depending upon the size and shape of articles.

Shoe-Shaped Funnel or Hopper.

A funnel may require to be of the shape shown in Fig. 163, which, on inspection, will be seen that its surface is composed of half an oblique cylinder for the front, two right-angled triangles for the sides, and half of a right cylinder (a round pipe whose ends are square to the centre line) for the back.

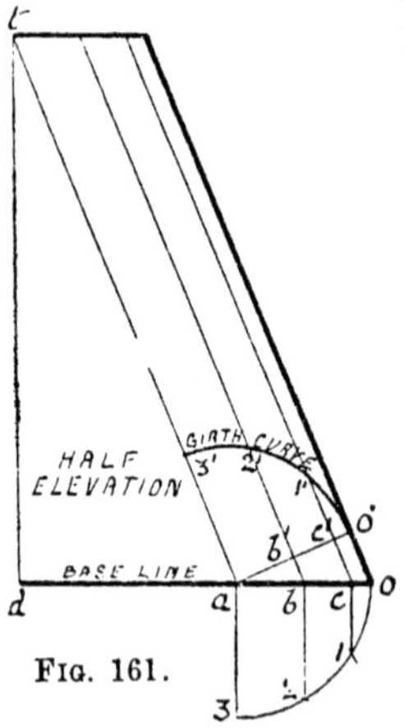

FIG. 161.

The striking out of the pattern is illustrated by Fig. 164. An elevation is drawn, and a semicircle described on the base, this being divided into six equal parts, and numbered as in the figure. Perpendiculars are run up from points

3, 4, and 5, and a line drawn through 3′ parallel to 6 *t*. On to this line the points 6, 5′, and 4′ are projected by running lines square to 3′ *a*.

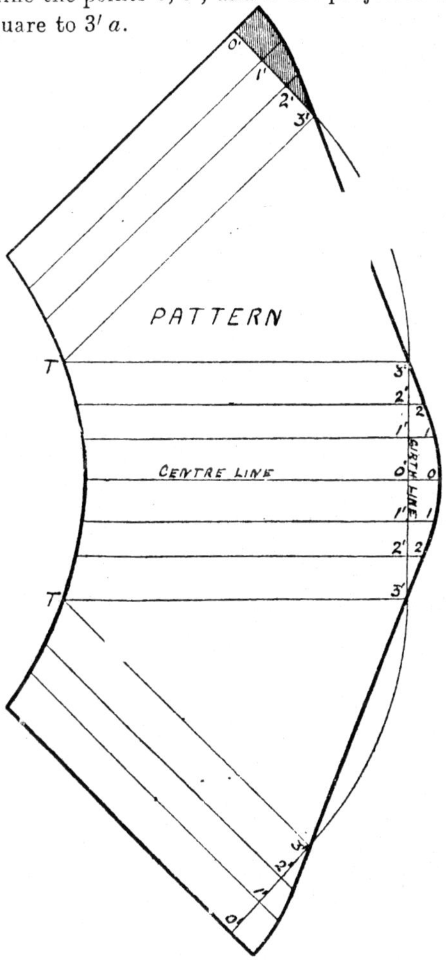

FIG. 162.

The pattern is obtained by drawing in a centre line 6 6, and marking it off equal in length to 6 *t* from the elevation. Then the distances 6″ 5″, 5″ 4″, and 4″ 3′ are measured from the elevation and stepped along the centre line of pattern, as indicated. Through these points lines are drawn square to the centre line. Now set the compasses to

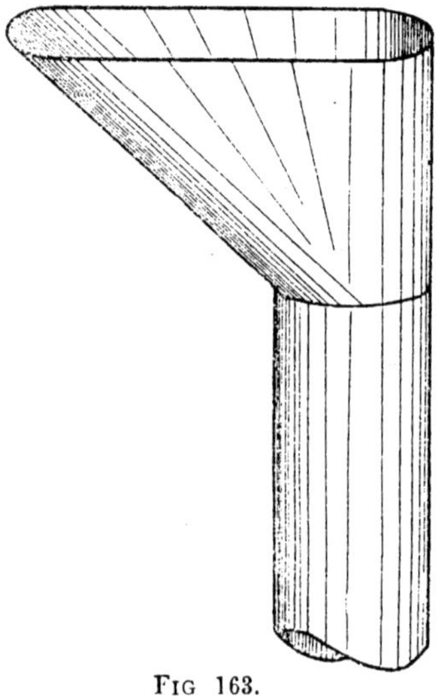

FIG 163.

the length of one of the small arcs, say, 3 to 4, on the semicircle, and commencing at point 6 on the pattern, mark off points 5, 4, and 3 by stepping from one line to the other, as seen. The right-angled triangle A B 3 is now marked out by making A B equal *a b* and 3 B equal to 3′ *b* from the elevation. The last portion of the pattern on each end is for the straight pipe part, and this will, of course, be equal in length to the quarter-circle 3 to 0.

Allowance for wiring, grooving, or other form of jointing can be added to the pattern as required.

The typical examples shown in this and the last chapter will, it is hoped, illustrate what has been said about articles whose surfaces are compounded of the surfaces of two or

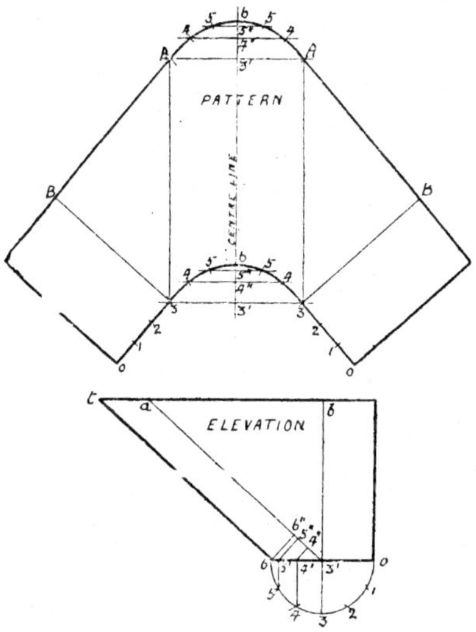

FIG. 164.

more solids. In Chapter XXV. such objects as tall-boy chimney-pots and ventilator bases are dealt with.

CHAPTER XXI.

ELLIPTICAL WORK.

THERE are many objects of elliptical shape that require to be made up out of sheet and plate metal. It is, therefore, essential that workmen in these trades should know one or two practical methods for describing ellipses. Whilst there are a great number of different ways in which an ellipse may be set out, there are only two that are of much use for workshop purposes. We shall now proceed to describe these two methods.

Construction of Ellipse.

One way of describing the ellipse is that known as the "string method," and is illustrated by Fig. 165. The diameters A B and C D are first set out at right angles, as there shown, and the points F′, F (called the foci) obtained by setting the compasses at half the long diameter and using C or D as centre, and cutting A B in F′, F. A pin or nail is now stuck in each of the points A and F, and a piece of string brought round the two nails, as shown, and tied. The nail is then drawn out of A and fixed in F′, as seen in the lower figure. The string is stretched tight by holding a scriber or pencil, as at P, and at the same time the ellipse described by moving the pencil all round, as shown. To get an accurate result, string that has very little stretch should be used.

If it is desired to mark an ellipse on a plate or sheet, where there will be difficulty in fixing the pins, a good plan is to clamp a batten on to the plate and drive the pins into this.

There is an important property of the ellipse which is worth while remembering, and that is: "The sum of the

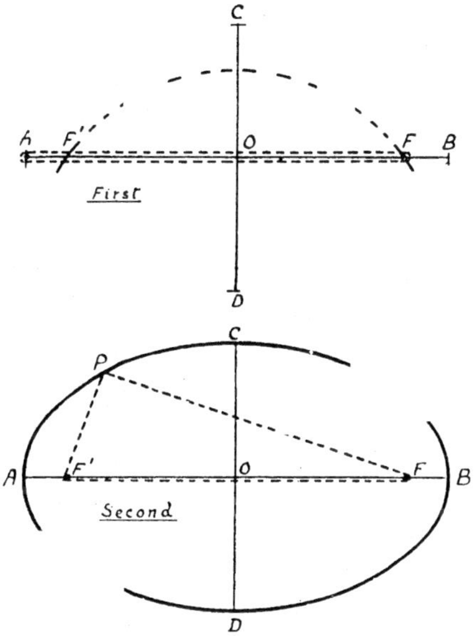

FIG. 165.

distances of the foci from any point on the ellipse is a constant quantity, and is equal to the long diameter." It is, indeed, from this property that we are enabled to construct the ellipse by the string method; for if the lengths P F′, P F be added together, they will, for any position of P, be equal to A B. Knowing this peculiarity of the ellipse, the ingenious reader should be enabled to devise one or two other simple methods for its construction.

The string method is most adaptable for large ellipses, and for smaller ones what is known as the "trammel method" is most suitable. This latter will now be explained.

A trammel (Fig. 166) need be nothing more than a strip of cardboard, wood, or sheet metal. Half the long diameter, P E, and half the short diameter, P F, of the ellipse must first be set along from the end of the trammel. And then to construct the ellipse, two lines at right angles are

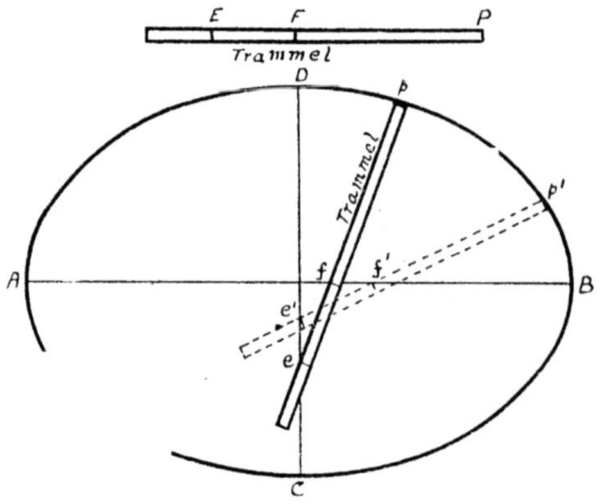

FIG. 166.

drawn, and on these the trammel placed, the points *e f* being respectively on the lines D C and A B. The trammel is now moved into different positions, points on the ellipse being obtained by marking the end P. When a sufficient number of points are obtained they can be connected with an even curve, and so the whole ellipse described. The important thing to notice is that the point *e* must always slide on the line D C at the same time as point *f* is moving along line A B. Two positions of the trammel are shown on Fig. 166.

It is not a difficult matter to make a trammel with two adjustable pegs and wood or metal cross-shaped slides (to lie along the ellipse diameters), and with this simple apparatus construct ellipses in a similar manner to that in which circles are described by compasses.

Circumference of Ellipse.

The circumference of an ellipse can be found by adding together the semi-diameters and multiplying by $3\frac{1}{7}$ Thus in Fig. 165, suppose A B = 24 in. and C D = 18 in.; then the circumference equals—

$$(12 + 9) \times 3\tfrac{1}{7} = \frac{21 \times 22}{7} = 66 \text{ in.}$$

The rule, as given, is only approximately correct, but is good enough for workshop practice when the ellipse is not very flat. Calculated by a rule giving more accurate results, the circumference should be about $\frac{3}{8}$ in. longer than above. Unfortunately, however, very accurate results are difficult in manipulation. In practice, the simplest way of obtaining the length of the circumference is to bend a thin wire along a quarter of the ellipse, as set out, and multiply this length by 4.

Area of Ellipse.

The area of an ellipse can be calculated by multiplying the semi-diameters together, and this product by $3\frac{1}{7}$. Thus, for an ellipse having diameters 24 in. and 18 in. the area equals—

$$12 \times 9 \times 3\tfrac{1}{7} = \frac{12 \times 9 \times 22}{7} = 339\tfrac{3}{7} \text{ sq. in}$$

By applying the above calculations to what has been stated in Chapter XII., the cubic contents, or number of gallons, that an elliptically conical vessel will hold can be obtained.

Elliptical Cone or Cap.

Just as we may have a circular cone, either right or

oblique, so in the same way we may have an elliptical cone. A sketch of a cone whose base is an ellipse, and whose axis is perpendicular, is shown in Fig. 167. A cap may be of this shape, or an object may be formed by some part of an elliptical cone surface. We shall now give a few examples of pattern-marking for this class of work.

In Fig. 168 the method employed to set out the pattern for a complete and also for a frustum of an elliptical cone is shown. A half-elevation of the cone *c* 0 *t* is drawn, and also a quarter of the base ellipse. This latter is divided into four equal parts, and taking *c* as centre, the points 1, 2, etc., are swung on to the base line *c* 0. The points 1′ 2′, etc., are then joined to the apex *t*. To mark the pattern out the compasses are set respectively to *t* 0, *t* 1′, *t* 2′, etc., and the arcs of circles, as shown, described from the point T. Then, fixing the compasses to the length of one of the parts on the quarter-ellipse, and commencing at 4 on the pattern, the points 3, 2, 1, and 0 are stepped from one arc to the other, the points then being joined to form an even curve. To form a complete cone, two parts like 0 T 0 would have to be cut out.

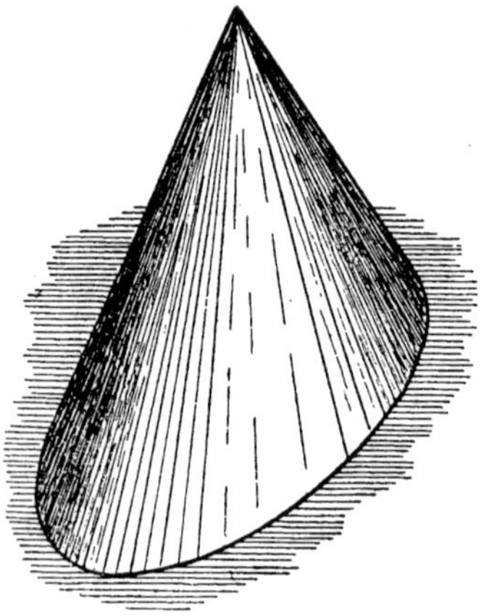

FIG. 167.

For an article made up like the shape of a frustum of a cone, the inner portion of the cone pattern would have to be cut away. Thus, suppose *a b* represents the half top of the article, then the lengths of lines from *t* down to where they cross *a b* will give the lengths of lines to mark the points to form the curve B D B. Thus, T B equals *t b*, and T D equals *t d*, and so on for the other lines.

It should be noted in setting out the shapes of tapered elliptical articles that only three dimensions for top and bottom can be worked to. In the present case we have the length and breadth of the bottom and the length of the top only. If required for shaping, or other purposes, the width of the top can be measured from *a d*, the length of this line giving half the width of the top. It should also be remembered that articles of the above description are not equal tapering, the ends having a greater overhang than the sides.

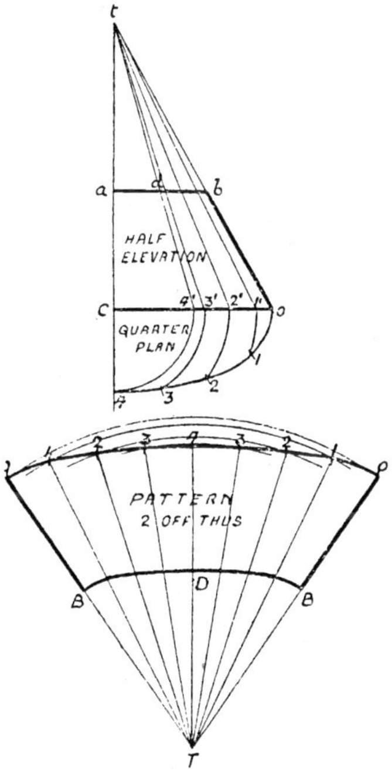

Fig. 168.

Elliptical Coal-Bucket.

There are many different kinds of elliptical coal-buckets, one of the commonest being that known as a "Waterloo," a sketcli of which is shown in Fig. 169. To set the pattern out for the body of this is generally considered a somewhat difficult task. With careful consideration, however, and some understanding of the principles of development, the reader should find the difficulties disappear.

In the example as set out in Fig. 170 it is assumed that the back and the front of the bucket have the same taper:

hence the body will come out as a portion of an elliptical cone. The elevation is drawn as shown, and the end lines produced to meet in T. The centre line T *t* is drawn square to 6 0, and produced to 3′. The semi-ellipse is described and divided into six equal parts, perpendiculars being dropped from each division point on to 6 0. Through the feet of these perpendiculars lines are drawn from T and produced to the top curve of the elevation, thus obtaining the points 1′, 2′, 3′, etc. The points 1, 2, 3, etc., on the semi-ellipse are then swung about *t* on to the line 6 0, these latter points being joined to T, and the lines produced upwards to meet the horizontals drawn through the points on the top curve of the elevation. Thus, 5″ is obtained by connecting T to 5^0 and producing to meet the horizontal line drawn through 5′. In the same way the other points 1″, 2″, etc., are fixed.

FIG. 169.

Now for the pattern. The curve 0 0 (Fig. 170) is obtained in exactly the same way as that on Fig. 168, the length of lines A 0, A 1, etc., being measured from T up to the base line 6 0. Thus A 5 equals T 5^0, and so for other corresponding lines. The pattern construction lines are then drawn from A through each point and produced outwards. These radial lines are cut off to their proper lengths by taking corresponding lengths from the elevation. Thus A *a* = T 0″, A *b* = T 1″, A *c* = T 2″, and so for the rest of the points.

Allowance for wiring is added to the top end of pattern, for throwing off and knocking up on the bottom, and grooving on the sides.

The pattern for the foot is laid out exactly as in Fig. 168, the point *c* on the elevation (Fig. 170) representing the apex of the elliptical cone. For the inner curve, the lengths are measured from *c* down to the line 6 0, and for the outer curve down to the bottom line. Thus, C 5 and C *h* on the pattern are respectively the same length as *c* 5^0 and *c h* on the elevation.

Allowance is made for wiring on the outer part, a single edge on the inner, and for grooving or riveting on the ends. Details of the methods of attaching the bottom and foot to body are also shown on Fig. 170. If made of black sheet iron the bottom edge is annealed by running around in the fire, after which it is (1st) carefully thrown off by stretching, and annealed again. The flange is then levelled with a mallet and (2nd) edged over. The bottom (3rd) is slipped in, and

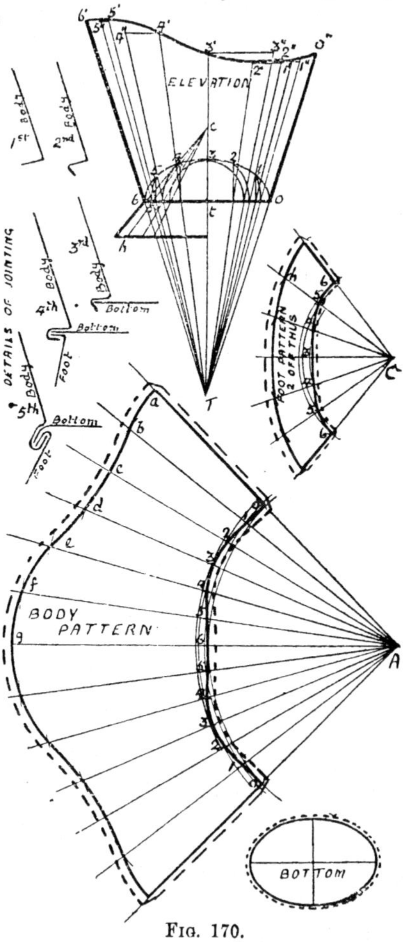

FIG. 170.

after a single edge has been turned on the foot this, also, is put in and (4th) paned down, the latter operation being best performed on a bick-iron and then run around on a hatchet-stake. The final operation (5th) is the doubling-over, or knocking up, as shown in Fig. 171. A special knocking-up hammer is used, and the bench-stake being either a head, as shown, the back end of a side-stake, or the end of a bench-bar.

Whilst the illustration (Fig. 171) shows the knocking-up process on a coal-bucket, this method of attachment, it may be pointed out, is very commonly applied to a large number of sheet-metal articles.

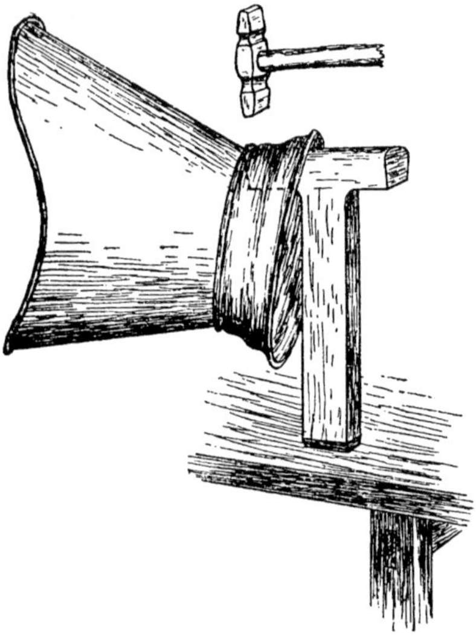

FIG. 171.

Oblique Elliptical Cone.

An article may take the shape of a portion of an elliptical cone of the above description—that is, one whose centre line is not square to the base.

The setting out of the pattern for an object of this character can be done in a similar manner to that shown in connection with the oblique cone (Chapter XVIII.). A side-elevation is first drawn (Fig. 172) and the half-plan of the top. The line T *t* is drawn up square to 0 6 produced, and then with *t* as centre and *t* 1, *t* 2, etc., as radii, the points are turned down on to 0 6, giving the points 1′, 2′, etc. These latter points are then joined to T. Now, using T as centre and T 0, T 1′, as radii, the arcs of circles are swept out. The compasses are next fixed to the length of one of the six parts of the semi-ellipse, and, commencing at 0 on the pattern, the points 1, 2, 3, etc., are stepped from arc to arc. The inner curve is obtained by drawing the radial lines on the pattern and cutting these by the arcs run around from the points where the corresponding lines intersect the bottom line *a d*.

The ends of the frustum will, of course, be similar in shape, and if it is desired to obtain the width of the ellipse at the bottom, this can be done by drawing *c b′* square to T *c*, and making *c b′* equal to *c* 3, joining *b′* to T, and then drawing *c′ b″* square to T *c*; then *c′ b″* will be the half width of the ellipse at the bottom of the frustum.

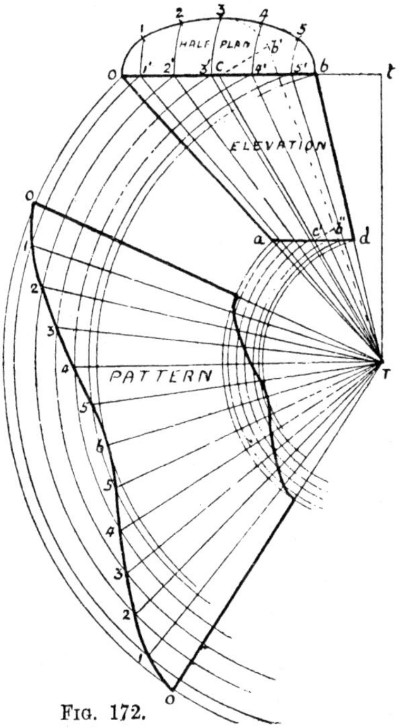

Fig. 172.

Overhanging Coal-Bucket.

A coal-bucket whose body can be set out on the assumption that it is part of an oblique elliptical cone is shown in the elevation (Fig. 173).

The back and front are produced to meet in T, and a perpendicular, T *t*, run up to meet the line 0 6. Using *t* as centre, the points on the semi-ellipse are swung down on to the line 0 6, and the lines T 1′, T 1″, T 2′, T 2″, etc., drawn as in Fig. 170. The pattern is then set out as before explained.

Elliptical Round Coal-Vase.

A coal-vase sometimes follows the shape shown in Fig. 174, the top being elliptical and the bottom round. If the top were made oval instead of elliptical, the pattern might be set out by one of the methods shown in Chapter XIX.

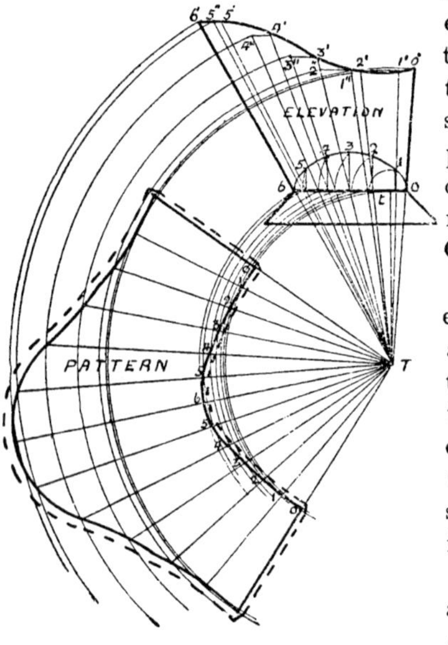

FIG. 173.

In this case, however, the method of triangulation will have to be used. It will at the same time further explain its application to articles of this description that are irregular in shape.

An elevation and a quarter plan are drawn as shown, the quarter ellipse and quarter circle each being divided into three equal parts. The points on the plan are connected up, the lines thus representing the plans of the six triangles that make up a quarter of the complete body surface. To set out the pattern we shall require to get the true lengths of all the lines shown in plan. The first line of the pattern to set down should be 3 D, this being made equal in length to 3′ *d* in the elevation. To obtain the true length of the required second line (3 C), set 3 *c* along the base lines from 3, thus marking the point *c*′; then 3′ *c*′ will give the length of 3 C. The small arc through C will be drawn by using point 3 as centre and the length 3′ *c*″ as radius. The compasses are now set to the length of one of the arcs on the quarter circles, and with D as centre the point C is cut. The line 2 *c* is now set along the base line from 2′, thus fixing *c*″, this latter point being joined to 2″. The line 2″ *c*″ gives the length of 2 C on the pattern. As before, an arc is now described shown passing through 2, using 2″ *c*″ as radius and C as centre, this being cut by using 3 as centre and the length of one of the parts

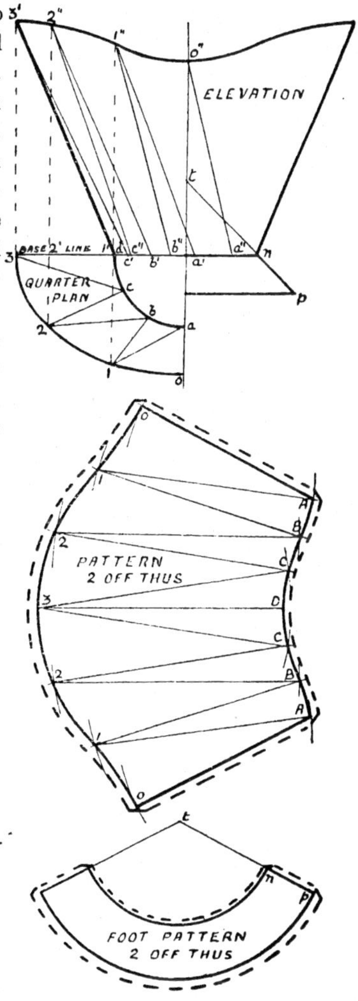

Fig. 174.

on the quarter ellipse as radius. Thus point 2 on the pattern is determined. The point *b'* is fixed by making 2' *b'* equal to 2 *b*, then 2'' *b'* will give the length of 2 B for the pattern. The length 1' *b''* is made equal to 1 *b*, line 1'' *b''* then giving the length of 1 B. The line 1' *a'* is marked along equal to 1 *a*, thus giving 1'' *a'* the length required for 1 A. And then the finishing line 0 A is obtained by setting 0 *a* along the base line from the foot of the centre line to give *a''*, and measuring off 0'' *a''*. In this way the twelve triangles that build up the half pattern are set out. The points are joined up, and allowances put on as before.

The foot being a frustum of a right circular cone, the pattern will be marked by making the radii *t n* and *t p* on the pattern the same length as the letters denote in the elevation.

The methods of jointing can be the same as before, or the bottom can be knocked up on to the body, and the foot slipped over and riveted.

Before passing from the above example in the use of the method of triangulation, it is perhaps as well to point out that the true lengths of lines can be obtained by drawing a pair of lines at right angles, and setting along these the respective distances from plan and elevation; those from the plan being measured along the horizontal, and those from the elevation up the vertical, the slant lines then giving the true lengths for the pattern.

CHAPTER XXII.

ROOFING WORK.

Galvanised Sheets and Gutter Angles.

GALVANISED corrugated sheet iron has an extensive application in roofing work. It is comparatively cheap, and when properly galvanised fairly durable. There is much dispute as to the length of time it will last. No definite "life," however, can be assigned to a galvanised iron roof, except all the conditions are fully known, and these are most difficult to determine, the length of time it will last depending upon the quality of iron and galvanising, thickness of sheet, and the kind of atmosphere the roof is placed in. In the sulphurous atmosphere of a manufacturing town it is probable that galvanised iron will not last more than one-quarter the time that it will in a pure country air. And, again, it will last longer in a dry atmosphere than in a moist one.

Galvanised iron is iron coated with zinc, and this latter metal has the distinct advantage of forming an oxide on its surface that is not dissolved by ordinary rainwater. If the water, however, becomes by any means acid, as it does in the neighbourhood of towns by dissolving the acid fumes, then this protecting film of oxide is eaten away, the coating of zinc soon disappearing and the sheet iron rusting into holes. When the galvanised iron begins to show signs of deterioration, it is a good plan to at once paint it, and to follow this up periodically. A good paint to use is one of a metallic oxide character. Common tar should not be used, as this is not a good medium for protecting sheet iron.

In fixing corrugated sheets it is usual to have a side lap of one corrugation, as shown in Fig. 175 (*a*), and this should be arranged the same way up as in the sketch, and not upside down, as one occasionally finds sheets erected. In the latter case, the joints are almost sure to leak. A safer joint is (*b*), in which two corrugations are lapped over. This makes a much better job, but adds somewhat to the cost, both in labour and material. The end laps of

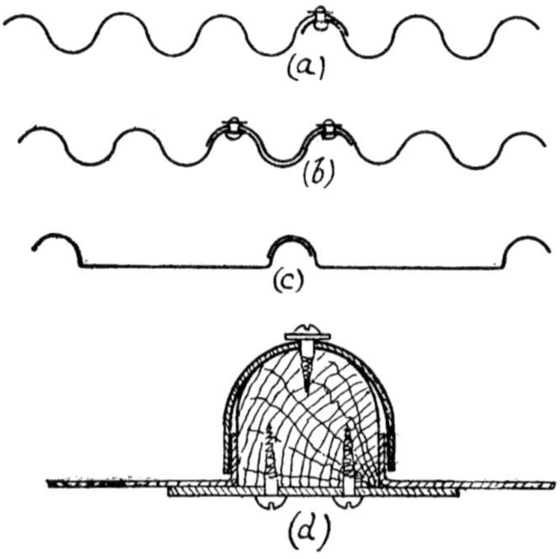

FIG. 175.

sheets run about 6 in., sometimes less, and occasionally more. The longer lap is always preferable, especially if the roof is flat. Where much snow is likely to lodge a large end lap is the safest, so as to avoid as much as possible the backing-up of the water. It is usual to fix a washer (made out of about 16 gauge iron) on the rivet before hammering down and snapping. This is to avoid leakage around the rivet. Sometimes rubber washers are used in addition, and on very exceptional occasions canvas packing is placed between the joints. It should be remembered that, however carefully galvanised iron is fastened to timber, the holes in the sheets are bound to pull a little and get loose on account of the difference in expansion and contraction, due to changes of temperature, between metal and wood. On a wholly iron structure this is not so bad; but even in this the intensity of the sun's heat-rays will cause a greater expansion in the galvanised sheet iron than in the framework underneath.

Tiles are sometimes formed out of galvanised sheet iron, and also out of sheet zinc, as shown in Fig. 175 (*c*). They are lapped over and nailed or riveted the same as corrugated iron. Sheet zinc for roofing purposes unfortunately has a high degree of expansion and contraction for changes of temperature, and, therefore, should never be fastened together in long lengths. A method that can be followed to overcome this difficulty is explained by Fig. 175 (*d*). The timber section represents a rafter or roll, to the bottom of which a flat plate is secured. The sheets have the edges turned up on each side, and dropping in between the rolls rest on the flat plate. The caps can be made up in short lengths (say, about 3 ft.), and fitted over roll and edge of sheet, and lapped over each other. No nails or screws should pass through the joints, so as to allow perfect freedom for expansion and contraction.

In the case of curved sheets of either corrugated iron or zinc there is not so much trouble with expansion and contraction, as the change of length is taken up in increased or decreased curvature.

Roofing Fittings.

The roofing sheet metal worker is called upon to make mouldings, gutters, ventilators, finials, downspouts, and pipe bends of all descriptions, and in addition much intricate work in the ornamental line. We will take, first of all, a few cases of pattern-cutting for

Moulding or Gutter Angles.

The commonest form of a gutter angle is perhaps of a square elbow for a half-round gutter (Fig. 176). It may be made out of thin galvanised sheet, say 24 to 20 gauge, having a bead or flange along the edge and a soldered joint, or, as in the sketch, made out of 16 or 14 gauge black iron, riveted at the joint and galvanised or painted afterwards. The setting out of the pattern, which is explained by Fig. 177, is a simple matter. A semicircle is described, as shown, and divided into six equal parts, and the girth line of the pattern made the same length as the semicircle either by calculation or marking along six lengths, each equal to one of the parts on the semicircle. From each of the division points on the girth line a perpendicular is run up, and from the points on the semicircle lines are drawn parallel to the girth line. The intersection of corresponding lines will give points on the pattern curve. Thus, where the line drawn up from 2 on the girth line intersects

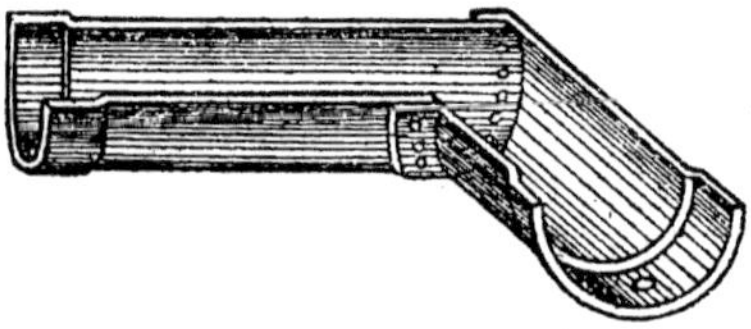

FIG. 176.

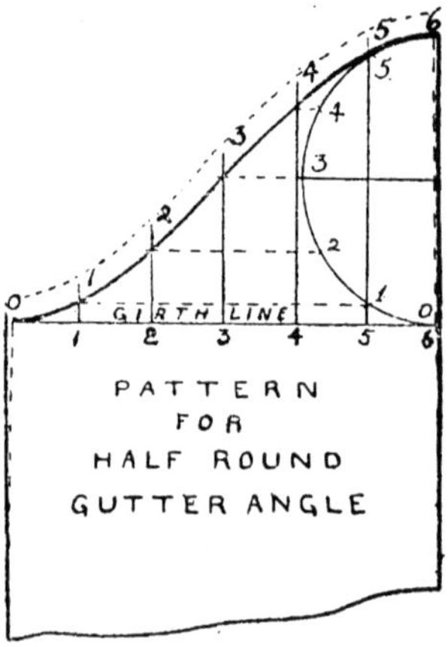

FIG 177.

the line drawn through 2 on the semicircle will give point 2 on the pattern curve. In the same manner all the other points can be determined. A free curve being drawn through the points, the net pattern is complete. The lap for flanging is added, as shown by the dotted line. The arm of the gutter angle which fits inside will, of course, not require any lap. It will also be an advantage to have the girth of this arm a little less than the other, the side lines of the pattern being cut slightly tapered, as shown, by the two dotted lines running along the sides of the pattern.

A good deal of care is necessary in the flanging, this being best carried out in the case of thick gauges by throwing over when hot. The holes for rivets should be punched in the plate after flanging, the holes for the inside arm being marked from these and punched by the use of a burr.

Obtuse Gutter Angle.

To set out the pattern for a gutter or moulding angle which is required to fit on or into a greater angle than a right angle, will demand a somewhat different method to that shown in the last case. Thus, suppose an elbow is wanted to fit on to an angle of 130^0, as in Fig. 178, then some such method as that illustrated by Fig. 179 will have to be used. The shape of the section is first set out and a line E F drawn across the top.

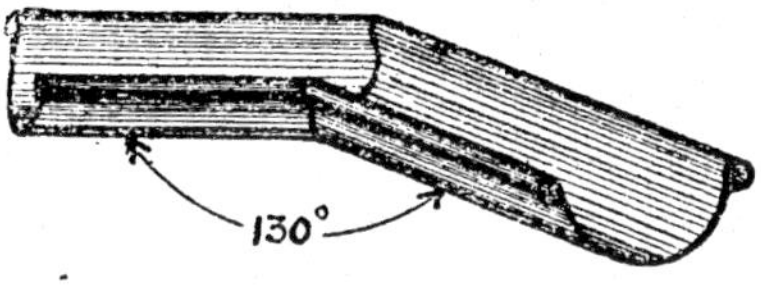

FIG. 178.

Two parallel lines C D and A B are drawn perpendicular to E F, these representing the width of gutter. Across these a line E A is drawn, called the joint line in the figure, making an angle of $\frac{130}{2} = 65^\circ$ with A B. The gutter section is divided up into any number of parts, such as 0 to 1, 1 to 2, 2 to 3, etc., and lengths equal to these set out to obtain the full girth of the gutter. It will be seen that the gutter section is divided into fourteen parts, hence the total girth of the section as laid out on the girth line of pattern will run from 0 to 14, as shown by the numbers. Through each division point on the section lines are drawn square to E F, and produced until they cut the joint line. From the points where these lines meet the joint line, dotted

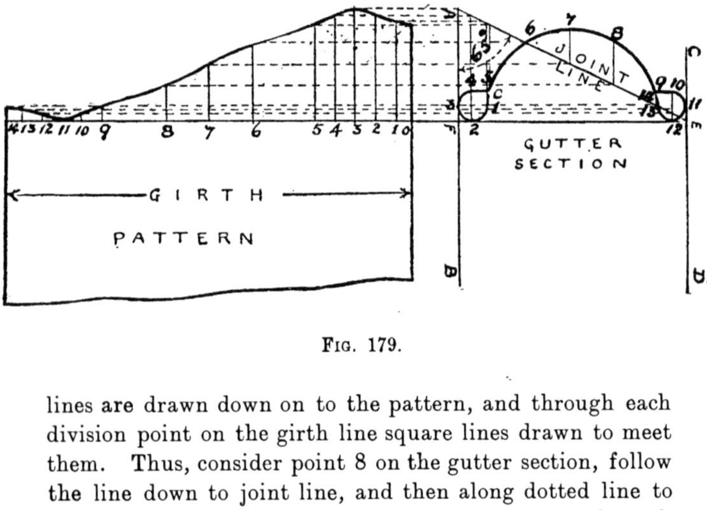

FIG. 179.

lines are drawn down on to the pattern, and through each division point on the girth line square lines drawn to meet them. Thus, consider point 8 on the gutter section, follow the line down to joint line, and then along dotted line to pattern, where it will intersect with the line drawn through point 8 on the girth line. This gives a point on the curve to form the cut of pattern. In the same way all the other points can be determined, and thus the pattern completed. No lap will in this case be needed, the edges of metal being butted together and soldered.

In bending sheets or plates for angles of moulding, care

must be taken to bend them into pairs of right and left hand.

Where the gutter or moulding has many bends, it must be formed to the exact shape of section, or else it will be most difficult to fit the corner joint together.

Square Angle for O.G. Gutter.

Fig. 180 shows a sketch of an internal angle for an O.G. gutter, and patterns for both internal and external angle-pieces. A section of the gutter is set out as shown on the pattern for an external angle. This is then divided into seven parts—0 to 1, 1 to 2, 2 to 3, etc.—and these lengths measured and set out to give the girth of gutter or width of pattern. From the points on the section, lines are drawn down, and from the corresponding points on the girth line, lines are drawn across. Where these meet give points on the pattern curve, as will be seen. The points are joined up, and thus the cut of pattern obtained. In joining up it should be remembered that where the line on the section is straight, the corresponding part on the pattern will also be straight. Thus 5 to 6 is seen to be straight on the section; hence on the pattern curve the line joining these

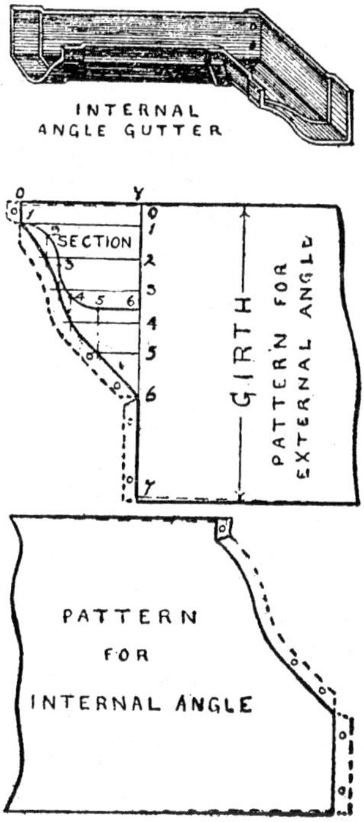

Fig. 180.

two points will also be straight. For heavy sheet iron the laps will be as shown by the dotted lines, and as in former cases flanged over when hot. For light galvanised sheet iron laps will be allowed on the straight parts of cut only; the edges of the curved part butting together, and being soldered from the inside. Laps will, of course, only be needed on one arm of the elbow.

The pattern for an internal elbow can be struck out as above, or, which is much better, when the pattern for the external angle is cut out can be marked off it as shown in the lower figure of Fig. 180. The cut of the end of pattern will be exactly the same as in the external angle, but used in the reversed manner. Laps will be as shown in the figure. Holes can be punched in the laps that will remain straight after the plate is bent, as these will not interfere with the part that has to be flanged over. The pattern for the inside arm should be slightly tapered, as in the half-round gutter angle, and this is shown by the side dotted lines.

It might be as well to here explain that an external angle-piece is an elbow which is supposed to fit *on* a corner, and that an internal angle-piece is an elbow which is made to fit *into* a corner.

Obtuse or Acute Elbow for O.G. Gutter.

An obtuse elbow is one whose arms are extended to an angle which is greater than a right angle (90^0), and an acute elbow one whose arms are opened out less than a right angle.

The method here given will apply to either case, and, indeed, might have been used for the square elbow instead of that shown in Fig.180; but for that particular angle-piece the method illustrated by Fig. 181 will not be so good as the one previously explained.

As this problem of jointing together two pieces of gutter or moulding to form a mitre or bevel joint is important, we will fully explain it by means of Fig. 181. To take a concrete case, let us suppose that the arms of the elbow make an angle of 100° with each other.

The exact shape of the section must first be set out. The double curve of this being drawn by dividing the straight

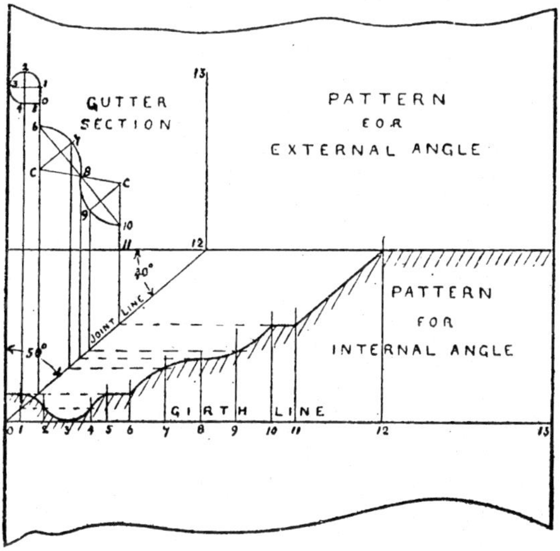

Fig. 181.

line 6 to 10 into four equal parts. Through two of the points, as shown, draw perpendicular lines. Produce line 11 to 10 up, and line 5 to 6 down to meet these lines, thus obtaining points C C, the centres of the arcs. Before drawing the curves in, it is as well to join the two centres by the line C C, and where this crosses the line 6 to 10, will be the meeting point of the two arcs. (Particular notice should be taken of this construction, as double curves are often required in sheet metal work.) Now to set out the patterns. First a plan of the joint line must be drawn, and as the angle of the elbow is 100° the joint line will make $\frac{100}{2} = 50°$ with the outside line, as shown. For construction purposes, however, it will be easier to set the joint-line angle from the back of the section, and a general rule for obtaining this angle will be: "Deduct half the elbow angle from 90°." Thus, in this case the angle will be—

$$90 - \frac{100}{2} = 40°,$$

and this will be set out as shown in Fig. 181.

The section is divided into parts 0, 1, 2, 3, 4, etc., up to 13, and lines drawn down through each point on to the joint line. From the end of the joint line a girth line is drawn, as indicated, and the exact lengths of 0 to 1, 1 to 2, 2 to 3, etc., from the section set along this line. Lines are now drawn up through each of these points square to the girth line, and where they intersect the corresponding dotted line will give a point on the pattern curve. Thus, for instance, the dotted line which is drawn from the joint line at the foot of the line drawn down through point 9 on the section will intersect the line drawn up from point 9 on the girth line. So with each other pair of lines. If the pattern curve be carefully cut along, the upper portion of the figure will give a pattern for an external angle or elbow, and the lower part a pattern for an internal angle.

It should not be forgotten that, whilst the bending up of this class of work is simple, the highest degree of accuracy in striking out the patterns and in forming the moulding or guttering to the exact shape of section is essential if the parts are to fit together properly. All sheet metal work of an ornamental character, if it is to look well, must be made as neatly as possible, having neither lumps nor hollows nor superfluous solder about the joints.

Valley Gutter Elbow.

To mark out the shape of sheet to form a right-angle elbow for a square valley gutter (Fig. 182) is an easy matter. The girth is first laid out (Fig. 183) by setting along the width of bottom, depth of sides, and breadth of flanges, the cut for the mitre being formed by making *a b* on the pattern equal in length to the line A B

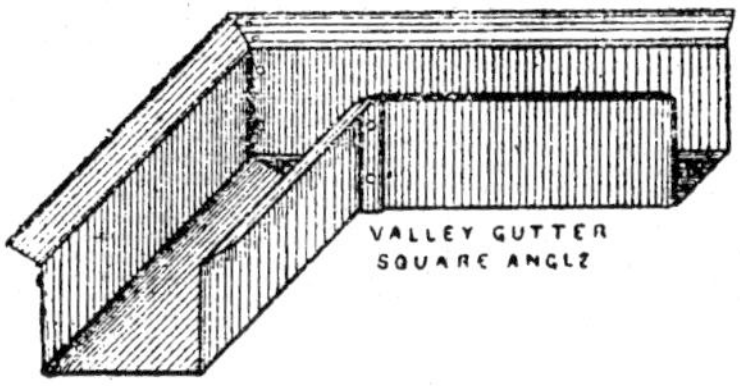

Fig. 182.

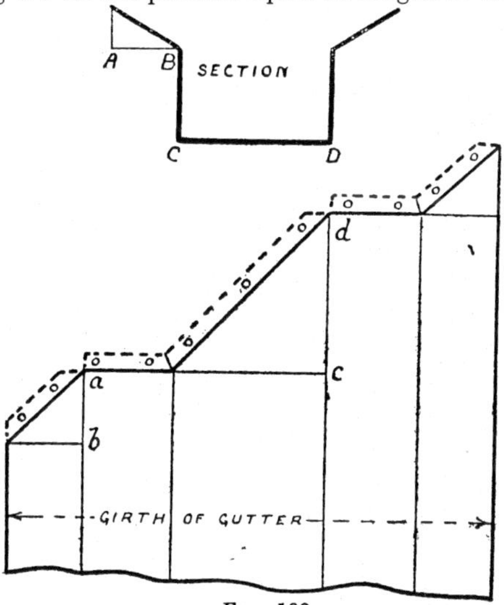

Fig. 183.

on the section, and *c d* equal to C D. The flanges or laps for riveting are added on as shown. After the joint is riveted, it should be carefully soldered along to prevent leakage.

Special Method for Square Elbows.

Before leaving gutter or moulding angles it will be as well to call attention to a special method that can be applied to square elbows, in the striking out of patterns to form the cut for any shaped section.

The shape of moulding is first set out (Fig. 184) and divided up into any convenient number of parts. Lines

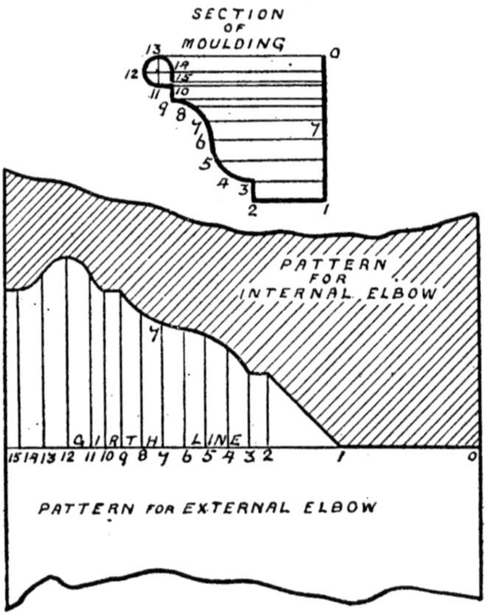

FIG. 184.

square to the back are drawn across through each division point. The girth line of pattern is marked down and lines drawn up square through each division point of this, these lines being cut off equal in length to the corresponding line on the section. Thus, to take one line only, the line **7 7** on the pattern will be the same length as **7 7** on the section; similarly the lengths of the other lines can be set off.

In practical sheet metal work it is most difficult to project lengths from one view or figure to the other; hence it is always the best plan to transfer the lengths with the compasses, as in the above case.

CHAPTER XXIII.

ROOFING WORK *(continued)*.

Cornices, Mouldings, and Ridge Caps.

In the previous chapter we dealt with the marking out of patterns for sheet metal moulding or gutters that form a plain mitred joint, and in this chapter we purpose explaining the way in which cornices, guttering, etc., may be jointed, and the patterns laid out when they meet at a double rake.

In Fig. 185 a sketch is shown of a cornice running along the eaves and up the gable of a building. By jointing the corner at right angles in the ordinary way, and then forming another joint to turn the cornice up the edge of the gable, the problem in this case becomes a comparatively easy one.

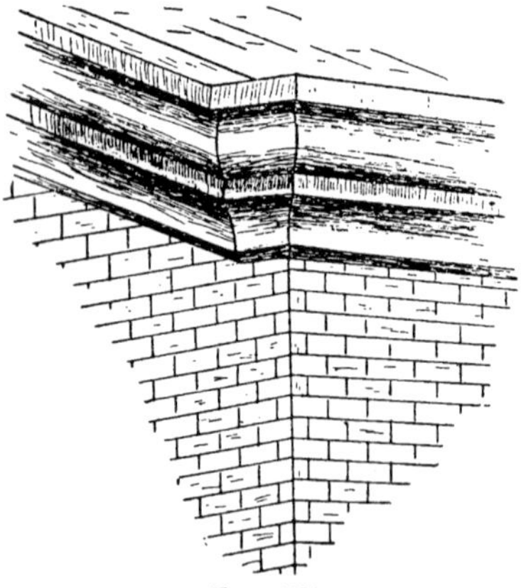

Fig. 185.

An elevation of the moulding is shown in Fig. 186, the roof being pitched at an angle of 30°. The section is divided up and numbered, the sum of the lengths 0 1, 1 2, etc., up to 9 giving the girth of the moulding.

Two pattern cuts will be required, one for the part of the cornice that is horizontal, and the other for the joint formed by the small horizontal piece at turn of corner and the part of cornice running up the gable edge. The girth of the section is first laid out for the widths of the patterns, and the cross lines drawn as seen in the figure. The cut for the corner mitre will be set out in the ordinary way by making the cross lines equal in length to the correspondingly numbered lines on the moulding section. Thus, 1 1′ = 0 1, 2 2′ = *a* 2, 3 3′ = *a* 3, and so on for the remaining lines.

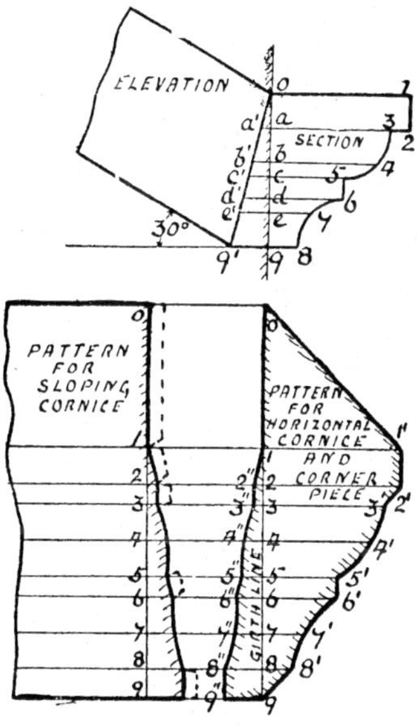

FIG. 186.

The cut for the gable cornice joint will be laid out by making 2 2″ and 3 3″ on the pattern each equal to *a a*′ on the elevation, 4 4″ equal to *b b*′, and so on for the remaining lengths. The cut on the pattern for the sloping piece of cornice will, of course, be the same as that on the left side of the corner-piece pattern, only reversed.

After cutting the sheet zinc or galvanised iron to the required shape, care must be taken to bend the pieces right and left hand, so that the edges will come together correctly.

Strong joints can be made by carefully soldering on the inside of cornice, and, if required particularly strong, laps can be left on the straight parts of the edge of one pattern (shown by the dotted lines), and these turned on to the insides of the moulding and soldered or riveted as desired.

Oblique Cornice Joint.

Instead of turning the moulding round the corner and up the gable by two joints, as in the last case, sections of cornices may be made that will come together in one joint at the corner. This particular method of jointing is illustrated by Fig. 187. The shape of one of the mouldings

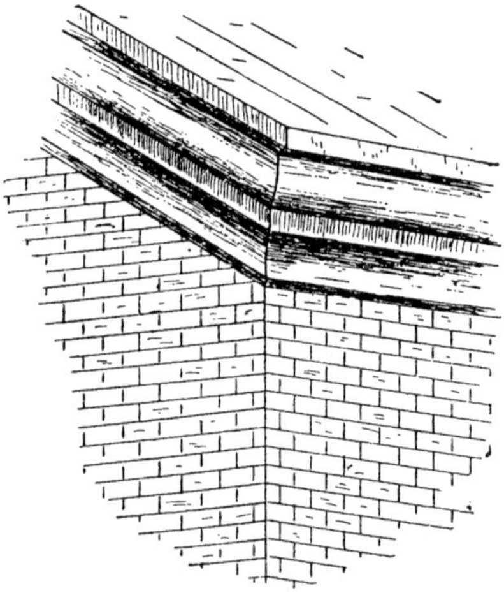

FIG. 187.

must first be fixed (in this case the one along the eaves), and the section for the other projected from it.

In Fig. 188 a section of the horizontal moulding is shown set out, and from this the shape of the gable cornice projected. The latter is obtained by drawing a line S S, at any part, square to the gable-slope line, through the numbered points on the eaves section, running up lines perpendicular to it, and cutting these off, to the left of S S, equal in length to the similar lines on the eaves section. Thus, $a' 1' = 0\ 1$, $a'' 2' = a\ 2$, $a^0 3' = a\ 3$, and so for the other pairs of corresponding lines. The points obtained are then joined up (as shown by the dotted lines), the resulting figure being the shape to which the moulding for the gable must be made.

In marking the pattern it should be remembered that the cut for the eaves moulding will come out as in the last case—that is, the same as in an ordinary flat square-mitred joint. The cut for the pattern of the sloping cornice will be obtained by first setting down a girth line equal in length to the girth of the projected section—that is, by making $0'' 1'' = 0' 1'$, $1'' 2'' = 1' 2'$, $2'' 3'' = 2' 3'$, and so on. The lengths of the construction lines on the pattern are measured from the elevation—that is, $0'' 0^0$ is made the same length as $0' 0$, and $1'' 1^0 = a' 1$, $2'' 2^0 = a'' 2$, $3'' 3^0 = a^0 3$, $4'' 4^0 = b' 4$, and in the same way for the remaining lines.

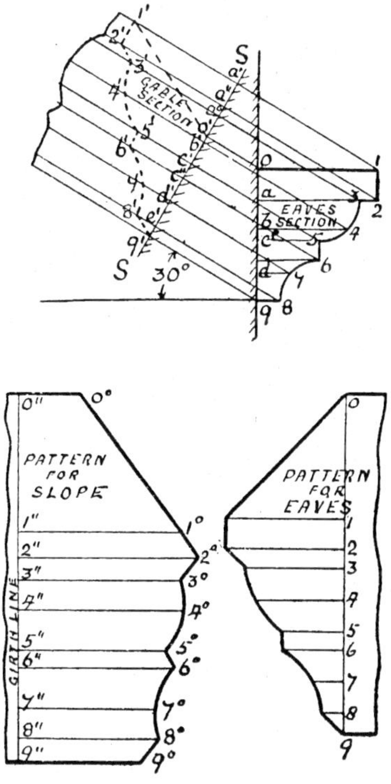

FIG. 188.

For the joint to be made properly and with ease, care

must be taken that the setting out is done accurately, and that the sheet metal is bent to the exact shape of the respective sections.

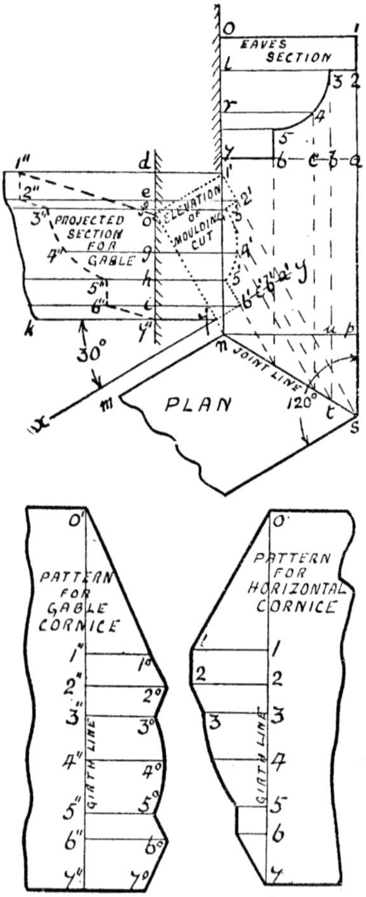

Fig. 189.

Double-Rake Moulding Joint.

Where the gable-end of a building is not square to the sides, but is inclined, the problem of connecting the two mouldings with a single joint becomes more difficult than in the last case. It represents, perhaps, one of the most complicated cases of sheet-metal cornice jointing it is possible to have. However, if the reader carefully follows each step in the setting out as shown, he should, even without a very extended knowledge of geometry, be able to accomplish the task of striking out a pattern.

One example of this class of jointing is shown in Fig. 189, in which the gable-end of the building makes an angle of 120° with the sides, whilst the pitch of the roof is 30°.

The shape of the section is first set out, and a plan drawn showing the required angle of 120°. From each numbered point on the section projectors are run down to the joint line, the line $n\ p$ then being drawn square across.

The pattern for the cut on the horizontal cornice can now be set out. First, lay down the girth line, as shown, by making it equal in length to the sum of the numbered parts on the section. Then through each point draw lines square across, and cut these off equal to the lengths of the lines between $n\ p$ and the joint line. Thus, 1 1 and 2 2 are each equal to $p\ s$, and 3 3 will be the same length as $u\ t$, the other lengths passing through 4, 5 and 6 being cut off in the same manner.

Before the pattern for the gable-cornice can be laid out, the length of its construction lines must be obtained, this being done by drawing a side elevation of the inclined cornice. Draw $x\ y$ parallel to $m\ n$, and then from the latter line run up a prependicular from n to intersect $x\ y$ in $7'$. Now draw the line $7'\ k$ at the required angle of 30°. Through each point on the joint line run up projectors, and cut these off, above $x\ y$, to the heights of the corresponding lines drawn above 7 a up to the eaves section—that is, $a'\ 1' = a\ 1$; $a'\ 2' = a\ 2$, $b'\ 3' = b\ 3$, and so on for the rest of the lines. If the points as found are joined up, it will be seen that the figure (shown marked out by small dots) will be the elevation of the moulding-cut. From this, the projected section for the gable-cornice can be obtained. Draw the line $7''\ d$ perpendicular to $k\ 7'$, and passing through $0'$; then mark off $d\ 1''$ and $e\ 2''$ each equal to 0 1. Afterwards, $f\ 3''$ should be made equal to l 3, the length $g\ 4''$ the same as r 4, also $h\ 5''$ and $i\ 6''$ each equal to 7 6. Joining the points up, the figure (shown by long dots) will give the shape of cornice-section, for the gable, that will join on to the given eaves section.

For the pattern of the gable-cornice cut, the girth line

will be measured from the projected section, and it will be seen that the same numbers are used in both section and pattern. The lengths of the construction lines on the pattern are cut off equal to those on the right hand of 7″ *d* measured up to the points 1′, 2′, 3′, etc. Thus, on the pattern, $1'' \, 1^0 = d \, 1'$, $2'' \, 2^0 = e \, 2'$, $3'' \, 3^0 = f \, 3'$, and so on for the remaining lines.

In bending into shape, the gable-cornice pattern will, of course, be bent to the projected section, whilst the pattern for the horizontal moulding will be shaped to the eaves section.

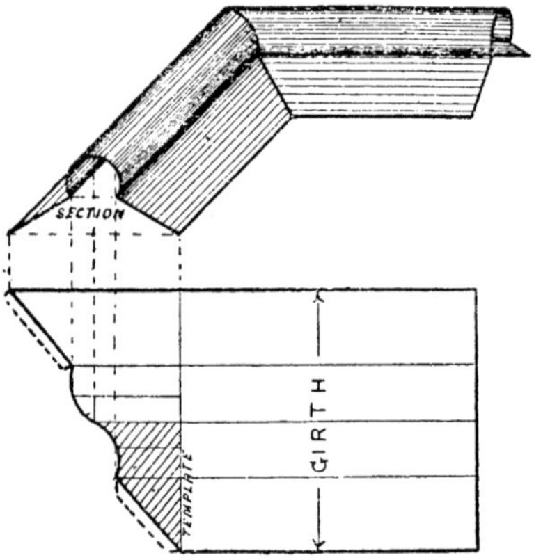

FIG. 190.

Although the setting-out in the last three cases has special reference to sheet-metal roofing work, it should be borne in mind that the principles will apply to all kinds of work where moulding or beading has to be fixed in a similar manner.

Ridge-Cap Elbow.

The setting-out as shown in Fig. 190 needs little explanation. The girth as taken from the section is first laid out, and construction lines drawn through each division point, these being cut off by the respective distances as shown projected down. In practice, it might be here remarked, this method of projecting lengths is hardly permissible, on account of the liability of error and inconvenience of drawing long lines parallel. The lengths should be measured from the section shape and transferred directly to the pattern. It will be sufficient, in practice, to mark out a piece like the shaded portion on the pattern, as this can be used as a template to strike out the remaining part by reversal.

Ridge-Cap Tee-Piece.

On comparing the shaded parts of the patterns in Figs. 190 and 191, it will be seen that they are exactly the same; hence the template for the elbow can also be used to mark out the patterns for the tee-piece. The template can first be

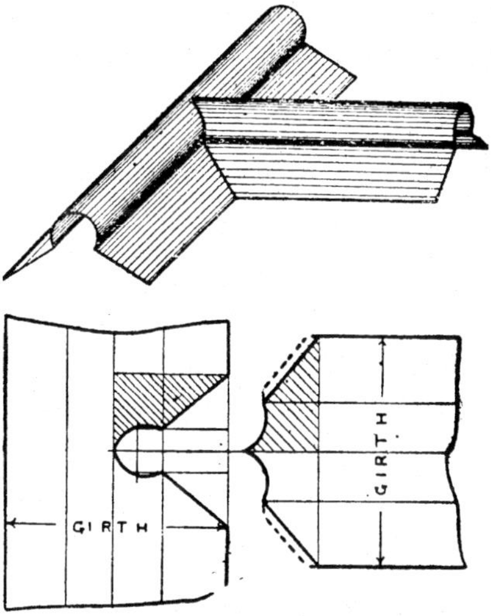

FIG. 191.

fixed in the position of the shaded parts (Fig. 191), and then reversed, to scribe out the other sides. The patterns, of course, could also be struck out by direct measurement, as in the case of the elbows.

CHAPTER XXIV.

ROOFING WORK *(continued).*

Domes, Finials, and Downspout-Heads.

Dome-Covering.

In cutting out the shape of the segments for a dome-covering (Fig. 192), no great skill is required. All the setting-out necessary is shown in Fig. 193. A half-section of the dome is drawn, and divided up into convenient parts, and numbered as seen. The dome being octagonal, the angle that the plan of the joint line will make with the base line will be—

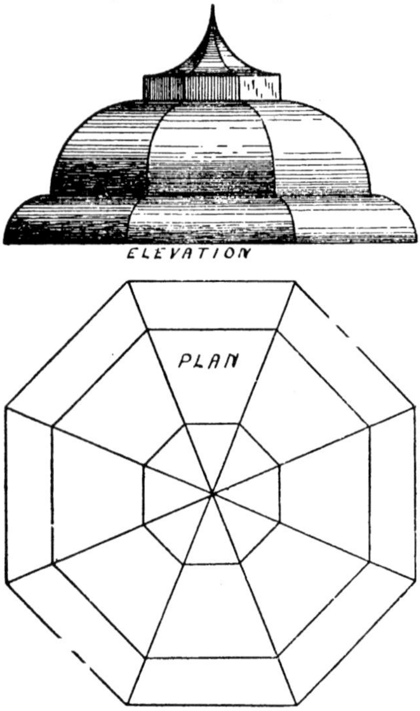

FIG. 192.

$$\frac{360}{8 \times 2} = 22\tfrac{1}{2}^\circ.$$

When the joint line is drawn in at the required angle, the figure below the base line, it should be noticed, will represent the plan of half a segment.

Lines are run down from the division points across the base to the joint line. The girth line is laid out in the usual way, and construction lines drawn across, these being cut off equal to the length of the corresponding lines drawn between the base and joint lines. Thus, to give one example, line 8 8° on the pattern is the same length as 8′ 8″ on the plan, and so with the rest of the lines.

If there is a ridge-roll at the joints, then the width of this must be allowed for in marking out the pattern, one-half the width of roll being set along inside the plan of joint line, which will give the required reduction.

The number of pieces of sheet metal required to make up a segment will, of course, depend upon its size. There should be no trouble, however, to determine this when once the shape of a complete segment is marked out.

If a roll or ridge-cap has to be bent to the shape of a joint, then this can be accomplished by first marking out the exact shape of a joint. This is shown set out at the bottom of Fig. 193.

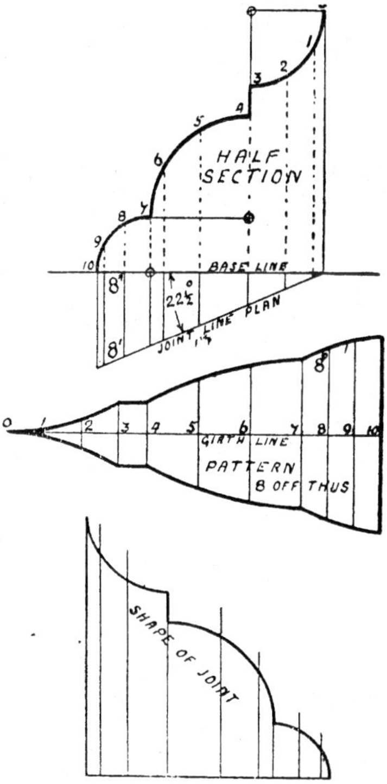

Fig. 193.

The joint-line plan and its division points are laid down, and construction lines run up from the intermediate points, the lengths of these being cut off equal to the heights of the corresponding line on the half-section. It should be observed that the three curves of the joint shape come out as quarters of ellipses, and, if desired, can be marked out by the trammel method, as explained in Chapter XXI.

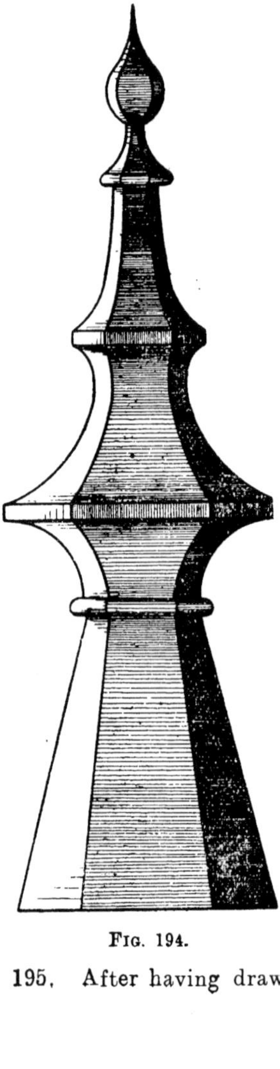

FIG. 194.

Roof Finial.

There can, of course, be a multitude of designs for a sheet-metal finial, all depending upon the taste of the designer, the limit of cost, and the kind of building that the finial is to be fixed upon. For a high building it should be remembered that small details of ornament on the finial are a waste of time and money, as they are, of course, not noticed from the ground.

A very simple form of hexagonal finial is shown in Fig. 194. It can be made out of either copper, brass, zinc, or galvanised sheet iron, the latter two metals being the ones usually chosen.

The half-sectional elevation and the pattern for one of the strips are shown set out in Fig. 195. After having drawn in the shape of the section as seen, the curves that form the outline are divided up into a convenient number of parts; there being seventeen in the present case. Lines are then drawn at points *a*, *b*, *c*, and *d*, making angles of—

$$\frac{360}{\text{twice number of sides}} = \frac{360}{12} = 30^\circ$$

with the cross lines; these really being plan views of one of the joint lines. Lines are then drawn down through each point parallel to the centre line on to one of the 30^0 lines.

For the strip pattern the girth line is first stretched out, its total length being made up by adding together the lengths of the different parts, as numbered on the outline in the sectional elevation. Lines are then drawn across the girth line through each numbered point, and the lengths of these cut off equal to the corresponding lines in the elevation between base and joint lines. Thus, for example, the lines 6 6′, 8 8′, and 9 9′ on the pattern will be respectively equal to the lines 6′ 6″, 8 8′, and 9′ 9″ on the elevation. In exactly the same way all the other lines required for marking the width of the pattern at the different parts can be measured from the elevation.

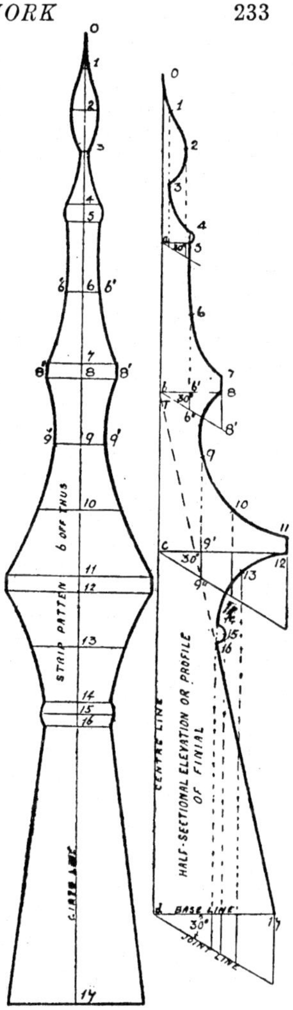

FIG. 195.

It will be noticed that

there are four 30° lines in the elevation, the object of the three top ones, of course, being to avoid having to run the dotted lines for the widths all the way down to the base line.

Instead of having the whole strip in one piece, it can, if necessary, be divided up into any number of parts, depending upon the size of the finial. After the pieces are connected, the sections can then be jointed together.

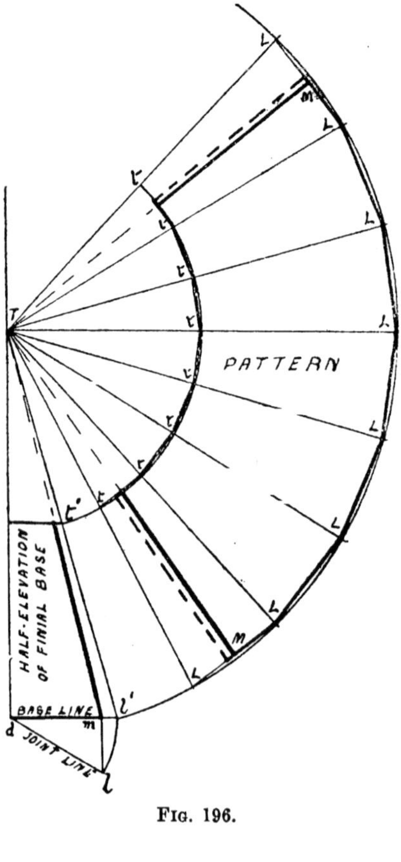

FIG. 196.

Sometimes the base part of the finial is made separately, which, it can be seen in this case, will come out as a frustum of a hexagonal pyramid. The apex of a complete pyramid can be found by producing the line 17 to 16 (Fig. 195) up to meet the centre line at T. The setting out of this frustum is shown separately, and to a smaller scale, in Fig. 196. The point l is swung around d to l', and joined to the apex T. Then taking T as centre and T l' as radius, the arc is described as shown. The compasses are now set to twice the length $l\ m$, and the points L stepped around the arc. It is a good plan to mark off one side more than is required (in this case seven), and then bisect the two end parts to obtain the seam exactly up the middle of a side. These joint lines are marked T M in the figure. The line T t' will give the radius for cutting off the points to form the inside part of pattern shown by the lines $t\ t$.

In making up the finial the strips should first be bent to the required shape of section—that is, the centre line of the strip should be formed to the outline in the elevation (Fig. 195). The strips are then all tacked together with solder, and after carefully testing the finial to see that it is symmetrical and without twist, the joints soldered up, as much of this being done from the inside as possible.

Downspout Head.

A downspout or hopper-head lends itself admirably for treatment by ornamental work in sheet zinc or galvanised sheet iron. In this case, as in the last, the shape of moulding chosen may be of any section to suit the individual taste and estimated outlay. A simple design is that shown in Fig. 197; its form in plan being that of the five equal sides of an octagon.

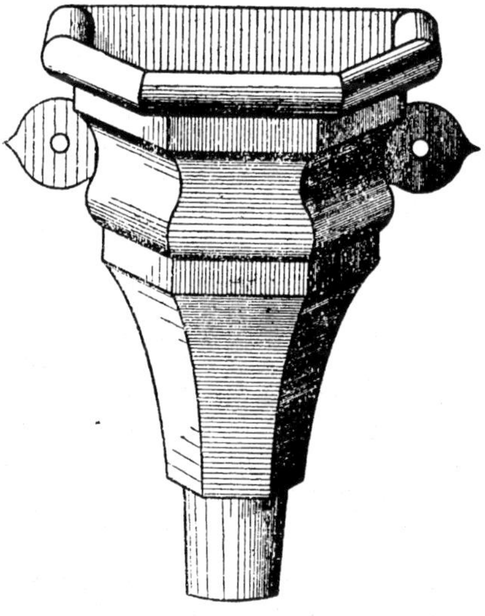

Fig. 197.

Before the pattern for a strip can be marked out, the shape of the moulding must be determined, as shown in

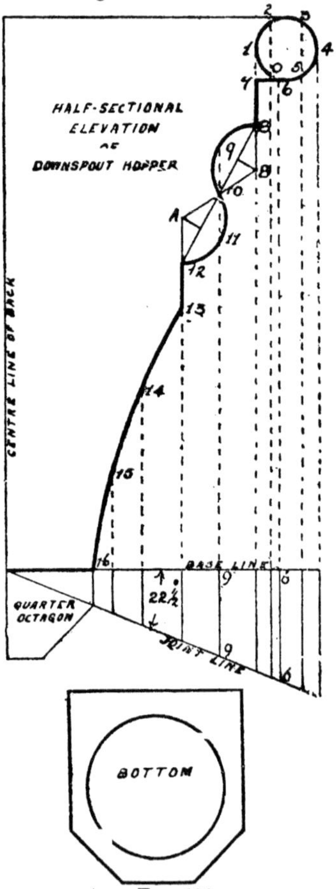

FIG. 198.

Fig. 198. The centres for the double bend or O.G. part can be found by joining points 8 and 12, and dividing the line into four equal parts, then drawing lines square through the two end divisions until they meet the lines drawn up from 12 and down through 8 in points A and B. If A be joined to B, then the point where the two arcs run into each other will be at the point marked 10. Having drawn in the outline, it must then be divided up conveniently for measuring (say, in this case, sixteen parts), and a base and joint line drawn. As previously mentioned, the angle that the joint line should make with the base line will be—

$$\frac{360}{\text{twice number of polygon sides}} = \frac{360}{16} = 22\tfrac{1}{2}^\circ$$

Perpendiculars are drawn through each point down to the base line, and then produced across to the joint line.

For the pattern of a front strip the girth line is first set down, its total length being obtained by carefully measuring off the parts between the numbers in the elevation and marking along as shown in the pattern (Fig. 199). Lines at right angles to the girth line are drawn through each point, and these cut off equal to the length of the corresponding line measured between the base and joint lines in the elevation. Thus, to give two examples, which should make it clear, the lines 6 6′ and 9 9′ on the pattern will be respectively equal to lines 6 6 and 9 9 as seen between the base and joint lines on the elevation. In the same way, all the other lengths for the pattern can be measured and marked off. It should be noticed that several of the lines are the same length; also that the lines 7 to 8 and 12 to 13 on the elevation being straight, the corresponding part on the outline of pattern will also be straight. Having found all the points for the cut, they are carefully joined up, and the strip pattern is complete.

The width of the side strips at the top will be made the same as that of the front strip, the back line being drawn parallel to the girth line.

It will be readily seen that the three side curves of the patterns are all the same: hence in practice it will only be

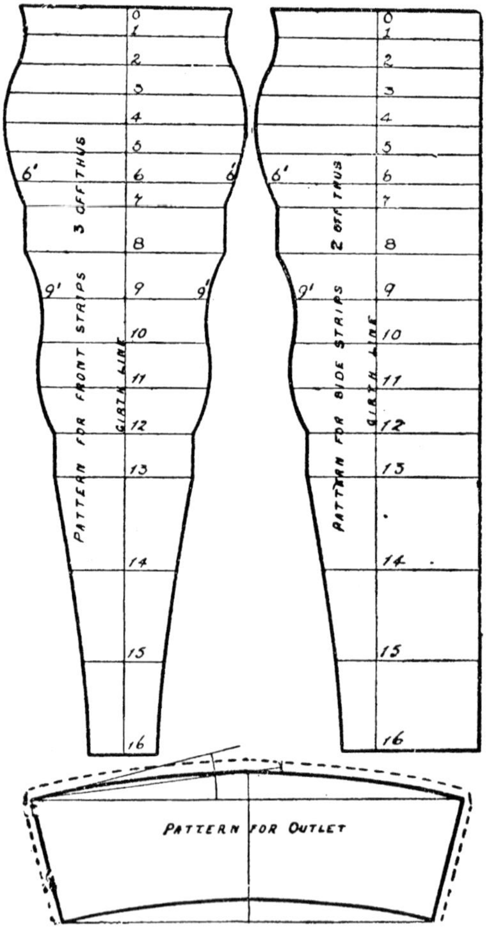

FIG. 199.

necessary to mark out one curve, the others being scribed from this.

The strips will be shaped and tacked together as in the last case, the complete soldering-up being done along the joints on the inside.

The shape of the back of the hopper can be determined from the elevation, or marked off directly after the five strips are soldered together. Lugs should be left on each side of the back, as shown in Fig. 197.

The shape of the bottom piece is shown in Fig. 198, and that for the outlet in Fig. 199, the pattern for the outlet being marked out by one of the methods explained in previous chapters.

CHAPTER XXV.

VENTILATOR AND CHIMNEY-POT BASES, HOPPERS, ETC.

A VERY common form of base for a ventilator or cowl is that shown in Fig. 200, and known as a "tall-boy" base.

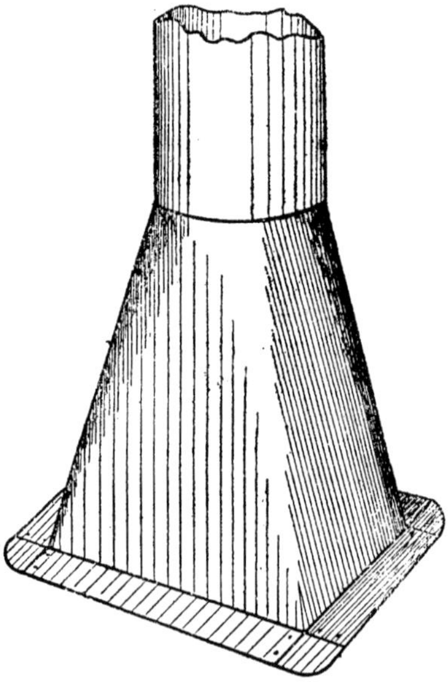

FIG. 200.

It is either square or rectangular at the bottom, and

circular at the top. On examining the plan in Fig. 201, it will be seen that the curved part of the article *t* 0 3 will be exactly a quarter of an oblique cone whose apex may be considered to be at *t*. Four of these equal segments will, of course, make up the curved portion of the surface; the remaining parts being flat triangles.

In making the pattern, a half-elevation and quarter-plan is first drawn (Fig. 201). The quarter-circle in plan is divided into three equal parts, and the division points joined up to *t*. The point *b* is now swung around *d* as centre on to the base line, and connected up to *e*. To get the true lengths of the lines, of which 0 *t*, 1 *t*, and 2 *t* are the plans, their lengths are set along the base line from 0 and joined up to *e*; the respective true lengths, therefore, being 0′ *e*, 1′ *e*, and 2′ *e*. The middle line B 3 of the pattern will be made equal in length to the line so named in the elevation; then T T, drawn at right angles, and the lines B T cut

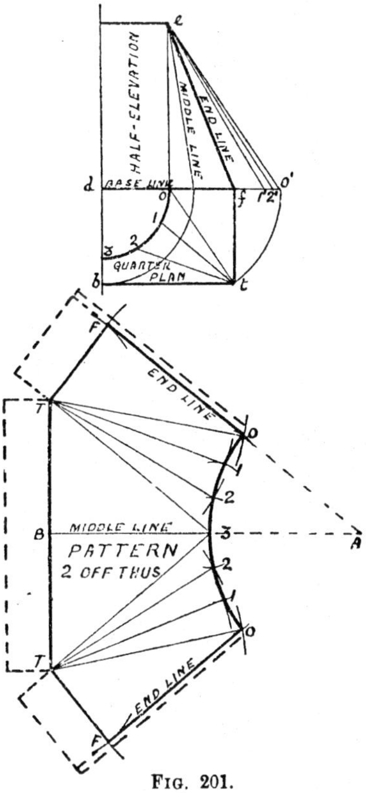

FIG. 201.

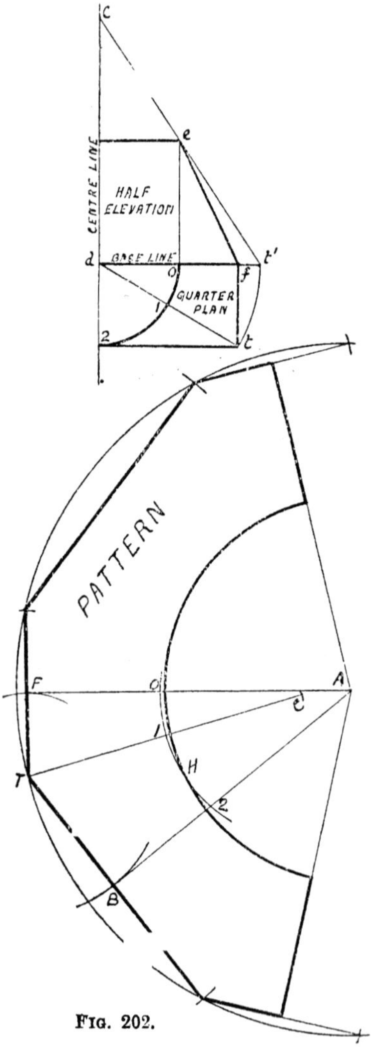

FIG. 202.

off equal to *b t* from the plan. With centre T and radii respectively equal to *e* 2′, *e* 1′, and *e* 0′, arcs of circles are described. Then opening the compasses to the length of one of the arcs (say 3 to 2) in the quarter-plan, and commencing at point 3 on the pattern, the points 2, 1, and 0 are obtained by cutting the first drawn arcs. The compasses are now set to the length of the end line on the elevation, and with 0 on the pattern as centre, an arc drawn (shown passing through F); this being cut by another arc which is described from point T, with radius equal to *t f*; and so the point F is obtained. An even curve joining up the points 0, 1, 2, etc., is drawn, and allowances put on the sides for grooving, and on the bottom for the flange, and the pattern is complete.

After having obtained one-half the pattern, it will be quite accurate enough for practical work to produce the lines B 3 and F 0 until they meet in A, and use this as a centre in a similar way to that explained in connection with Fig. 202. If the tall-boy base is square at the bottom, then a portion of the pattern for one-eighth of the surface will be all that is required to obtain the centre A; the line down the middle of the corner meeting the line along the centre of a side.

After the base has been formed into shape, grooved-up, and the bottom flange bent over, corner plates are riveted on to the flanges as shown in Fig. 200.

Another method for marking out the pattern, which is quite good enough for ordinary practice, is shown in Fig. 202. The line *d t* in the plan is swung around *d* on to the base line, the point t^1 then being joined up to *e*, and produced to meet the centre line in *c*. The compasses are set to *c e* on the elevation, and taking a point C on a line like C T in the pattern, the arc 0 to 2 is drawn. The lengths 1 to 0 and 1 to 2 are now made equal to the lengths of the correspondingly numbered arcs in the quarter-plan. The line C T is measured off equal to the line *c* t^1 in the elevation. The compasses are next fixed respectively to the radii *t* 2 and *t f*, and the arcs at B and F described. Lines are then drawn touching these arcs, and passing through the points 0 and 2 to meet in A. This gives what may be called an approximate centre for the describing of the pattern. Taking A as centre and A T as radius, the part circle is drawn; then commencing at T the sides and ends are stepped along as shown. The end lines of the pattern are best obtained by marking full end lengths on the arc, bisecting them, and then joining up to A as shown in the figure. The inside curve of pattern is marked out by taking A as centre and A H as radius (this being the average between A 0 and A 2), and running around to meet the seam lines. Allowances as required must, of course, be put on as in the last method. It will be observed that in this case the pattern (Fig. 202) is developed for the whole surface.

Hopper or Hood with Flat Back.

A form of hopper to fit against a wall (Fig. 203) having a square or rectangular top, and a circular bottom, can have its pattern set out in the same way as the tall-boy base. All the necessary marking out is shown in Fig. 204. To obtain the pattern lines a half-elevation and a half-plan are first drawn, the lines *e* 5, *e* 4, *e* 3, *b* 3, etc., being set along the top line from *d* and joined up to point 6. The pattern is struck out by commencing with line 6 F, which is made equal in length to 6 *f′* from the elevation. E E is then drawn square to 6 F, and F E cut off equal to *e f*.

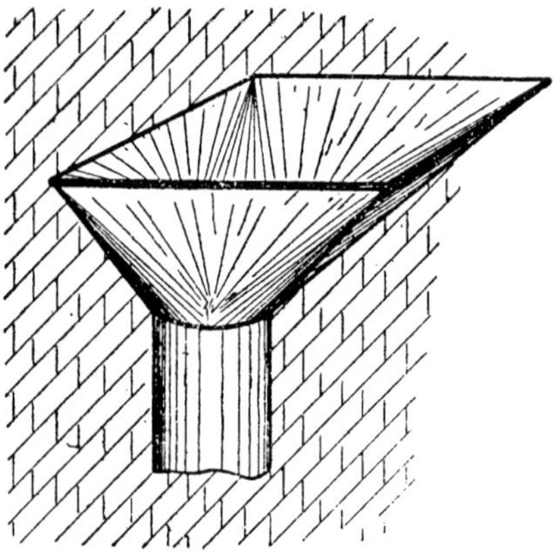

FIG 203.

Now with E as centre and radii respectively equal to 6 5′, 6 4′, and 6 3′, describe arcs of circles (as seen passing through points 5, 4, and 3). Then with the compasses set to a distance equal to one of the arcs in plan, say, 3 to 4, and commencing at 6, cut the first drawn arcs, thus determining points 5, 4, and 3. Again, with E as centre and *e b* as radius, draw an arc (shown passing through B), and with 3 as centre and 6 3″ as radius, cut this and so fix the point B. Having found B, now use this as a centre, and with radii 6 2′, 6 1′, and 6 0′, draw the remaining arcs, cutting these as before, and thus determining points 2, 1, and 0. Then taking 0 as centre, and radii equal to 6 *d* from the elevation, describe an arc (shown passing through A), and cut this by using B as a centre, and *b a* as radius, thus fixing the point A. Join the points 0, 1, 2, etc., by an even curve, and the other lettered points by straight lines, and the net pattern is complete.

The above pattern has purposely been set out without showing any construction lines on its figure, as all that is required in workshop practice is to get the correct outline, and by as few lines as possible. It should also be noticed that there is really no need to draw the construction lines as shown in the plan and elevation, all that is wanted being the exact distance between the points, such as 5 and *e* for setting along the

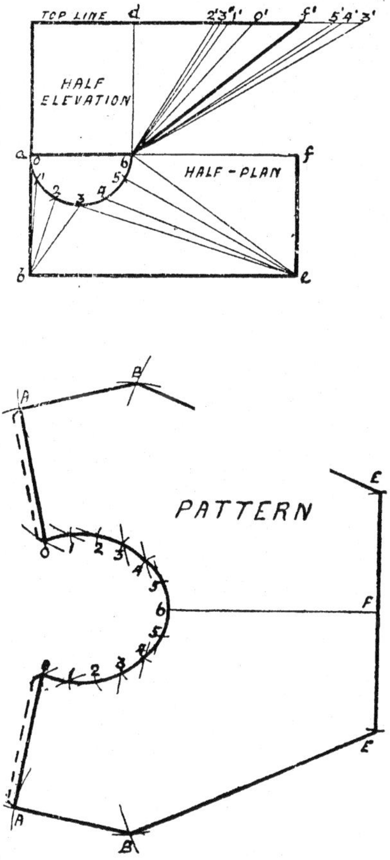

FIG. 204.

top line, and the distance between the points 6 and 5′ for obtaining points on the pattern outline. (These remarks, it might be here observed, apply to all classes of patterns.) Any allowance required for seaming, wiring, or beading must, of course, be added to the net pattern.

Article with Square Top and Round Base.

An article or part of an article may have a round base and a square or rectangular top, as seen in Fig. 205. Its pattern can be developed by treating the curved portions of the surface as parts of oblique cones, and the flat parts as triangles.

The setting out of the pattern is explained by Fig. 206, in which a quarter-plan and half-elevation is shown. The quarter-circle is divided up into four equal parts, and the lines *b* 1 and *b* 2 set along the base line from point *t* and joined up to *t*′. A middle line T 0 of the pattern is laid down and made equal in length to *t*′ 0 from the elevation. A line is now drawn through T square to the line T 0, and T B cut off equal to *t b* from the plan. Now, using B as centre and radii respectively equal to *t*′ 1′ and *t*′ 2′, arcs of circles are drawn as shown. Then setting the compasses to the length of one of the arcs in plan, say, 1 to 2, and commencing at 0, the points 1, 2, etc., are

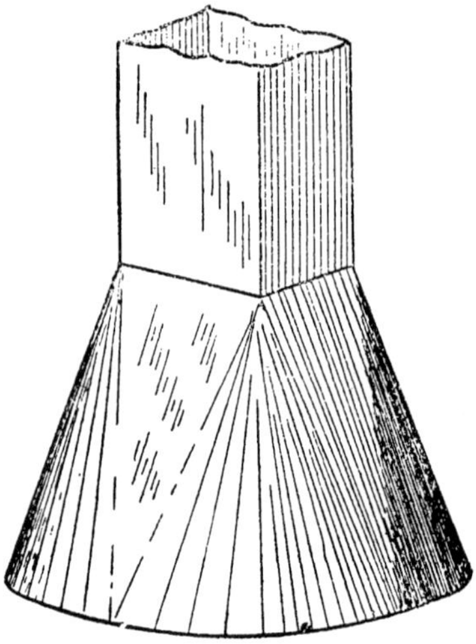

FIG 205.

marked. The triangle T B 0 is next set out, its construction being simple, and a repetition of the first part of the pattern. An approximate centre in this case can be found, as with the pattern in Fig. 201, by simply producing the lines 0 T and 2 B until they meet, as shown by the dotted lines.

If the top of the article is rectangular or polygonal in form, its pattern can be struck out as above, but in these cases a greater number of lines would have to be used. Also, if the centre of the top does not come vertically over the centre of the bottom, the pattern can be readily marked out with the same method, the only modification being the same as that applied to Fig. 204.

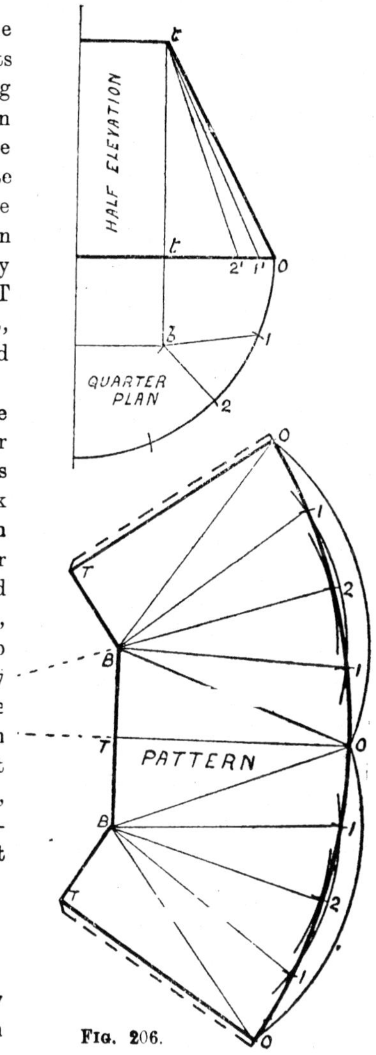

FIG. 206.

Ventilator Base of Pyramid Shape.

A ventilator base **may be of the form shown in**

Fig. 207, which, it will be seen, amounts, geometrically, to the fitting of a cylinder on to a square pyramid concentrically.

The pattern cuts, both for the pyramid and the pipe surfaces, are shown struck out in Fig. 208. A half-elevation is drawn, and a line making 45° with the base line set down, this being cut off by the line 0 2, which is drawn square to the base line. The arc 0′ 2′ is now described, and it will thus be seen that the figure 0′ 2′ 2 0 can be taken as representing one-eighth of the complete plan of the ventilator base. Line 0 0″ is produced to meet the centre line in *c*. The line 0 2 is bisected, and lines *d* 1, *d* 2 swung on to the base line about *d*, and the points 1^0 and 2^0 joined up to *c*. For the pattern, the compasses are opened out to the length *c* 2^0, and a circle described as shown. Five sides are now stepped around the circle, each side, 2 2, being equal in length to twice the line 0 2 from the plan. The last two sides are now bisected, so that to make up the complete pattern there will be three full sides and two half-sides. Each side is now divided into four equal parts, and from the division points lines drawn to the centre C. The compasses are next set respectively to the radii *c* 0″, *c* 1″, and *c* 2″, and the three arcs drawn on the pattern to cut the radial lines. Where these arcs intersect the correspondingly numbered line will give a point on the curve. The points are then joined up, such

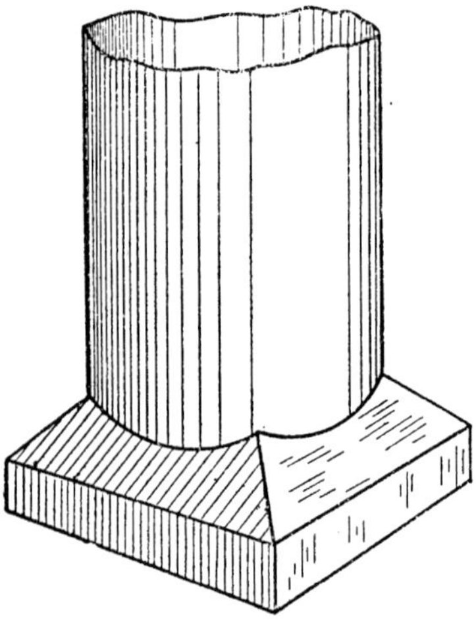

FIG. 207.

as 0″, 1″, and 2″, with even curves, and the net pattern is complete. (It is as well to remember that the inner curves are parts of an ellipse, as in many cases they can be marked out by a much simpler method.) Allowance is then put on the pattern for the base flange, corner laps, and side seams.

If it is desired to run the seam down a corner instead of the middle of the side, as shown, then, of course, four full sides would have to be marked out, and not three full and two half-sides as on Fig. 208.

For the cut on the bottom of pipe a girth line is first stretched out by setting along alternately the lengths of arcs 0′ 1′ and

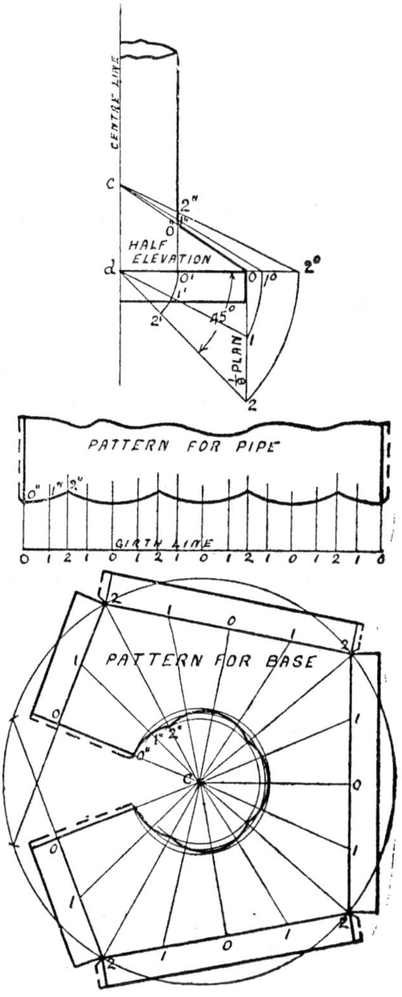

FIG. 208.

1′ 2′. Lines square to the girth line are then run up from each point, and these cut off respectively equal 0′ 0″, 0′ 1″, and 0′ 2″ from the elevation. The new-found points are then joined up to form the curve. It is as well to remember,

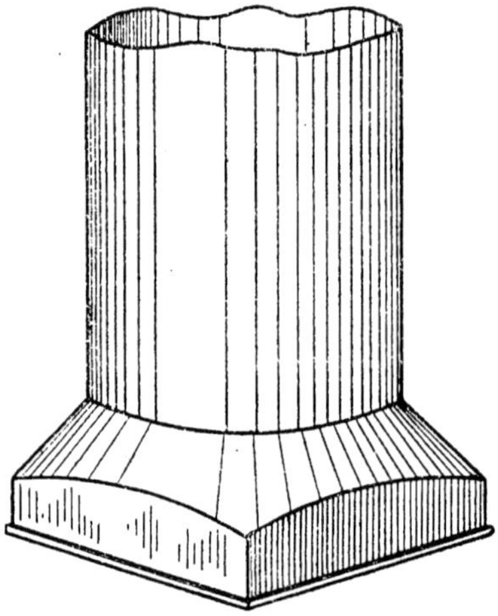

FIG 209.

and it will act as a test for the accuracy of the setting out, that the lengths of the curves 0″ 1″ and 1″ 2″ should be the same both on the pipe and base patterns.

Ventilator with Conical-Square Base.

Sometimes a ventilator base follows the design shown in Fig. 209, which it is not difficult to imagine represents the intersection of a round pipe and cone for the top, and a cone and square pipe for the bottom.

The patterns can be struck out in the way shown in Fig. 210. The half-elevation and part plan are drawn as in the last case. The arc 0 3 is divided into three equal parts, and the division points joined to *d*. Lines *d* 1′, *d* 2′, and *d* 3′ are turned around on to the base line, and perpendiculars run up to meet the line 0 *c* in points 3″, 2″, etc. The pattern for the conical part is obtained by fixing the compasses to the length *c* 0 describing the arc as shown, and setting along it twelve lengths each equal to the length of one of the corresponding arcs in the plan. After the radial lines are drawn in, the compasses are set respectively to the lenaths *c* 0″, *c* 1″, etc., in the elevation, and the arcs on the pattern drawn. Where these cut, the same numbered line will give points on the curve, which can be joined up as shown. The inner curve of the pattern is, of course, marked out by using a radius equal to *c t* from the elevation.

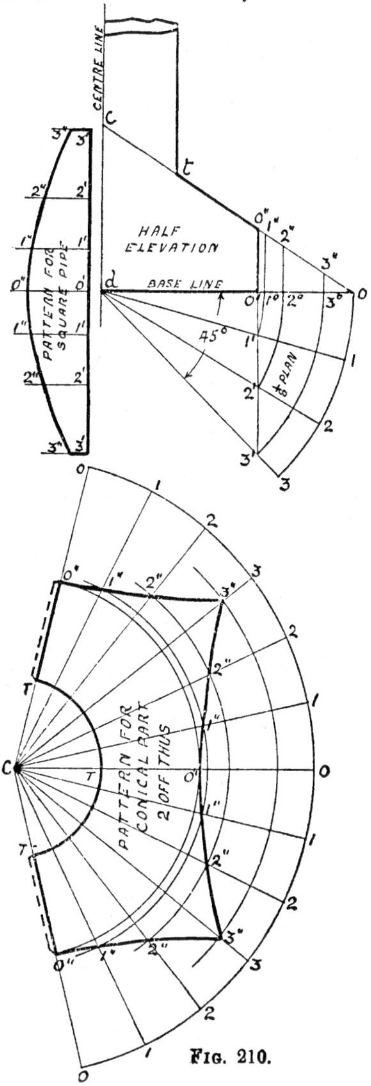

FIG. 210.

The pattern for one side of the square base is shown set at the top of Fig. 210. Here the line 3′ 3′ is made twice the length of 0′ 3′ from the plan, the division points being the same. Lines are drawn square through each point, and cut off respectively equal to 0′ 0″, 1° 1″, 2° 2″, and 3° 3″ from the elevation. The resulting curve is (to those who understand geometry) a hyperbola, and may be set out by other methods common to that curve. None, however, are simpler than the one shown.

Very many different kinds of bases for ventilators are made; but sufficient has perhaps been shown to explain the general principles involved in the marking out of the patterns for flat-bottomed bases. The above can easily be modified to cover the setting out for bases resting on the ridge and sides of a roof.

CHAPTER XXVI.

SHIP VENTILATORS, ETC.

VENTILATORS for ships are made in many shapes, forms, and sizes, one of the commonest kind being that shown in Fig. 211. It is usually made of iron, and occasionally of copper or brass.

Several methods are in vogue for marking out the plate patterns, according to the practice of the particular locality. As previously stated, in work of this character that has to be hollowed or stretched, it is impossible to set out the patterns that they will work out dead true to shape. The most that can be hoped for is to get as good an approximation as possible, and at the same time take care that the pattern is slightly on the full side.

In Fig. 212 a side elevation of a bell-mouthed ventilator is shown. Its body is formed of four pieces, two cheeks and the throat and the back parts, the bell-mouth being made up in three pieces.

For the cheek pattern the simplest plan is to take the elevation of the body as the shape of the pattern; the only modification required being that shown by the dotted line, *e l k*. The point *l* can be obtained by making the line *m n* equal in length to half the mouth diameter, that is, *r e*. Drawing *n t* to touch the semicircle on *a f*; cutting off *t p* equal to *t n*, and then making *r l* equal to *o p*, as shown by the construction lines. An arc of a circle is then drawn through the points *e l* and *k*.

Before attempting to strike out the patterns for the back

and throat pieces the elevation of the two joint lines must first be drawn. Divide the curves *a e* and *f k* each into, say, three equal parts, and on the four lines that join the division points describe semicircles, as shown. Now mark the middle points of the semicircles, such as *s* and *u*, and set around the arc on each side a length equal to half the diameter of the respective semicircles. Thus the arc *s* 4′ will be equal in length to *r e*, and the arc *u* 3′ equal to *v d*, and so for each of the other two semicircles. Perpendicular lines, such as 4′ 4 and 3′ 3, are then drawn from the points on the semicircle to their diameters, and thus points on the joint lines obtained. These are connected with an even curve, as shown by the dotted lines, which will then give an elevation of the two side seams.

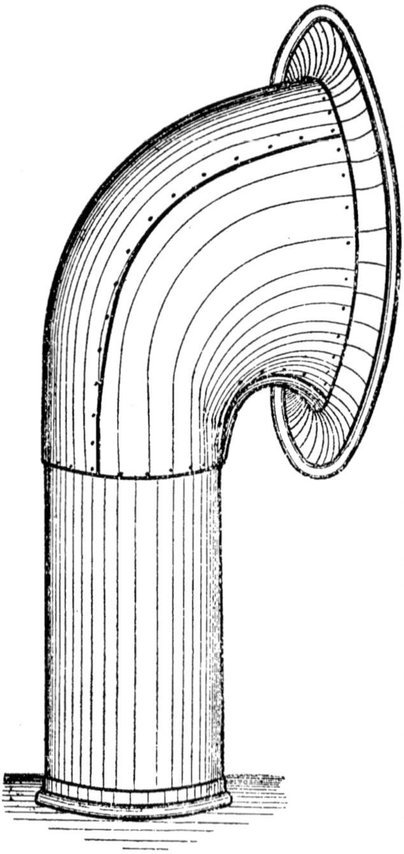

FIG. 211.

For the back pattern, a centre line A E is marked down, equal in length to the respective parts of the curve *a e* in the elevation. The line D 3 is set off equal in length to the arc *d* 3′, and the line B 2 equal to arc *b* 2′. In the same way the other points, 1 and 4, are found. To get the lengths of the side curves on the pattern, a line, *c* 0, is first drawn across the two curves in the elevation, as shown; the point *c* being the middle of the back curve, and the line *c* 0 being drawn by the eye, to make as near as possible equal angles with the two curves. The parts of the joint line 0 1 and 0 4 are carefully measured along, and their lengths set above and below the point 0 on the outside line of the back pattern. In this way the points 1 and 4 are obtained. Arcs of circles, 4 E 4 and 1 A 1, are then described to form the ends of the pattern. It should be remembered that when working up the plate for the back, that the centre will lengthen a little and the sides contract some-

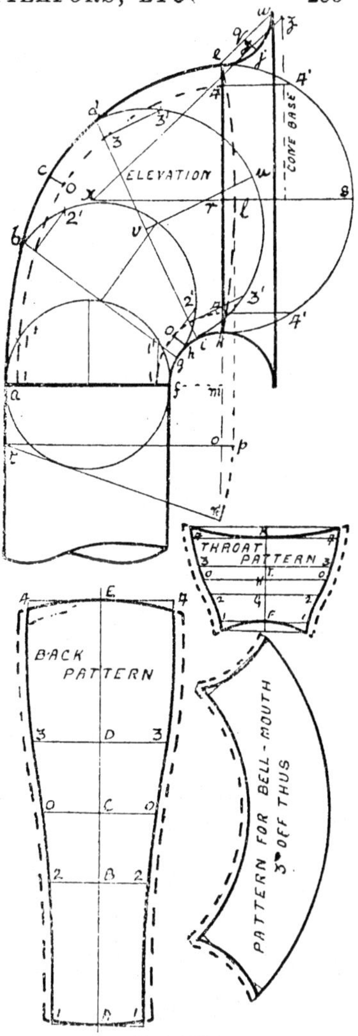

FIG. 212

what: hence the side-curves of the pattern should be made slightly longer than measured from the elevation. This is best allowed for by making the arcs 4 E 4 and 1 A 1 somewhat flatter than they would be if drawn exactly through the three points as found.

The throat pattern can be set out in identically the same manner as that for the back, and so that the reader may the more readily follow the construction, the same numbers for the outside curves have been chosen. In this the centre line, F K, on the pattern is the same length as *f k* on the elevation, and the lines, F 1, G 2, etc., equal in length to the arcs *f* 1′, *g* 2′, etc. The outside curves of the pattern will be the same length as the throat seam line; 0 1 and 0 4 on the pattern being made equal in length to 0 1 and 0 4 on the elevation. In working up the throat plate the outer edges will, of course, have to be stretched: hence they will lengthen somewhat, so that it is as well to keep F K the same length as *f k* on the elevation, but to draw the arcs 4 K 4 and 1 F 1 on the pattern slightly flatter, and, consequently, reduce the lengths of the side curves somewhat. The exact amounts to allow on or take off, as the case may be, are matters of experience, or of difficult calculation, the main thing being to keep on the right side, as, in any case, some small allowance must be made for trimming.

If for a large head, the bell-mouth will be made in several pieces; in the present case three have been chosen. The pattern will come out as part of the surface of a cone. The first thing, then, is to find the slant height of the cone, and thus the radius for the pattern. Join *e* to *w*, and from the middle point *q* draw the perpendicular *q j*. Make *j y* one-third of *q j*, and through the point *y* draw the lines *z x* parallel to *w e*, to meet the axis of the supposed cone in *x*. The line *j z* is made equal in length to the arc *j w*, and thus the slant height of the cone is determined. The line *x z* is now used as the radius for the outer curve of the bell-mouth pattern; the length of this curve being made equal to one-third the circumference of the cone base circle. In working up the piece of sheet for the bell-mouth, it will be found that the draw at the ends will not be uniform; consequently it will be necessary to allow a little on the ends of the pattern, as shown by the arcs

The allowances for the joints are added to the back and throat pieces, and also to the inner side of the bell-mouth pattern.

The beading around the bell-mouth is formed either by wiring or split-tube, as shown in Fig. 213.

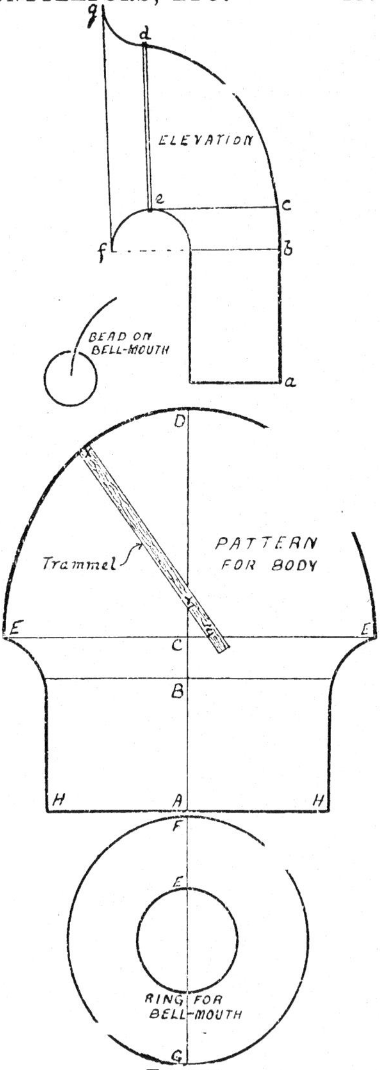

FIG. 213.

Small Ventilator Heads.

The body of a small head may be worked up from one piece, in much the same manner as a copper kettle-spout. In Fig. 213 an elevation of a small head is shown. To mark out the pattern the centre line, A D, is drawn the same length as the curve *a d*, the lengths of the intermediate parts also corresponding to those of the elevation. The line A H is made equal to half the circumference of the ventilator shaft-pipe, and C E cut off equal to one and a half times the length of *c e* on the elevation. The outline of the top part of the pattern comes out as a semi-ellipse; and this can, perhaps, be best marked out by what is known as the "trammel method." On a strip of hoop-iron or a wooden lath mark from the end a distance equal to the semi-axes, or diameters, of the ellipse; in this case C D and C E respectively; thus obtaining two points like Y and Z, as shown on the sketch of trammel in Fig. 213. Fix the trammel in several successive positions, always keeping the points Y and Z on the lines C D and E E respectively, and mark the position of X; thus points for the required ellipse will be obtained, and when joined up with an even curve, will give the boundary of the top portion of the pattern, as shown. A curve is now run from E to join on to the pipe portion of the pattern.

For the bell-mouth, a ring in this case will be best; its diameter, F G, being equal to *f g* in the elevation. The width of ring F E will have to be somewhat larger than the length of arc *f e*, to allow for draw. This width can be calculated, but it will be sufficiently accurate to make F E equal to about one and a quarter times *f e*. The bell-mouth can be fixed to the body by making a knocked-up joint.

An enlarged view, showing the method of fixing the bead, which is usually split-tubing, is also shown in Fig. 213.

Irregular Circular-ended Tapering Article.

A ship's ventilator may also be constructed in segments, as shown in Fig. 215. In order that the method adopted in obtaining the shape of the segment patterns may be clearly understood, it will be an advantage to first go carefully over the setting out of the pattern for an irregular article whose ends are circular, and not parallel. An elevation of such an article is shown in Fig. 214. This class of object gives good scope for illustrating the use of the method of triangulation in obtaining surface developments, and should, therefore, be taken particular notice of, as by this method any article whose surface is developable can have its pattern set out.

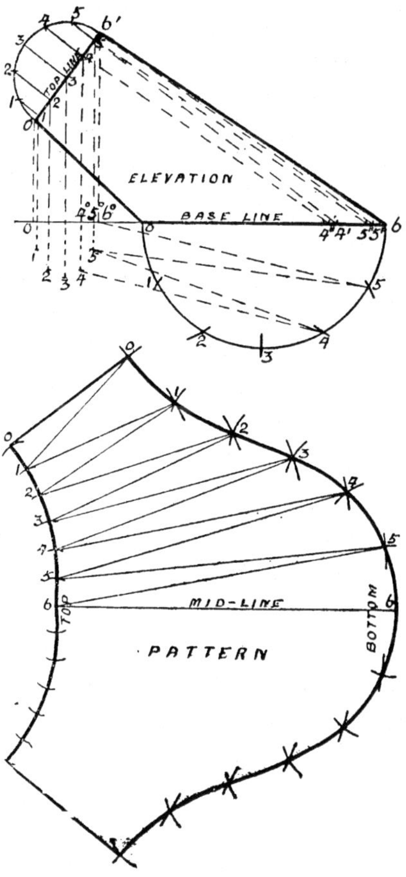

FIG. 214.

Imagine the circles that form the top and bottom of the article in Fig. 214 divided respectively into twelve equal parts, and that corresponding points be joined; then, on each quadrilateral so formed a diagonal drawn. It will thus be seen that the surface of the article would be divided into twenty-four triangles. The pattern is then built up, as it were, by getting the true shape of each of these triangles and adding them together, as shown in the one half of the pattern in Fig. 214.

Let us now go over the construction. From the numbered points on the top line projectors are run down to, and across, the base line; their distance below this being cut off equal in length to the corresponding line on the top semicircle. Thus the dotted lines 4 4^0 and 5 5^0 will be respectively equal to the perpendiculars drawn through points 4 and 5 on the semicircle down to the top line, and so on for the other lines. If the points 0, 1, 2, etc., be joined up, it will be seen that the half-plan of top becomes a semi-ellipse. There is no need in practice to draw in the ellipse; all that is wanted being the plans of the points.

For the pattern the mid-line 6 6 is first laid down, being made equal in length to the line 6 6′ from the elevation. Now, to obtain the true length of the diagonal for line 6 5 on the pattern, measure from 6^0 on the ellipse to 5 on the bottom semicircle, setting this distance along the base line from 6^0, and so obtaining point 5′. The length of the dotted line 5′ 6′ from the elevation is now measured off and used as radius from point 6 at the top end of the pattern, and a small arc drawn (shown passing through 5 at the bottom end of the pattern). The compasses are now set to the length of one of the six arcs on the base semicircle, and with point 6 at the bottom end of the pattern as centre, a small arc is drawn to intersect the first arc, and thus fix the point 5. The dotted line 5 5 from the plan is now set along the base line from 5^0, and the point 5″ marked. The line from 5″ on the base line to 5 on the top line is measured off, and used as a radius from point 5 at the bottom end of the pattern to describe the small arc passing through point 5 at the top end. This arc is cut by setting the compasses to a radius equal to the length of one of the six arcs

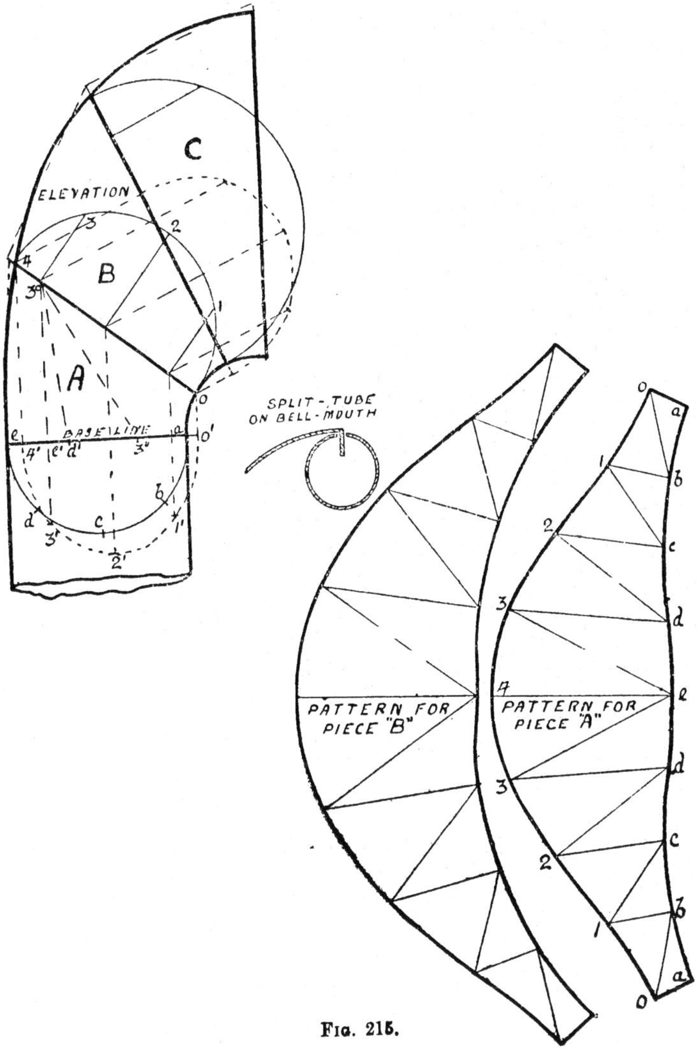
C
ELEVATION
B
A
BASE LINE
SPLIT-TUBE ON BELL-MOUTH
PATTERN FOR PIECE "B"
PATTERN FOR PIECE "A"

FIG. 215.

on the top semicircle, and using point 6 at the top of the pattern as centre. Thus point 5 at the top end of the pattern is determined. In the same way the lengths of all the other lines can be found. Thus 5^0 $4'$ on the base line equals 5 4 on the plan, and the line 5 4 on the pattern will equal the dotted line drawn from $4'$ on the base line to 5 on the top line; the distance 4^0 $4''$ will equal 4 4 on the plan, and line 4 4 on the pattern equals $4''$ 4 on the elevation, and so on for the remaining lines.

It is well to remember for practical purposes that there is no need to draw any of the dotted lines on the plan or elevation, or any of the construction lines on the pattern. All that is wanted being the fixed points, such as those obtained on the lower half of the pattern by the intersection of arcs.

The above method has been explained at some length, on account of its great importance. The reader should, therefore, find no difficulty in following its application to a

Ventilator Head in Segments.

An elevation which illustrates this method of constructing a head is shown in Fig. 215; the body being divided into three segments, A, B, and C.

The patterns for two of the parts are shown set out, and after what has been said in connection with Fig. 214, it will be sufficient to briefly indicate the method of obtaining one pattern, say, for segment A. The back and throat curves are each divided into three equal parts, and the joint lines drawn. On these semicircles are described and divided into, say, four equal parts. Now to deal with the lines required for the pattern of segment A. From each of the division points of the semicircle, which is described on 0 4, drop perpendiculars on to that line, and from the feet of these perpendiculars run lines down square to *a e* and across it, cutting them off equal in length to the corresponding lines on the semicircle. There is no need to join the joints so found; but if they are connected together it will be seen that they form a semi-ellipse as in Fig. 214. The mid-line 4 *e* of the pattern will, of course, be made the same length as 4 *e* in the elevation. Now fix the compasses to *e* 3′ and set this distance along the base line from *e*′ and so determine the point 3″. The line 3″ 3° will give the required length of *e* 3 for the pattern. Next measure *d* 3′ and set this along the base line from *e*′, and so obtain the point *d*′. The line *d*′ 3° will then be the length required for the line *d* 3 on the pattern. In the same manner the lengths of all the other lines required to construct the eight different triangles on the half-pattern can be obtained. The lengths *e d*, 4 3, etc., will be taken from the lengths of one of the parts on the respective semicircles, as in Fig. 214.

All the construction lines actually required to strike out the pattern for the piece B are shown on the elevation; but as the marking out is only a repetition of that already gone over for segment A, there is no need for any further description. The pattern for C is not shown; but this will, of course, come out in the same manner as for the other segments.

If the backs of each segment are to be left straight (which is sometimes done in very common heads), as in Fig. 214. then the patterns as laid out in Fig. 215 will be quite correct; but if they are to be hollowed to the required curve, then it will give a more accurate result to produce the joint lines out to the dotted lines of the backs, and describe the semicircles on these, thus making allowance for the drawing-in of the backs of the segments somewhat in hollowing.

As pointed out in connection with the pattern for Fig. 214, it should also be noted that there is no need to draw a single construction line on the pattern, all that is wanted being the points obtained by the intersection of the arcs. The construction lines are simply shown to illustrate the principle of setting out the pattern.

No allowance for jointing has been made, as this can be added according to requirements.

The method of fixing the split-tube to form the bead around the mouth is also shown in an enlarged view on Fig. 215.

Conical Ventilators, Etc.

The patterns for a ship's rib-head ventilator, instead of being set out by the method of triangulation, as just explained, can often be more conveniently developed by treating each segment of the ventilator as part of a cone. Before describing how this can be done, however, it will be necessary to first explain how a cone and a cylindrical pipe can be made so as to fit together exactly. To do this it will perhaps be the best plan to go over the striking out of the patterns for a simple cowl made up in the way of a

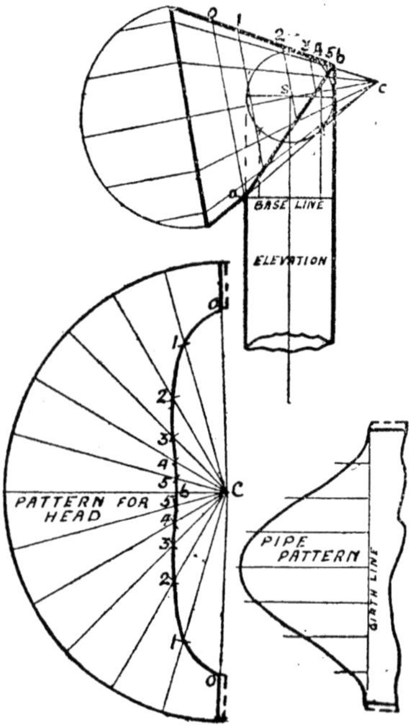

Fig. 216.

Cone and Pipe Connection.

The most important thing to take notice of

in jointing together a cone and pipe is to so arrange them that the elliptic cut on the cone shall be exactly the same shape and size as the cut on the pipe. This is done by imagining that both cone and pipe are tangential to a common sphere. In practice it amounts simply to drawing in a circle, equal to the diameter of the pipe, and then fitting the cone so that its outside lines shall touch the circle. Thus, in Fig. 216 a circle is described from centre, *s*, having a radius equal to that of the pipe, and then drawing the pipe and cone to touch this in any position, as required. The points where the outside lines of cone and pipe intersect will give the ends, *a* and 6, of the joint line. This construction should be most carefully gone over, on account of its great importance, it having to be used in all cases where conical and cylindrical pipes are required to be connected together in this way.

The pattern for the round pipe will be set out in the usual manner, lines above the base line being measured off to determine the length of those to the left of the girth line on the pattern.

The construction lines for the conical pattern will be obtained in the ordinary way by running radial lines, as shown, and from where these cross the joint line, *a* 6, drawing lines square to the axis on to the outside of cone. After having set out the pattern for the complete cone the points, 0, 1, 2, etc., are obtained by cutting off the lines, C 0, C 1, etc., equal to *c* 0, *c* 1, *c* 2, etc., from the elevation.

No allowance for connecting pipe and cone is shown, this being put on according to the method of jointing followed.

Ship's Rib-Head Ventilator.

An elevation of one form of this is shown in

Fig. 217, the head being made up in four pieces. So that each segment shall come out as a portion of a cone, it will be necessary to construct circles on the centre line, as shown by those described from centres *n*, *o*, and *p*. The shape of segment A is formed by drawing a cone, apex *a*, to touch the sphere whose centre is at *n*. The joint lines for the parts B, C, and D are determined by drawing a

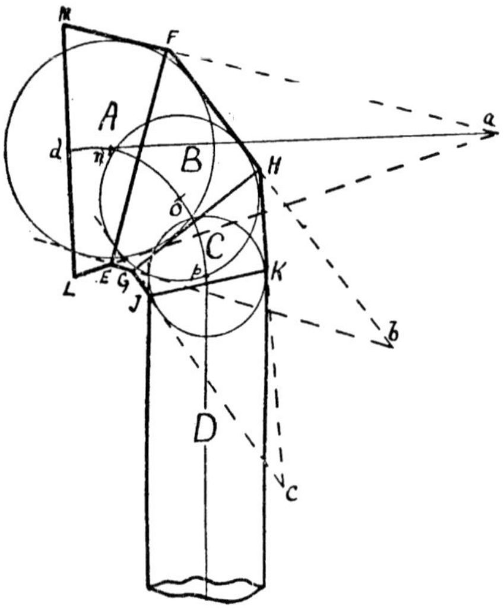

FIG. 217.

cone to touch spheres *n* and *o*; a cone, apex *c*, to touch spheres *o* and *p*, and the pipe, D, to touch the sphere *p*. Where the outside lines of the respective pairs of cones intersect will give the ends of the joint lines. Thus E F is the intersection of cones whose apexes are *a* and *b*, G H that of *b* and *c*, and J K the joint between cone *c* and the cylindrical pipe, D. To have a circular mouth, the line, L M, must be square to the cone axis, *d a*.

The back and throat of a ventilator may be curved as shown by the dotted lines in Fig. 218, and the shape of the segments still obtained in the same manner as above.

The pattern for one segment only—that of B—is shown set out in Fig. 218. The complete cone is first constructed by making *b* 0 equal to *b* 4, and drawing in a base line, 0 4. Upon this base line a semicircle is drawn and divided into four parts. From each division point on the semicircle a perpendicular is dropped on to the diameter, and then radial lines run on to the apex *b*. Where these radial lines intersect the joint lines, 4 *d* and *f e*, lines square to the cone axis are run on to the outside of the cone. Thus, all the construction lines required for the pattern are determined. The complete cone pattern is first marked out in the usual way by taking *b* 0 as radius and stepping eight lengths, each equal to one of the arcs on the semicircle along the girth line. These points are now joined up to *b*, and the radial lines so drawn cut off by using the lengths already found on the side *b* 0 of the cone. Thus, to follow the fixing of one pattern point, the line 3 3^0 is drawn on the semicircle, 3^0 joined to *b*, and $3'$ $3''$ drawn parallel to the

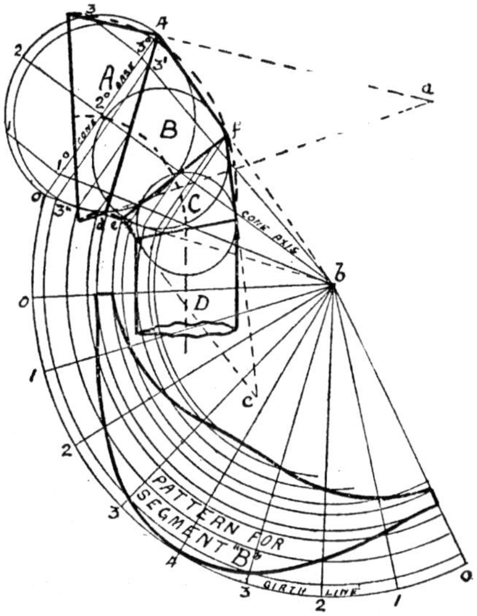

FIG. 218.

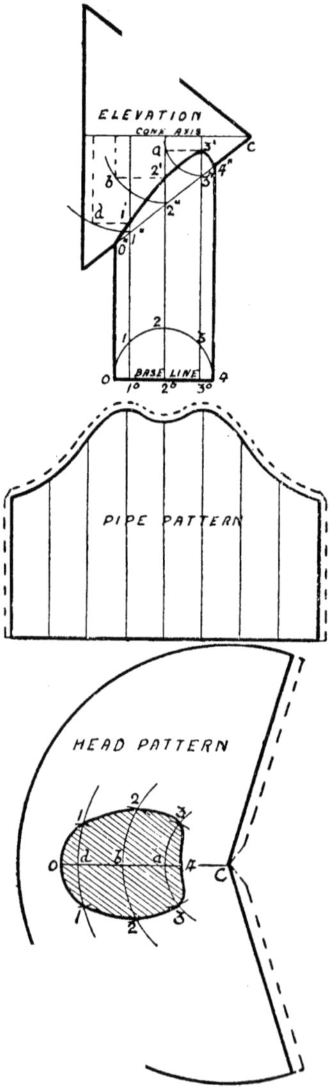

cone base, or square to the axis; the distance *b* 3″ is then swung around on to the radial line, *b* 3, and in this way a point on the pattern curve is obtained. In a similar manner the other points can be determined.

The patterns for the segments A and C are not shown set out, as these, of course, can be developed in exactly the same way as explained above in connection with segment B. The pattern for the pipe D can be marked out as in an ordinary elbow, or similar to the straight pipe in Fig. 216.

A simple and cheap form of cowl can be designed by fitting a

Round Pipe on Cone.

In this case, where the two centre lines are at right angles, the patterns come out in an easy manner. An elevation of such a cylindrical pipe and cone fitting together is shown in **Fig. 219.** So that the construction lines for the patterns may be obtained, the usual method of drawing an elevation of the joint line must first be gone over. Describe a semicircle on the base line and divide it into four equal parts, running lines up through each division point to the cone axis, as shown. Now, from the points where these lines intersect the axis, draw the arcs 1″ *d*, 2″ *b*, and 3″ *a*, of indefinite length. Then measure the lengths of the respective lines which are drawn across the semicircle, marking their distances along the cone axis, projecting down, and so obtaining the points *a b d*. Thus *d* 1′ will be equal in length to 1 1° on the semicircle, *b* 2′ equal to 2 2°, and the line *a* 3′ equal to 3 3°. By running lines back through *a*, *b*, and *c* parallel to the cone axis, points on the joint curve will be determined.

The pattern for the cylindrical portion will be struck out, as before explained, by measuring off the construction lines between base line and joint curve.

The hole in the head pattern can be drawn by marking C O equal to *c* 0″ on the elevation, and then describing the three arcs to the respective radii: C *d* equal to *c* 1″, C *b* equal to *c* 2″, and C *a* equal to *c* 3″; the point 4 being fixed by making C 4 equal to *c* 4″. The lengths of the arcs, *d* 1″, *b* 2″, and *a* 3″, on the elevation are then carefully measured, and their lengths set respectively along *d* 1, *b* 2, and *a* 3 on the pattern. The points found are then joined up, and so the shape of the hole obtained.

For stock patterns, or where a number have to be marked off the one pattern, a greater degree of accuracy will be ensured by having more construction lines, such as dividing the semicircle into six or eight parts, instead of four, as in the present example.

The dotted line on the top of the pipe pattern shows the necessary allowance for a flange for riveting on to the cone.

Pipe on Cone Obliquely.

When the centre line of the pipe is inclined to that of the cone (Fig. 220), then the determining of the joint curve is a more difficult matter. The only real difference, however, between the construction in this problem and the last is in the arcs *d* 1″, *b* 2″, and *a* 3″; for whereas in the former case they were arcs of circles, in this example they come out as parts of ellipses. The difference of construction, then, lies in obtaining the shapes of the elliptic arcs. To do this all that is necessary is to first get the two diameters of the respective ellipses, and then set the small arcs

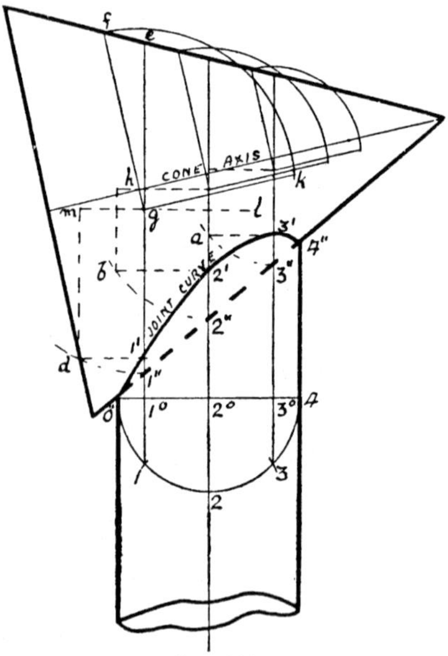

FIG. 220.

out by the method shown in connection with Fig. 213. It will perhaps be sufficient to explain how to get the diameters of the ellipse of which the arc *d* 1″ is a part, as the method will be the same for each arc. Draw the line 1 *e* parallel to the centre line of the pipe, and bisect 1″ *e* in *g*. Draw *g f* square to the axis of the cone, and on *h f* describe a quarter-circle, producing it a little beyond the point where it meets the cone axis. Now draw a line through *g* parallel to the centre line of the cone, to meet the quarter-circle produced in *k*. Then the line *g k* will be half the small diameter, and the line *g e* half the large diameter of the ellipse. These two lengths are set along a trammel, as previously explained, and the two points slid along the lines *g* 1 and *g l*, thus obtaining the curve *d* 1″. Now mark off *g m* equal to 1^0 1, and draw a line down parallel to *g* 1, and so fix the point *d*. The line *d* 1′ is then drawn square to *d m* or *g* 1, and thus the point 1′ on the joint curve is found. In the same way the other points 2′ and 3′, can be determined.

The striking out of the patterns is not shown, as this part of the work will be done in an exactly similar manner to that illustrated by Fig. 219.

Lobster-Back Cowl.

The construction of a lobster-back cowl (Fig. 221) follows somewhat similar lines to that of a quarter-bend, made up in segments, as shown in Chapter IV. In some cases where the throat part is curved the setting out of the patterns will be exactly the same as the quarter-bend; but in the present case, where the mouth and bottom pieces of the cowl meet, and are square to each other, some modification of the pattern is required.

A side elevation of the cowl, exhibiting the arrangement of the seg-

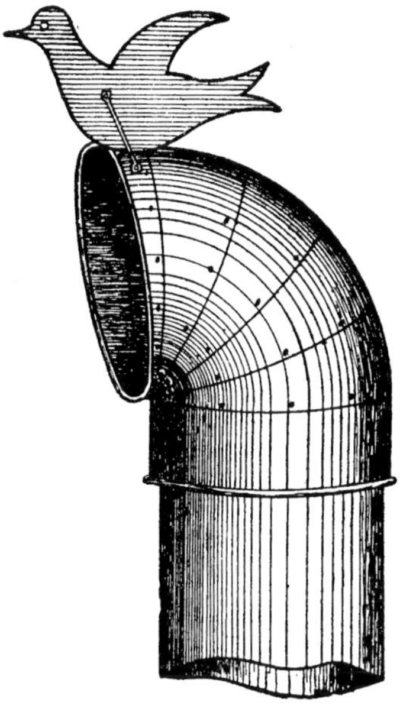

FIG 221.

ments, is shown in Fig. 222. The curved part of the back is formed of a quadrant of a circle, and is usually, as in this case, divided into four equal segments. The mouth and bottom pieces are respectively produced in to the dotted lines *a b* and *a c*, being themselves connected along the line *o a*. The four back segments are thus, as it were, joined on to these.

The construction lines for the mouthpiece are obtained by describing a semicircle on the line 0 6, dividing this into six equal parts, and running lines square to the diameter and across to the dotted line *a c*. The pattern is set out by first marking down the girth line equal in length to twelve times one of the small arcs on the semicircle, drawing lines square across, and cutting them off equal in length to the construction lines measured on either side of 0 6 in the elevation. Thus, to show one line, the parts $4' \, 4^0$ and $4^0 \, 4''$ on the pattern are equal to the same figured lines in the elevation.

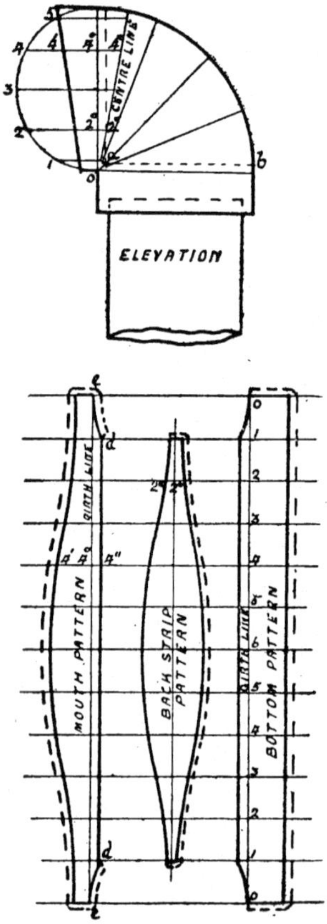

FIG. 222.

The bottom pattern will be marked out in the same way as the mouth pattern, and it should be noticed that the ends of both of these are cut away, so as to form the square throat, when bent and fixed in position. A small lap is allowed on to the end cuts, *d e*, of the mouth pattern, to cover for turning inside the bottom piece and riveting, if required.

The pattern for a back strip can be marked out from the same construction lines as previously used; but here the girth line will be shorter, being equal in length to ten only of the small arcs on the semicircle; this shortness being arranged so as to avoid all the strips meeting in a point under the throat. The width of the different parts of the pattern-strip will be measured from the line 0 6 in the elevation up to the centre line of the segment. Thus the distance $2^0\ 2''$ on the pattern will be the same as $2^0\ 2''$ on the elevation, and so for all corresponding lines. A lap will be allowed on one side of the strip, and also on both ends, so that they may be brought round and riveted on to either the mouth or bottom piece. Allowances have also been put on mouth and bottom pieces, to cover for wiring on the outer edges.

The wind-vane can be cut out any shape to suit the individual taste, a lug being left on the bottom to turn over and rivet on to the cowl-head.

In order that the head may revolve, a spindle is rigidly fixed along the centre of the pipe-shaft, which should fit into a centre on the head. This arrangement, however, is so well known that there is no need to give further details.

CHAPTER XXVII.

HOLLOWED ARTICLES.

To obtain the exact shape of a plate or sheet in the flat, for an article whose surface has a double curvature, is generally almost impossible. In practice, however, very good approximations can nearly always be found; the degree of accuracy in working up into the finished article depending more or less upon the treatment that the metal receives at the hands of the workman. In all good work, especially that which has to be under pressure, such as steampipes, the object aimed at should be to keep the plate in the finished article the same thickness all over, or, at any rate, to mind that it is not unduly thinned at any particular part.

As one workman will stretch or draw the plate more than another, it is obviously impossible in this class of work to set out a pattern that will suit all manipulators. Two things should be kept in view in marking out patterns for hollowed work—one is to make sure that the plate is not too small, and the other to mind not to waste metal. It is an easy matter to get a plate large enough, and then to shear and cut away in working up until the right size of object is obtained; but this manifestly is a most expensive method to follow, especially in the dearer metals such as brass and copper.

Generally, in hollow work, a good guide to follow is to try to set out the net pattern so as to have the same area as the surface of the finished article. After this, allowance can be made, if required, for any undue contraction or draw, and also for trimming and jointing.

Spherical Bowl.

The simplest article of hollow work to obtain the pattern for is probably that of a bowl as seen in Fig. 223. The size of disc for this can be obtained in several ways, all based upon the assumption that the area of the circular pattern is equal to the area of the curved surface of the bowl. The above assumption is practically correct, and would be strictly so if the metal of the hollowed bowl were exactly the same thickness as the sheet from which it had been worked up. To keep the metal the same thickness is almost impossible in practice; but what area is lost by contraction is generally balanced by that gained in expansion.

If the bowl is in shape a hemisphere, then the radius of the disc can be calculated if we remember that "the area of the surface of a complete sphere is equal to the square of the diameter multiplied by 3·1416." That is, area of sphere surface $= d^2 \pi$, and as the area of the disc must be equal to half of the above, we have—

$$R^2 \pi = \frac{d^2 \pi}{2} \therefore R^2 = \frac{d^2}{2}$$

and—

$$R = \frac{d}{\sqrt{2}} = \frac{d}{1 \cdot 4}$$

where R equals radius of pattern disc. This gives us a handy rule, for in all cases where the bowl is half a sphere, the radius of the circular pattern will be found by dividing the diameter of the sphere by 1·4.

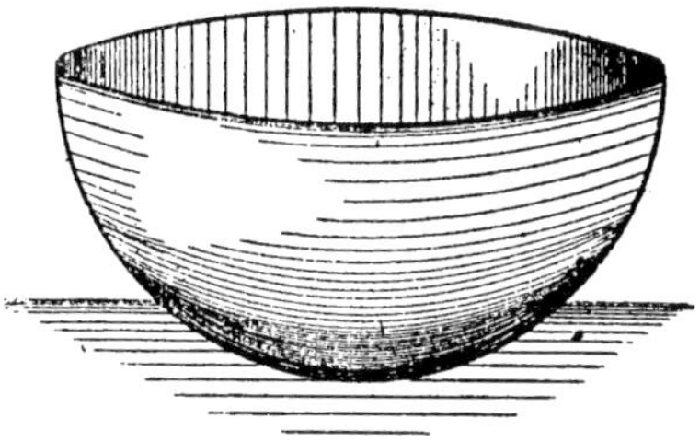

FIG. 223.

For those who are not good at calculating the same result can be obtained graphically by aid of the construction shown in Fig. 224. This construction, indeed, may be taken generally and applied to all hollowed work that comes out as any segmental portion of a sphere. Essentially the construction consists in setting out a right-angle triangle, one side

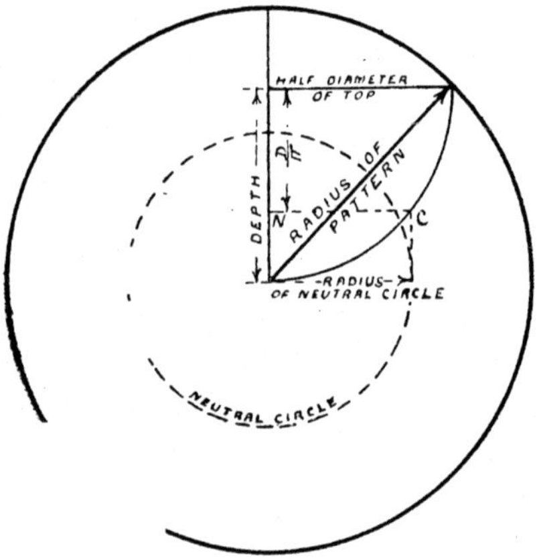

FIG. 224.

being equal to half the diameter of bowl, and the other equal to the depth; the hypotenuse or third side then giving the radius of the disc.

The general case is perhaps better explained by Fig. 225, the parallel lines representing the top of bowls, and the dotted lines with the numbered R giving the radius of the corresponding required disc. Thus the line R^1 will give the radius of a circular plate which will work up to a bowl whose top is represented by the line 1 1. In the same way a circular plate whose radius is equal to R^3 would form a bowl whose diameter at top would be equal to 3 3.

From Fig. 225 it is interesting to observe that the radius of a circle having the same area as a sphere will be equal to the diameter of the sphere. So that, if it were possible to work a sheet metal disc into a complete ball of the same

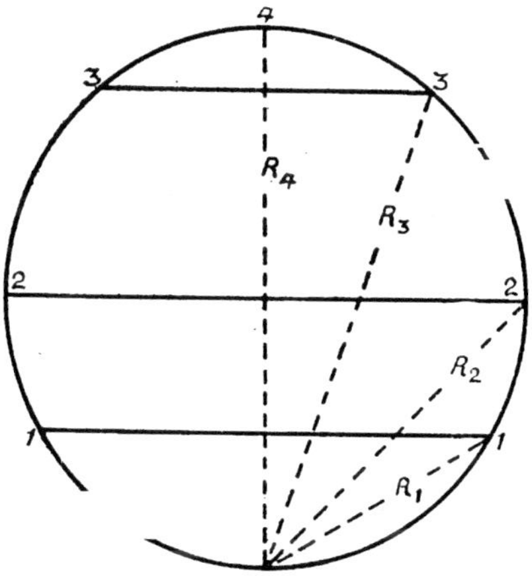

Fig. 225.

thickness metal, the diameter of the disc would be just twice that of the sphere.

In calculating the surface area of a sphere, or any part of a spherical surface, it is often handy to call to mind the relation that exists between the surfaces of the sphere and cylinder. "The area of a sphere, or the curved surface of a segment or zone of a sphere, is equal to that of the circumscribing cylinder or part of cylinder." Thus, in Fig. 226, the area of the segment of sphere A B C will be equal to the area of the curved surface of the part of cylinder *a c b b*, which is cut off by A C, produced as shown. In the same way the area of the curved surface of zone will equal the part of cylinder cut away by producing D E and F G.

This somewhat peculiar property of the sphere and cylinder occasionally comes in handy in setting out patterns for plumber's and coppersmith's work, and also in estimating weights of sheet metal required in work of this character. It is also of use when the end of a straight pipe is required to be worked to form part of a dome end, such as in coppers, steam-dome covers, and coal-scoops. Referring again to Fig. 226, suppose we require to work the end of a straight pipe into the shape *b* H F, or *b* L G, then the length of pipe wanted to form the domical part H F

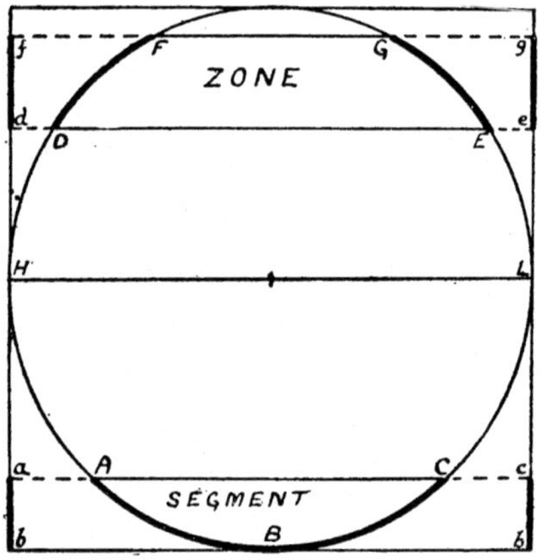

FIG. 226.

would be H *f*. It should be borne in mind that if the point *f* is to be worked around to F it will be necessary in the raising to so stretch the metal as to keep it a constant thickness.

Raising a Bowl.

In working up a bowl or any similar article or part of an article, the sheet may be either "raised" or "hollowed." The raising process is more particularly suitable to the softer metals, such as lead, pewter, copper, and brass, and is carried out, as shown in Fig. 227, the sheet metal being drawn over the head by working round course after course. The edges of the sheet will wrinkle a good deal, and particular care must be taken, especially in thin metal, that the sheet does not double over, or else the job will be ready for the scrap-heap. To avoid this, work around the bowl

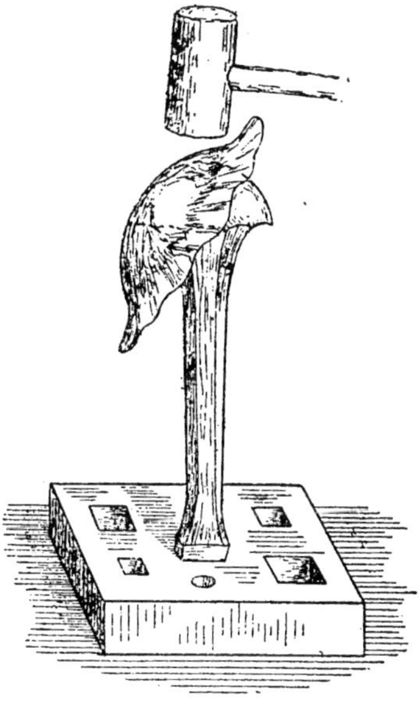

FIG. 227.

gently, and if unduly large wrinkles appear, work them out carefully to the edge of plate. If the job is of copper or brass, it should be annealed two or three times during the working up.

To obtain a smooth surface and to harden and stiffen the

metal the job should be finished off with the hammer. The planishing being done with a round flat or concave-faced hammer on a smooth bullet-head stake, as shown in Fig. 228. The blows should be carefully placed, commencing at the centre and gradually working out to the edge. The surface should not be struck twice in one place; but the hammer-marks should join on to each other. The greatest care must be taken that the sharp edge of the

FIG. 228.

hammer-face does not strike the surface, as it is almost impossible to obliterate marks of this character, and if left on, the appearance of the article is not by any means improved.

If the surface is to be polished it should be observed that it is free from scale and perfectly clean before the planishing takes place, as every particle of dirt on the surface will be driven into the metal by the hammering, and will be most difficult to remove in the polishing.

Hollowing a Bowl.

In hollowing, the sheet metal is hammered into a recess in either a block of wood, cast iron, or lead. For general work it is most convenient to have a wooden hollowing block as shown in Fig. 229. Recesses can then be readily sunk into the ends to suit the shape of the work in hand. A bullet-faced hollowing hammer is used, the sheet being first wrinkled around the edge, and the courses following each other up to the centre of the plate. The plate should not be hammered too much at the middle whilst the edges are stiff, as this will tend to unduly stretch, and thus thin, the metal at that part.

"Hollowing" is usually a much quicker process than "raising," but it has a tendency to thin the sheet in the middle part more than the latter method. The stiffer metals, such as steel, iron, and zinc, are generally treated by the former process.

Fig. 229.

Wrinkling Circle.

Whether the hemispherical bowl is raised or hollowed, it will be observed that the centre portion of the disc is stretched, whilst that part which is nearer to the edges will contract; there must, therefore, be a part of the disc or bowl which is neutral;

that is, neither stretches nor contracts. This will be a circle lying on the surface of the bowl, and shown by the line N C in Fig. 224. To determine the position of this circle, a distance equal to the diameter of the hemisphere divided by 3·1416 is set down from the top, and a line parallel to the top of bowl then drawn; this gives the diameter of the neutral circle. Taking a hemispherical bowl of 5 in. diameter, the distance down of the neutral circle will be—

$$\frac{d}{\pi} = \frac{5}{3{\cdot}1416} = 1{\cdot}6 \text{ in.}$$

In the working up, this circle should remain of constant diameter, and therefore gives us the boundary line where the wrinkles from the edge of disc should die away. In

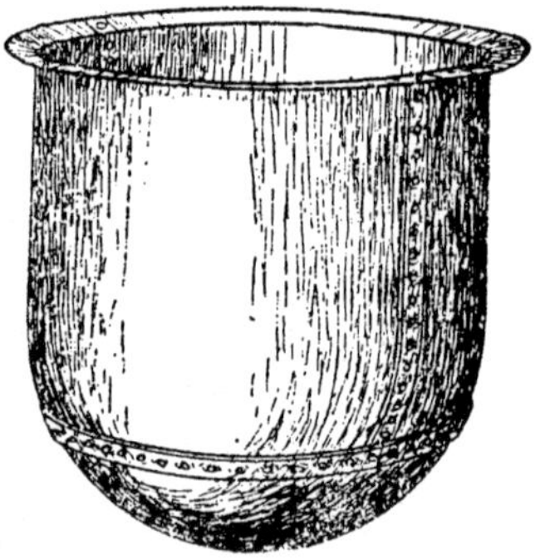

FIG. 230.

copper bottoms, and such like work, the wrinkling circle should be marked on the circular plate, and the wrinkles put in on the outside of this circle.

Patterns for Copper.

The setting out of the patterns for a copper (Fig. 230) conveniently comes in with our consideration of the sphere.

A diagram of the vessel is shown in Fig. 231, from which it will be seen how to obtain the sizes of bottom and sides. The radius for the bottom disc can be measured from the diagram, or can be calculated as follows:—

$$\text{Disc radius} = AC = \sqrt{AB^2 + BC^2}$$

The height of the plate for the body is determined by the property of the sphere and cylinder mentioned in connec-

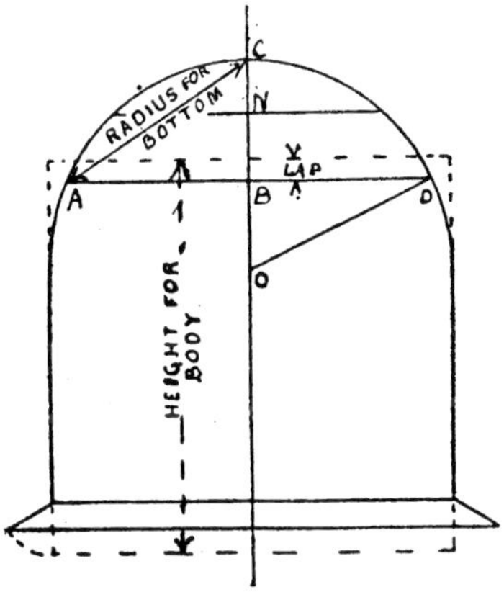

FIG. 231.

tion with Fig. 226. The body will be made up in either one, two, or more pieces, according to the diameter of the copper. The patterns are not shown set out, as their shapes being so simple there is no need for it.

After the body is roughly riveted together the bottom part can be worked over or razed in on a bench bar, as shown in Fig. 232. (It will, perhaps, not be here out of place to state that the bench bar in a general shop is probably the most useful tool to have, as it can be used for a score or more different purposes. If it has a square-tapered hole in the flat end, it can be made to carry various heads and do similar work to the horse in a floor-block.)

The stretching or throwing off of the brim is explained by Fig. 233. This can either be done on a head stake as shown, on an anvil, or on the edge of the flat end of bench bar. In stretching the brim the depth of flange should first be marked around the inside. When throwing the brim off on the stake, care should be taken that the more intense part of the hammer-blow falls near the edge of metal, as the greatest amount of stretch must, of course, take place on the outside of the brim.

The position of the wrinkling circle for the bottom, and

FIG. 232.

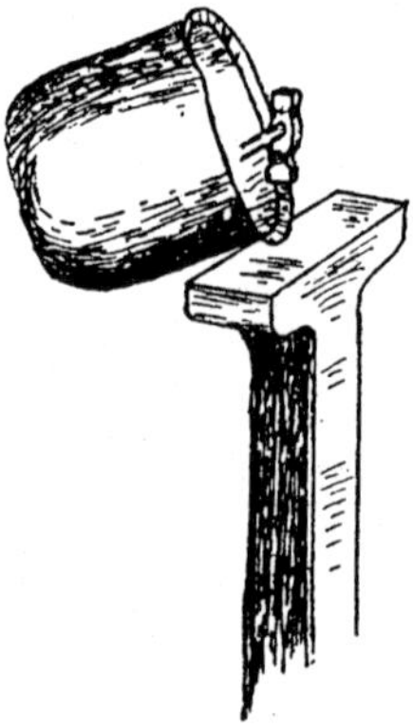

FIG. 233.

thus its diameter (Fig. 231) can be determined by the following rule: "To find the distance of the neutral circle from the bottom, deduct the chord from the arc, and multiply this difference by the radius and divide the product by the arc." That is—

$$CN = \frac{\text{radius } OD \times (\text{arc } ACD - \text{chord } AD)}{\text{arc } ACD}$$

Before proceeding to completely rivet up, the bottom should be attached with a few tacking rivets. In finishing the joints care should be taken that the seams are properly countersunk, and also that the tails of rivets are drawn up, so that the inside of the copper shall be completely smooth.

Sketches of hammers are shown in Fig. 234, the two on the left being types of hollowing or blocking hammers, the centre one a stretching or razing hammer, and the right-hand

FIG 234.

pair planishing hammers. It is, perhaps, hardly necessary to point out that there are hundreds of different shapes and sizes of the above, the form of hammer used depending upon the strength of material and kind of job in hand.

Capacity of a Copper.

Before proceeding to show how to calculate the number of gallons that a copper of the shape shown in Fig. 230 will hold, it will be necessary to explain how to find the volume of a sphere.

One of the simplest aids to the remembering of how to find the volume of a sphere is in the peculiar relation that exists between the volumes of a cone, sphere, and cylinder when their diameters and heights are equal. Imagine that Fig. 235 represents these three solids, then the relative volumes of cone, sphere, and cylinder will be as 1 is to 2 is to 3. So that, having found the volume of the cylinder, the sphere will be two-thirds, and the cone one-third of it. It will also be seen that the sphere is just twice the volume of the cone. The above relation facilitates the work in calculating the cubic contents of a vessel with hemispherical or conical ends.

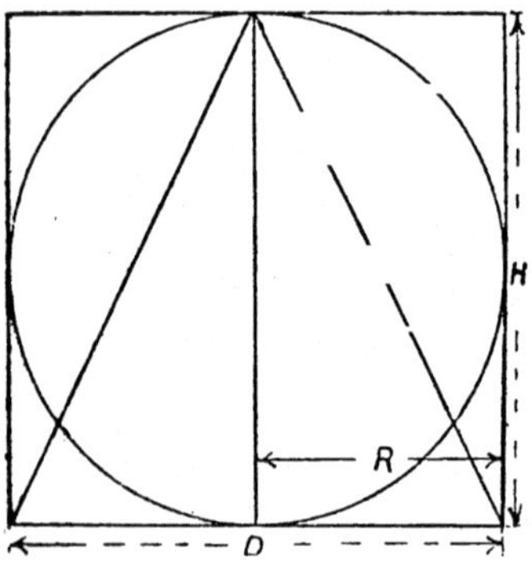

FIG. 235.

If a vessel has hemispherical ends, all that is necessary is to add two-thirds of the diameter on to the cylindrical portion, and calculate its volume.

If a vessel has a conical bottom or top pointing outwards, then its volume can be found by adding one-third the height of the cone on to the cylindrical portion, and calculating as before. If a vessel has a conical bottom pointing inwards, then, of course, one-third the height of cone would be deducted from the length.

It is sometimes convenient to use the ordinary mensuration rules, such as—

Volume of sphere—

$$= \frac{\pi d^3}{6} = \frac{\pi 4 r^3}{3} = r^3 \times 4{\cdot}1888$$

Volume of cone—

$$= \frac{\pi d^2 h}{12} = \frac{\pi r^2 h}{3} = r^2 h \times 1{\cdot}0472$$

Volume of cylinder—

$$= \frac{\pi d^2 h}{4} = \pi r^2 h = r^2 h \times 3{\cdot}1416$$

Where d = diameter, r = radius, and h = height.

In practice, it is handier to use some such rules as the following, which have been calculated on the basis of 277·274 c. in. to the gallon. Taking dimensions in feet—

Gallons in cylinder	= $r^2 h$	× 19·6
,, sphere	= r^3	× 26·11
,, cone	= $r^2 h$	× 6·53

If the dimensions are in inches, then the following multipliers must be used—

Gallons in cylinder	= $r^2 h$	× ·01133
,, sphere	= r^3	× ·01511
,, cone	= $r^2 h$	× ·00378

Taking one example to illustrate their use. Suppose we require to find the number of gallons in a hemispherical bowl of 20 in. diameter. Then—

$$\text{Gallons} = \frac{10 \times 10 \times 10 \times {\cdot}01511}{2} = 7{\cdot}55 = 7\tfrac{1}{2} \text{ (nearly)}.$$

After the above explanation, let us come back to the copper. Suppose it is 3 ft. diameter and 3 ft. 6 in. deep. Then deducting half the diameter from the depth, this will give us 2 ft. for the length of the cylindrical part. Adding two-thirds the depth of the hemispherical bottom on to the cylindrical portion, this will give us an equivalent cylinder of 3 ft. length. The cubical contents will, therefore, be—

$$\tfrac{3}{2} \times \tfrac{3}{2} \times 3 \times 19{\cdot}6 = 132{\cdot}3 = 132\tfrac{1}{4} \text{ gallons.}$$

CHAPTER XXVIII.

SOLID PANS, JUGS, EXPANSION BULBS, ETC.

Solid Round Pan.

A CIRCULAR pan or vessel, such as that shown in Fig. 236, can be raised or drawn out of the solid plate when such malleable metals as copper, brass, etc., are used. Articles of this description, when required in quantities, are drawn up out of blank discs, in two or three operations, by the aid of suitable dies in a hand or power press. When a few articles only are wanted, they are formed by hand, the disc being gradually worked over a bench-stake, by the use of mallet and hammer, until the required shape is obtained. Whether shaped by hand or machine, nearly all metals require annealing between each or every other operation.

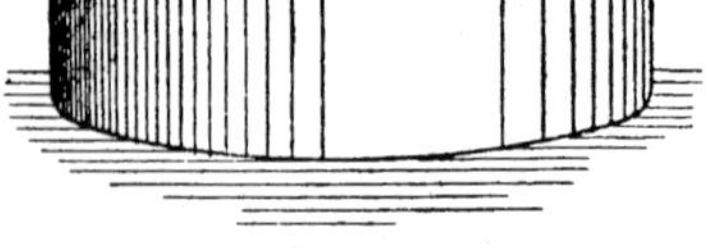

FIG 236

In calculating the size of disc the thing to be kept in mind is to have a circular blank of the same area as the combined area of the bottom and sides of the vessel. This can be calculated or found by graphic construction. We will show both methods. For calculating the radius of the disc, the following rule can be used: "Add the square of the radius of the pan bottom to twice the product of the radius and the depth, and extract the square root of the whole." Thus, in Fig. 237, suppose the diameter of the vessel is 2 ft., and its depth 1 ft. 6 in.; then the radius of the flat disc will be—

$$R = \sqrt{r^2 + 2rd} = \sqrt{144 + 24 \times 18} = \sqrt{576} = 24 \text{ in.}$$

A much simpler method, for those not good at calculations, is shown by the construction in Fig. 237. A B is the diameter of the circular pan, C the centre of the bottom, and B D the depth. Make B E equal to B D by turning the latter line down, as shown. Now describe a semicircle on A E, centre O, and produce B D to meet the semicircle in F. Join C to F, and the line C F will give the length of the radius for the blank disc. This latter method, it should be noted, will give the same result as obtained by calculation, the construction simply being a graphical method for working out the arithmetical problem as above.

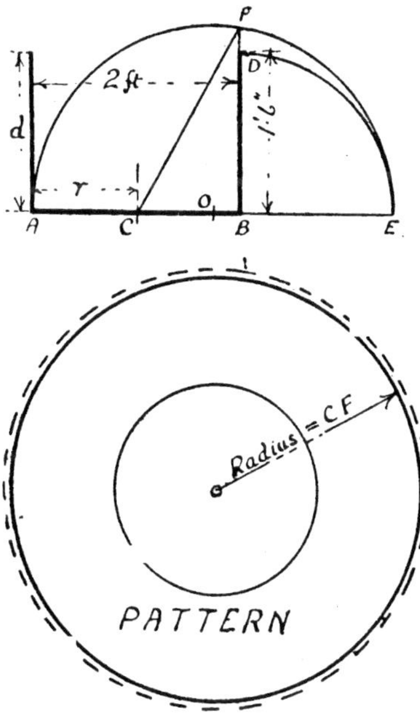

Fig. 237

In raising up the vessel care must be taken to keep the metal at as uniform a thickness as possible. If any trimming is required to be done, or wiring put on, then a small allowance should be added to the circular plate, as shown by the outside dotted circle.

Solid Round-Tapered Pan.

If the size of the disc is required for working up into a circular pan, having a flat bottom and tapered sides, then, with some little modification, the rules, as used in the former case, can be applied. First of all, the mean radius of the pan must be found, and this is done by adding the radius of the bottom to the radius of the top and then dividing by two. The rule for finding the radius of the blank will then read as follows: "Add the square of the bottom radius to twice the product of the mean radius and the slant depth, and extract the square root of the whole." Suppose a pan of this description (Fig. 238) is 2 ft. 6 in. in diameter at the top, 1 ft. 6 in. at the bottom, and 1 ft. 3 in. slant depth.

Let R = radius of disc.
,, r = mean radius of pan.
,, r_1 = radius of bottom.
,, r_2 = radius of top.
,, s = depth of slant side.

Then the mean radius will be—

$$r = \frac{r_1 + r_2}{2} = \frac{15 + 9}{2} = 12 \text{ in.}$$

The radius of the pattern disc will equal—

$$\text{R} = \sqrt{r_1{}^2 + 2\,r\,s} = \sqrt{81 + 24 \times 15} = \sqrt{441} = 21 \text{ in.}$$

The same result can be obtained graphically by the construction shown in Fig. 238. Bisect E F, and draw the line G H through the bisecting point parallel to A B. Draw H J parallel to C A, and thus cut off A J equal to

G H. Using A as centre, and A C as radius, turn A C down, and thus fix the point K. On K J describe a semi-circle, and draw the line A L square to A B to meet it. The length of the line drawn from L to F, the centre of the

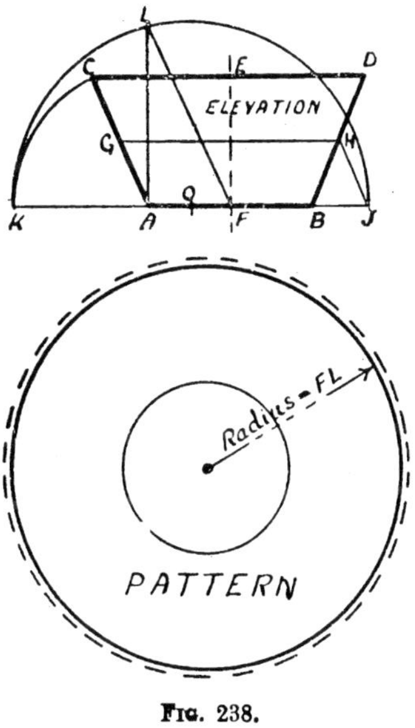

Fig. 238.

bottom, will give the radius for the blank disc. Any allowance required must, of course, be added on to the pattern, as in the last case.

Vessels with Double-Curved Surfaces.

The patterns for articles whose surfaces are of double curvature can be marked out very approximately by an adaption of the methods already explained. Before, however, the methods can be applied to this class of object, it will be necessary to give some preliminary explanation. Suppose it is required to get the size of a circular blank that will work up into a

Barrel-Shaped Vessel,

as shown by the section in Fig. 239. It should be remembered, as has already been stated, that the area of the pattern disc must be equal to the area of the bottom and body together of the vessel. The bottom being a circle, there will, of course, be no difficulty in finding its area. To calculate the area of the body-surface, however, is a more difficult task. We may consider the body as being a surface of revolution—that is, a surface swept out by the arc A D E moving at a constant distance from the centre line C F. It is manifest that if the arc revolves in this manner, the surface generated would be that as shown by the figure. Now, the area of a surface formed in this way is equal to the length of the generating

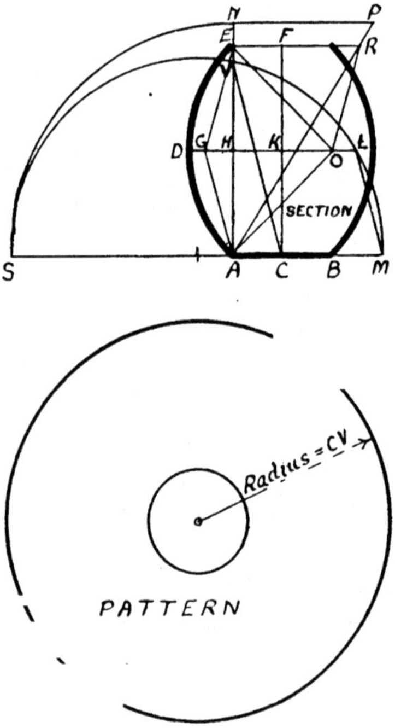

FIG. 239.

curve—in this case the arc—multiplied by the distance that its centre of gravity would travel in one complete revolution. The centre of gravity, it may be explained, can be looked upon as an imaginary point upon which the section curve would balance in any position. For an arc of a circle, the position of its centre of gravity can be calculated by the following rule: "Multiply the radius by the length of the chord, and divide by the length of the arc; the result giving the distance of the centre of gravity from the centre about which the arc has been described."

For the benefit of those readers who can manipulate figures, we will now explain how to obtain the radius of the pattern disc by calculation, and then afterwards show how the same result can be found by construction. The arc A D E, in Fig. 239, which is described from the centre, O, is a quarter of a circle, of radius $7\frac{1}{2}$ in.; hence its length will be—

$$\frac{15 \times 3{\cdot}1416}{4} = 11{\cdot}78 \text{ in.}$$

The length of the chord A E will be—

$$\sqrt{2 \times (7{\cdot}5)^2} = 10{\cdot}6 \text{ in.}$$

The distance O G, from the rule stated above, will then be—

$$\text{O G} = \frac{\text{Radius} \times \text{chord}}{\text{arc}}$$

$$= \frac{7{\cdot}5 \times 10{\cdot}6}{11{\cdot}78} = 6{\cdot}75 \text{ in.}$$

The distance of the centre of gravity, G, from the centre line C F will equal—

$$\text{O G} - \text{O H} + \text{A C}.$$

The diameter A B = 5 in., and in this case O H = H E = 5·3 in.; hence—

$$\text{K G} = 6{\cdot}75 - 5{\cdot}3 + 2{\cdot}5 = 3{\cdot}95 \text{ in.}$$

Having obtained the distance of the centre of gravity from the centre line, the radius of the pattern disc can now be calculated, as in connection with Fig. 238.

Let l = length of arc.
,, r_1 = radius of bottom.
,, r = distance of centre of gravity from centre line.

Then—

$$R = \sqrt{r_1^2 + 2\,r\,l} = \sqrt{(2\cdot5)^2 + 7\cdot9 \times 11\cdot78}$$
$$= \sqrt{99\cdot3} = 10 \text{ in. (nearly).}$$

Whilst, perhaps, somewhat uninteresting, the calculations, as explained above, are very important, and have wide application in finding the areas of surfaces of this character. They can also be applied to finding the capacity or cubic contents of vessels such as those shown in Figs. 239, 240, and 245.

Capacity of Barrel-Shaped Vessel.

Seeing that we have the dimensions in connection with Fig. 239, it will, perhaps, be better to explain how to find its volume before passing on. It should be remembered that whilst the calculations that follow apply to the vessel in Fig. 239, the same principle is applicable to all circular articles.

The distance of the centre of gravity of the segment A D E H A from O will equal—

$$\frac{\text{The cube of the chord}}{\text{12 times area of segment}}$$

$$= \frac{10\cdot6 \times 10\cdot6 \times 10\cdot6}{(11\cdot78 \times 3\cdot75 - 5\cdot3 \times 5\cdot3) \times 12} = 6\cdot17 \text{ in.}$$

Distance of centre of gravity of segment from centre line equals—

$$6\cdot17 - 5\cdot3 + 2\cdot5 = 3\cdot37 \text{ in.}$$

Then the volume of the vessel equals the volume of the centre cylindrical portion, together with the volume swept out by the segment revolving around the centre line—

$$\begin{aligned}
\text{Volume} &= (2{\cdot}5)^2 \times 3{\cdot}1416 \times 10{\cdot}6 \\
\text{,,}\quad &+ 2 \times 3{\cdot}37 \times 3{\cdot}1416 \times 16{\cdot}09 \\
\text{,,}\quad &= 208{\cdot}03 + 344{\cdot}59 \\
\text{,,}\quad &= 552{\cdot}62 \text{ cubic inches.}
\end{aligned}$$

To find the number of gallons the above would have to be divided by 277·274 (the number of cubic inches in a gallon). It will thus be seen that the vessel will hold just under two gallons.

Pattern for Barrel-Shaped Vessel by Construction.

To find the radius of the pattern disc, graphically, the line A N (Fig. 239) is made equal to the length of the arc A D E, and N P drawn square to it and equal to the radius O E. Line E R is then drawn parallel to N P, and the point R cut off by joining P to A. O G is then made equal to E R by joining R to O and drawing E G parallel to R O. The line K L is cut off equal to K G and L M drawn parallel to G A. Now, using A as centre, and A N as radius, the point S is marked off. A semicircle is next described on S M intersecting the line A N in V. The line C V is the length required for the radius of the pattern disc; and this, on measurement, will be found to be 10 in., as calculated before.

Circular Pan with Sides Curved Outwards.

The method shown above will apply to all kinds of different-shaped circular vessels, the only difference being in the finding of the centre of gravity of the side section. Perhaps one further example in the way of a pan with its sides curved outwards will make the construction followed plainer.

In Fig. 240 a section showing the shape of the pan is given. The line A N is cut off equal to the length of the arc A D E and N P drawn square to it, and made equal to the radius O A. The line A F, as shown, is marked off equal to the chord A E, and F R drawn parallel to N P, the point R being determined by joining P to A. Then the point G, which is the centre of gravity of the arc A D E, is fixed by making O G equal to F R. The point L, which is the corresponding centre of gravity for the right-hand arc, is found by drawing G L parallel to A B and making K L equal to K G. G is next joined to A and the lines L M drawn parallel to G A. After S has been determined by making A S equal to A N, a semicircle is described upon S M, so fixing the point V on the line A N produced. The length of the line C V will give the radius for the disc.

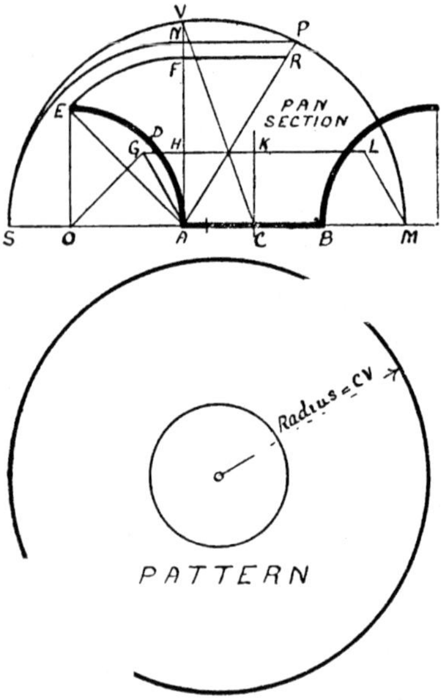

FIG. 240.

Copper Expansion Bulb.

A copper expansion bulb, or ball, as shown in Fig. 241, is sometimes fixed upon a length of steam or hot-water pipe, to allow for the varying length of the pipe due to changes of temperature. It is usually worked up from two circular discs of metal, the halves being fastened together with a brazed joint running around the bulb.

The setting out for the pattern disc is shown in Fig. 242. It is only necessary to mark out a quarter of the section shape, and then on this apply the construction used in Figs. 239 and 240. The point G can be taken as the centre of gravity of the curve (this being the point upon which a wire bent into the shape B G A would balance). A is joined to G, and produced to D, the line A D being made equal to the length of the double curve A G B. The line G F is next drawn parallel to A L and O F cut off equal to O G. The point H is then fixed by drawing F H parallel to G A. Line A D is turned up about A as centre to fix the point E, and on H E a semicircle described, cutting L J produced in K. The line A K gives the radius for the circular blank. After each half is worked up into the required shape, the centre circles are cut out to form the pipe inlet and outlet.

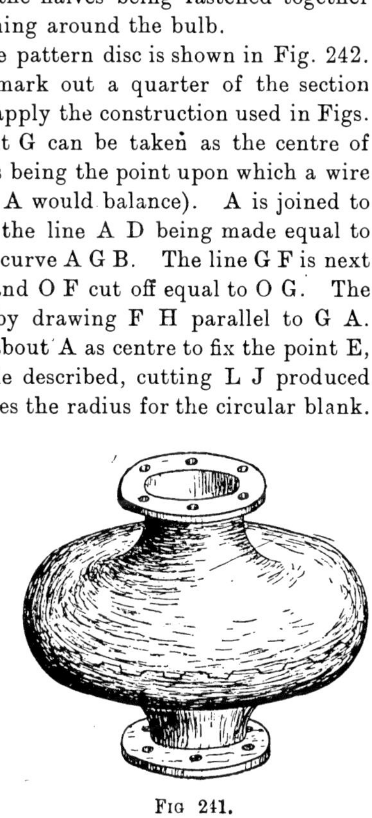

FIG 241.

We will now give a couple of examples of the application of the foregoing methods to the setting out of patterns for articles which can be worked up from a frustum of a cone. The first example is that of a

Steam Exhaust-Pipe Bell-Mouth,

as shown in Fig. 243. A heavy bead is worked on the pipe at the part where the bell-mouth runs into the straight pipe, and a split tube is fitted around the top edge of the outlet.

The setting out for the pattern is shown in Fig. 244. The bead is first of all allowed for by lengthening the pipes by the distance A B, which is equal to the length C D E measured around the bead. The position of the point G is found by the rule explained in connection with Fig. 239.

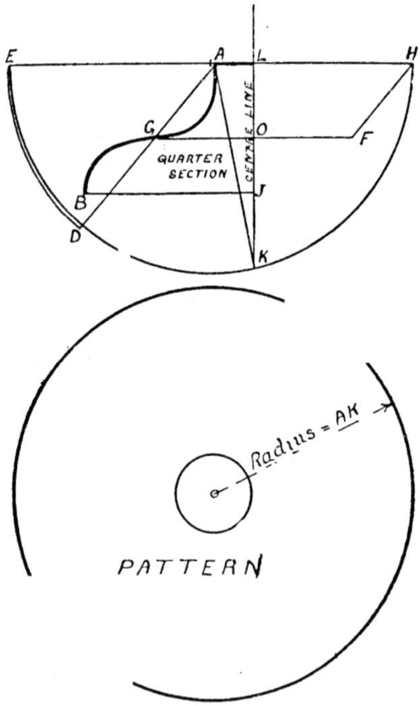

FIG. 242.

F is the middle point of B E. The points G and F are joined together and the length G J made equal to F B; then F H is drawn parallel to G J and made the same length as the arc K L. Now, if H be joined to J, the point W, where it intersects the line F G, will give the centre of gravity of the section shape. The line N M is now drawn in an average position, its length being marked off equal to the outline B K L O. M N is then produced to meet the centre line in T. The pattern is now laid out as that for a

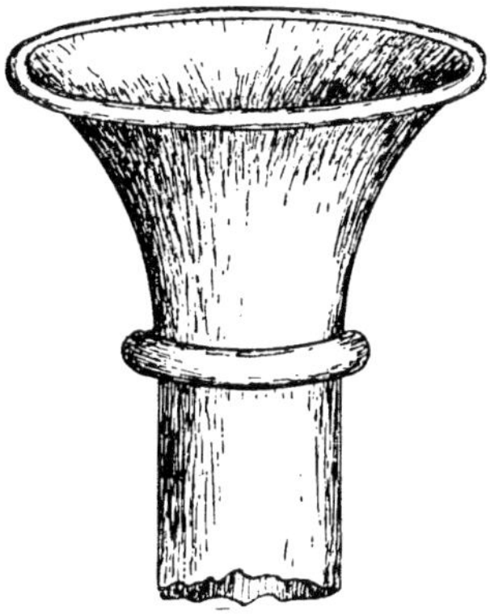

FIG. 243.

cone frustum, the lengths T M and T N being used as the radii, and the length of the arc R S P being made equal to four times the length of the quarter-circle on V M.

Copper Jug.

The second example is that of a jug, as shown in Fig. 245. The jug is made in four parts—the body, bottom, spout, and handle. The setting out of the patterns is shown in Fig. 246. A half-elevation, showing the spout portion, is first drawn. The line B D is marked in an average position on the outline, and the middle point C determined by drawing the line A C square to, and from the middle of the centre line. B D is made equal to the length of the double curve E F H, the body pattern then being struck out, as in Fig. 244.

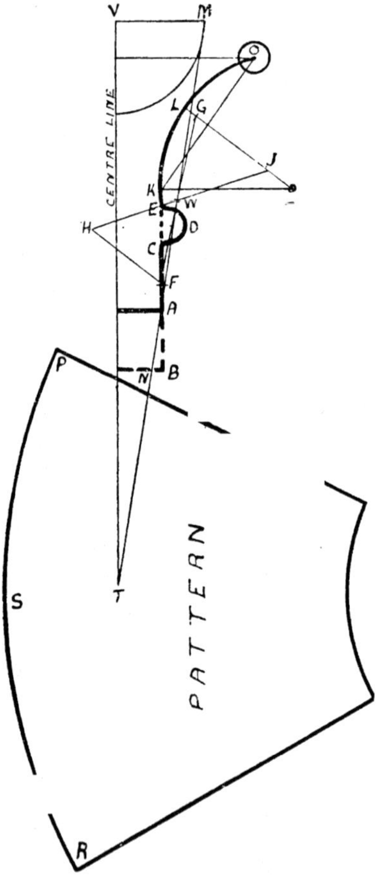

FIG 244.

The inner circle on the bottom pattern is the same diameter as the jug bottom, and for the outer circle the depth of the foot is added all round.

For the spout the half-section 0 1 2 3 is first marked **out**, this being divided up, as shown, and perpendiculars dropped on to the line H 3. Then, using centre P, the arcs *a b*, *c d* are swept around. The girth line 0′ 0′ on the pattern is made the same length each side its centre point as the

FIG. 245.

curve 0 to 3 on the spout. Construction lines are drawn through each division point, and these cut off, above and below the girth line, the same length as the arcs on the spout in elevation. Thus, $a' b' = a b$, $c' d' = c d$, and $e' f' = e f$, the parts, of course, being measured above and below the line H 3. The points obtained are connected up with curves, and so the pattern completed.

The hole for the spout will be cut in the body after it has been worked into shape. The handle can be made in the form of a tapered tube, loaded with lead, and bent into

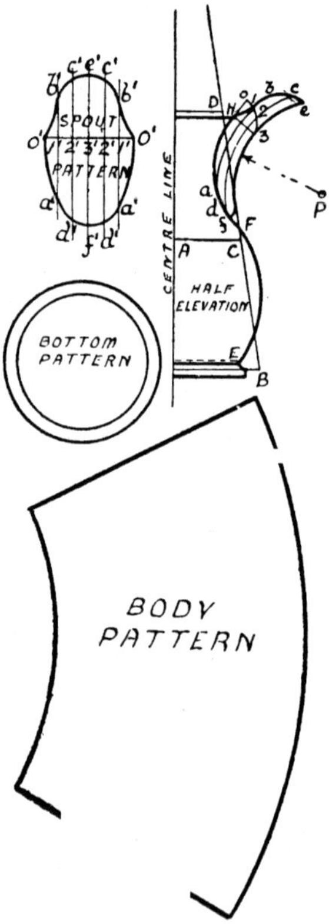

FIG 246.

shape; or it can be formed out of a bar of solid copper. It may be attached to the body by riveting.

The seam on the body can be brazed down to form the cone frustum.

After the body and bottom have been tinned on the

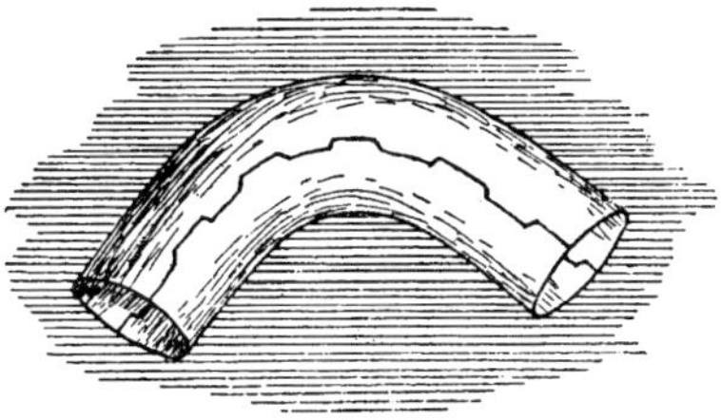

FIG. 247.

inside, the latter can be fixed in the former by soldering around the inside.

The surface of the jug may be polished and lacquered, or treated in any other way as desired.

CHAPTER XXIX.

WORKED-UP PIPE BENDS, BREECHES PIECES, ETC.

Pipe Bends.

SOLID drawn pipes, both of steel and copper, of diameters up to 6 in. or 7 in., can now by the aid of hydraulic or other bending machines be bent to form bends of various shapes, so that simple pipe bends made up out of sheet metal and brazed or riveted are not so common as formerly. Bends for pipes of large diameter, however, have to be made up, and also those for small pipes where no facilities exist for pipe bending. We shall, therefore, now consider a few typical cases of bends, tee-pieces, etc.

Quarter-Bend.

A quarter, or square, bend is usually made up in two pieces, the joints running along the back and throat, or along the two sides, as shown in Fig. 247. This latter method has several advantages over the former, one being that there is less waste in cutting out the plates, another that they are perhaps a little easier to shape, and a third the greater convenience in brazing side seams.

A method for obtaining the size of the plate is illustrated by Fig. 248. The exact shape of the quarter-bend is marked out as shown in the figure, and the joint line drawn in. Now before setting out the plates it will be as well to consider what happens when a plate is bent to form either the back or throat portion of the pipe. Consider the back piece first. As the plate is brought into shape by hollowing and razing, it will be observed that the back of the half-pipe strêtches and thus becomes longer, whilst the edge of the plate, which will form the joint, contracts, and thus becomes shorter. There must, therefore, be some line on the plate which neither gets longer nor shorter, and if we can obtain the length of this line, it will give us the length of the plate for back. The same reasoning also applies to the saddle or throat part of the bend; for, whilst the edges of the plate which form the joint line get longer in the working, the throat draws in, and thus the throat line becomes shorter. There must, therefore, also be a line on the saddle which remains of constant length. The posi-

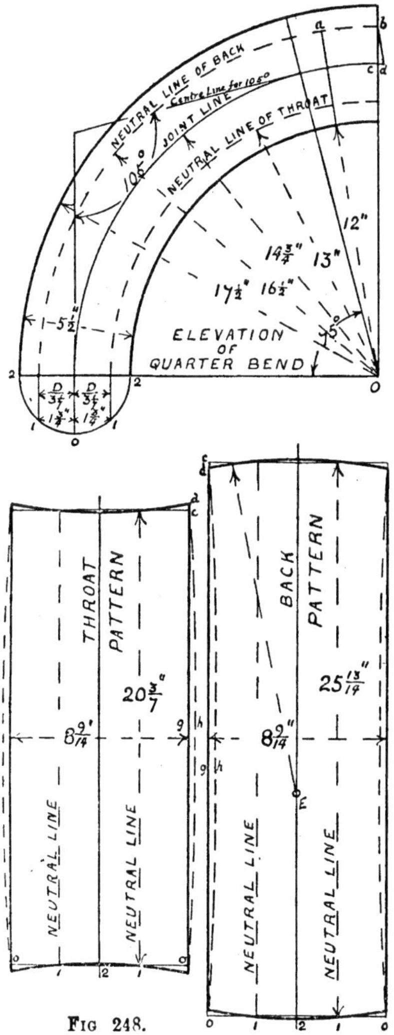

FIG 248.

tion of this neutral line on the back and throat pieces can be obtained from the following rule —

"Divide the diameter of the pipe by $3\frac{1}{7}$, and set this distance on each side of the joint." Thus, in the present case, the diameter of the pipe being $5\frac{1}{2}$ in., the distance of the neutral line from the centre line of pipe will be—

$$\frac{D}{3\frac{1}{7}} = \frac{D \times 7}{22} = \frac{52 \times 7}{22} = 1\frac{3}{4} \text{ in.}$$

This distance is set on each side of the joint line, and the neutral line drawn as shown in Fig. 248. The size of the plates then will be obtained by making their widths equal to half the circumference of the pipe, and their lengths equal to the lengths of the respective neutral lines. The widths of the patterns will be—

$$\frac{5\frac{1}{2} \times 3\frac{1}{7}}{2} = 8\frac{9}{14} \text{in.}$$

The length of the back piece—

$$\frac{16\frac{1}{2} \times 2 \times 3\frac{1}{7}}{4} = 25\frac{13}{14} \text{ in.}$$

and the length of the throat piece—

$$\frac{13 \times 2 \times 3\frac{1}{7}}{4} = 20\frac{3}{7} \text{ in.}$$

A somewhat peculiar fact should be noticed in connection with the lengths of the neutral lines, and that is, that the neutral line of the back is always, for a quarter-bend, exactly the diameter of the pipe longer than the neutral line of the throat. So that when the length of one line is determined the other can be obtained by adding or deducting the diameter of the pipe, as the case may be. In the present example we have—

$$25\frac{13}{14} - 20\frac{3}{7} = 5\frac{1}{2} \text{ in. (diameter of pipe).}$$

If there is any straight pipe on the end of the quarter-bend, then the length of this should be added on to the calculated length, and the ends of the pattern will be straight lines drawn square to the centre line. But if the bend has no straight portion, as in the present case (Fig. 248), then the ends of the pattern will need curving somewhat, as shown on the figure. Theoretically, no curvature at the ends should be necessary, as the area of sheet metal on the patterns, as calculated by the above rules, is exactly equal to the area of the pipe bend surface. Practically, however, it is impossible to draw metal evenly, and for some short distance from the ends, generally equal to the radius of the pipe, the sheet or plate will hardly be drawn at all. This difficulty is usually overcome by making each strip slightly longer than required and then trimming the ends off the pipe. If desired, though, the curvature of the ends can be approximately obtained in the following manner Draw the neutral lines on the patterns (Fig. 248) by making the distance 2 to 1 on each side of the centre line equal to the length of the arc 2 to 1 on the semicircle in the elevation. Now make *a b* on the back neutral line equal to the radius of the pipe. Join *a* to O, and draw *b d* parallel to *a* O. Then the length *c d* will be measured off and set on the pattern, as shown. There should be no trouble in finding the radius, so that an arc can be drawn passing through *d* and the end of the neutral lines. This radius is shown on the back pattern, marked E *d*. If required, the length of *c d* can be calculated from the following rule: "Square the diameter of the pipe and divide it by $6\frac{2}{7}$ times the radius of the back neutral line." That is in this case

$$c\,d = \frac{5\frac{1}{2} \times 5\frac{1}{2}}{6\frac{2}{7} \times 16\frac{1}{2}} = \frac{9}{32} \text{ in. (nearly)}.$$

In working a throat piece into shape, it will be found that the girth near the middle becomes, through the draw, somewhat less, and for the same reason the girth of the back will increase; consequently, when the two halves come together the joint line will be slightly out of the centre of the side of pipe. This can be modified if necessary by adding on to each side of the throat pattern and deducting from each side of the back pattern a length equal to "the square of the diameter of the pipe divided by seven times the radius of the throat," so that the camber *g h* of the side dotted curves will equal—

$$\frac{5\frac{1}{2} \times 5\frac{1}{2}}{7 \times 12} = \frac{11}{32} \text{ in. (nearly).}$$

This distance should be set out as shown by *g h* on the pattern, and an arc of a circle drawn, as seen by the dotted curves.

The patterns as marked out above will be the net size, and any allowance for trimming or jointing must be added on. If the side seams are to be riveted, then a proper allowance for lap must be made; but if brazed, by thinning the edges down to form a wedge-joint, then little or no allowance will be needed, as the width of lap will be worked down out of the sheet metal.

Bend Less or Greater than a Quarter.

A bend may require making to joint up two lines of piping that are not at right angles, or to fit on to two flange faces that are not square to each other.

Suppose the centre lines of the piping make an angle of 105°, as shown in the elevation, Fig. 248, then the flange faces will make an angle of—

$$180 - 105 = 75^\circ$$

with each other. This angle can be set out as shown in Fig. 248, and thus the shape of the bend determined. The lengths of the back and throat patterns can be found as explained in connection with the quarter-bend, or they can be calculated by the following general rule, which applies to all cases. Rule for length of back pattern: "Multiply the radius of the joint line by $6\frac{2}{7}$, add twice the diameter of pipe, multiply by the angle that the flange faces make with each other, and divide by 360." Rule for length of throat pattern: "Multiply the radius of the joint line by $6\frac{2}{7}$, deduct twice the diameter of pipe, multiply by the angle that the flange faces make with each other, and divide by 360." Thus for the 105^0 bend, as marked out in Fig. 248, the length of back will be—

$$\frac{(14\frac{3}{4} \times 6\frac{2}{7} + 2 \times 5\frac{1}{2})}{360} \times 75 = 21\frac{17}{28} \text{ in.}$$

And the length of throat will be—

$$\frac{(14\frac{3}{4} \times 6\frac{2}{7} - 2 \times 5\frac{1}{2})}{360} \times 75 = 17\frac{1}{42} \text{ in.}$$

The difference between the lengths of the back and throat patterns can be readily calculated, without using the above somewhat cumbrous rule. Thus: "Multiply four times the pipe diameter by the angle between the flange faces and divide by 360." So that in the above example the difference will be—

$$\frac{4 \times 5\frac{1}{2} \times 75}{360} = 4\frac{7}{12} \text{ in.}$$

In any kind of a bend, before proceeding to shape the plates, wires should be bent to the shape of throat, joint, and back curves, these being used as templates to which the two halves will be bent. If no special appliances are about in the shape of blocks, dies, etc., on which the parts can be worked, then the throat part can be manipulated on a heavy mandrel or tee-stake. The back can be hollowed on a hollowing block, and dressed into something like shape on a curved top tee-stake, bullet-head stake, or cod fixed on bar. Care must be taken that the plates are kept properly annealed as they pass through the various operations. After the halves are shaped to the templates, if required to be brazed, the edges should be thinned down and properly

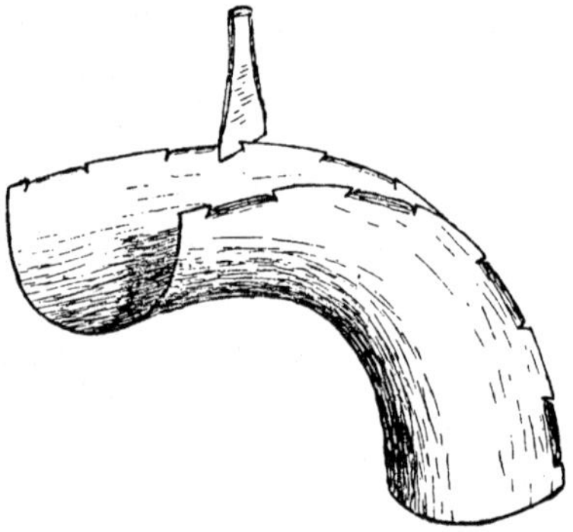

FIG. 249.

cleaned. The cramps are then cut on one half with a thin knife or chisel, which is held obliquely across the edge of the plate whilst being driven into the metal, as seen in Fig. 249.

The two halves are then fixed together and fastened with binding wire, and the cramps dressed down on a cod, as shown in Fig. 250. The bend will then be brazed and hammered up, as explained in Chapter XXXV.

Worked-up Breeches-Piece.

The methods applied to obtain the patterns in the last cases can with some little modification be used for all sorts of made-up bends. We will now explain the application to a three-way piece, as shown in Fig. 251. This kind of job can be made up in three pieces, the two side parts which form the waist and outside of legs, and one part which forms the inside of legs, or it can be formed of five pieces, three as above, together with a triangular gusset on each side.

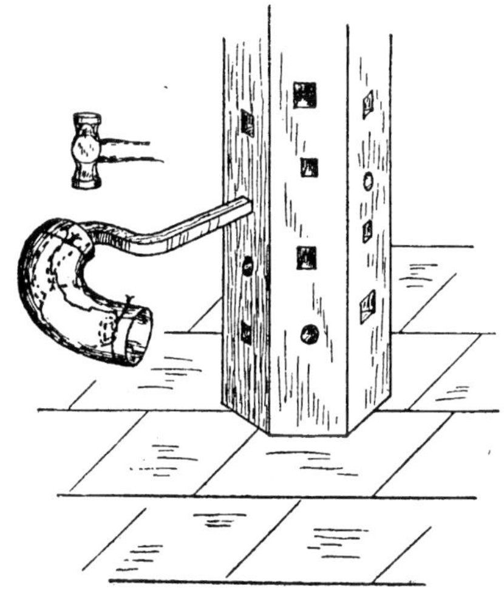

FIG. 250.

The patterns as set out in Fig. 252 are to build up the bend in three pieces. A half-side and half-end elevation of the bend is drawn, and it should be remembered in connection with this that the area of the waist pipe circle should be equal to the areas of the two leg-pipe circles added together. The leg-pipe being 4 in. diameter, the diameter of the waist-pipe will be equal to—

$$\sqrt{4 \times 4 \times 2} \doteq 5{\cdot}7 = 5\tfrac{3}{4} \text{ in. (nearly).}$$

Instead of bothering to calculate, the size of waist-pipe can readily be found by construction. Set out A B and B C at right angles (Fig. 253), each respectively equal to the radius of the leg-pipes, whether they are the same size or not; then A C will be the radius of the waist-pipe. In connection with this figure it is worth while noting that if A B, B C, C D, and D E are equal and drawn to form

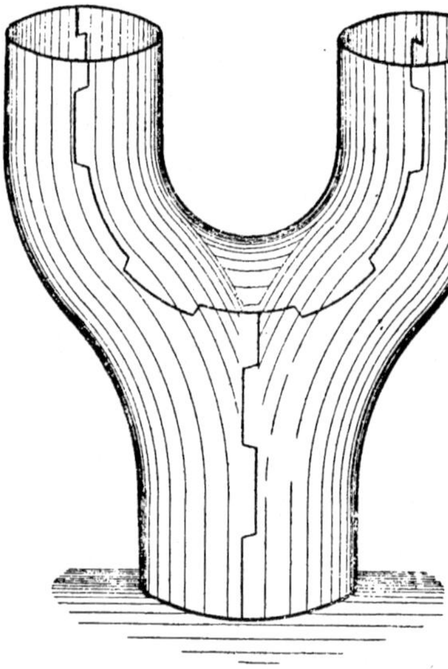

FIG. 251.

right-angled triangles, then the lines A B, A C, A D, and A E will give the radii of circles whose areas are as 1 is to 2 is to 3 is to 4.

To draw in the neutral lines (Fig. 252), their positions on the waist and leg-pipes are calculated as explained in connection with the quarter-bend.

For the waist and outside leg pattern, make the centre

line equal in length to the neutral line, and across its ends draw lines square, and cut these off equal to half the cir-

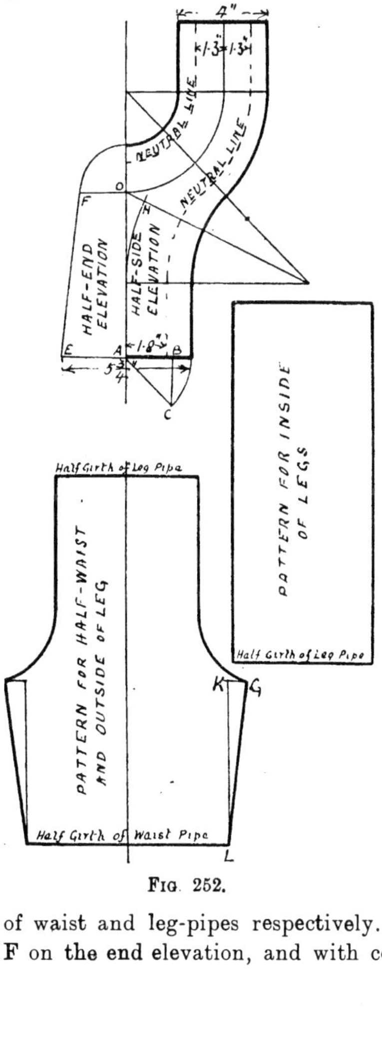

FIG. 252.

cumference of waist and leg-pipes respectively. Set compasses to E F on the end elevation, and with centre L on the pattern describe the arc K G, making it equal in length to the line O H on the elevation. Join L to G, and draw the side curve, and the net pattern is complete.

The pattern for the inside of legs will be equal in length

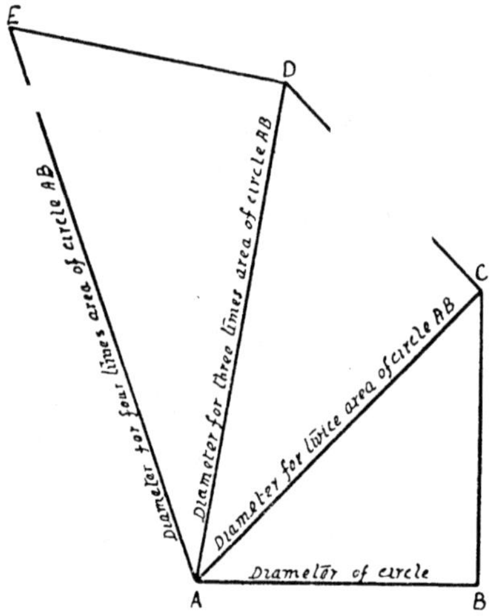

FIG. 253.

to the neutral line for that portion, and its width made equal to half the circumference of the leg-pipe.

Three-way Tee-Piece.

A sketch of this is shown in Fig. 254. The tee-piece may be of two shapes, one when the bulb is greater in diameter than the diameter of the large pipe as in the sketch, and the other when it is of the same diameter, and consequently the sides straight as in the half-end elevation, Fig. 255. We shall get the pattern out for this latter shape, as the setting out for the bottom portion, when the bulb is larger than the main pipe, will be dealt with in connection with a four-way piece.

A half-side and end elevation is shown in Fig. 255, and on examination it will be seen that the parts of the main and branch-pipes are cylindrical in shape; hence if the tee-piece were made up in three parts as in the last case, the pattern could be set out in the same way, or, as before mentioned, gusset-pieces could be inserted on each side. We shall, however, explain the method of working up from two pieces only, and jointed as in Fig. 254.

The neutral lines should be marked on the side elevation,

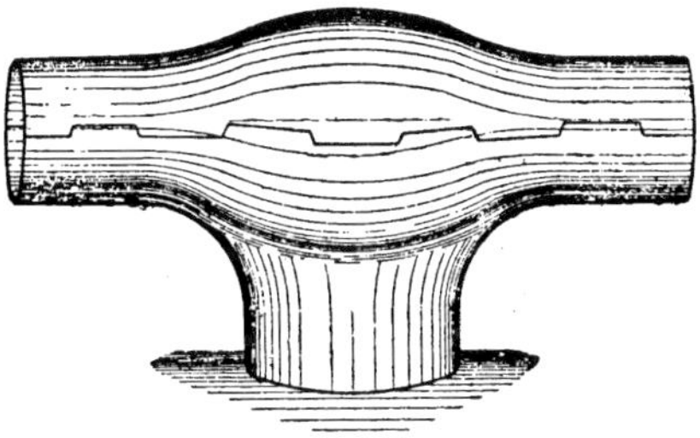

FIG. 254.

as shown. The pattern for the top piece can be set out by describing a circle, with radius equal to C D from the elevation, and then setting lengths along by marking off A B on the pattern equal in length to the neutral line A B in the elevation. The width of the ends is, of course, equal to half the circumference of the branch pipes. In cases like this, where the area of the main pipe circle is twice that of the branch pipe, it is worth noting that the diameter of the circle for the pattern of top part is equal to twice the diameter of the branch pipe.

The pattern for the bottom portion of the tee-piece is not so easily obtained. It is as well at the onset to keep in mind that the area of a pattern for an object which has to be worked up by hollowing or razing should be at least

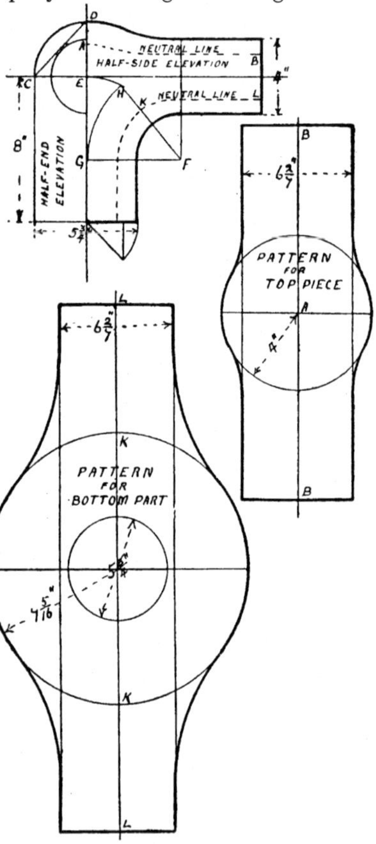

FIG. 255.

equal to the area of the surface of the finished article. This fact assists us considerably in calculating the sizes of the pieces of sheet metal required.

In the present example the diameter of the main pipe is 5·7 in., and the depth of the cylindrical part 8 in. What we require is to obtain a circle equal in area to the cylindrical surface plus the area of a 5·7 in. diameter circle. Put in the form of a rule, we have: "Radius of pattern circle is equal to the square root of the pipe diameter multiplied by the depth added to the square of the radius." Which, in this case, will work out—

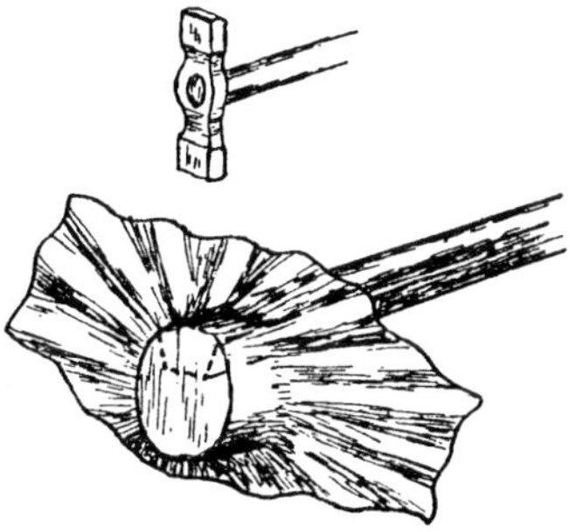

FIG. 256.

$$\text{Radius} = \sqrt{\left(\frac{5\cdot7}{2}\right)^2 + 5\cdot7 \times 8}$$

$$= \sqrt{53\cdot72} = 7\cdot32 = 7\tfrac{5}{16} \text{ in. (nearly).}$$

After describing the circle to the above radius, turn to the elevation (Fig. 255), and with centre F and radius F G. draw the arc G H; then with centre G and radius G E, mark

FIG. 257.

off H, and join up to F. The length of the neutral line K L will give the length to add on to the pattern circle, as shown.

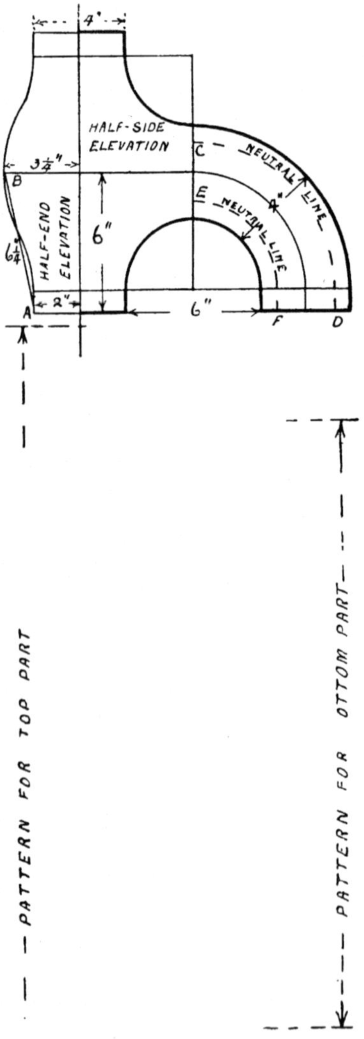

The bottom portion of the tee-piece can be raised as shown in Fig. 256, and when worked into the required shape, the disc cut out at the bottom of the main pipe.

Four-way Piece.

For the purpose of showing how to deal with a job that brings in conical work, we will conclude this chapter by going over the setting out of patterns for a four-way piece, each pipe being the same diameter, and the plates jointed as shown in Fig. 257.

The half-side and end elevations are shown in Fig. 258. It will be seen that A B on the end elevation represents the slant side of a frustum of a cone whose ends are 4 in. and 6½ in. respectively in diameter. Now, to get the size of the pattern circle we shall have to find the radius of a circle whose area is equal to the surface of the cone

FIG. 258.

frustum, together with the area of a 4 in. circle. To do this the following rule can be used: "To the sum of the end radii multiplied by the slant height add the square of the pipe radius and extract the square root of the whole." In the present example—

$$\text{Radius} = \sqrt{(3\tfrac{1}{4} + 2) \times 6\tfrac{1}{4} + 2 \times 2}$$
$$= \sqrt{36\tfrac{13}{16}} = 6\tfrac{1}{16} \text{ in. (nearly)}.$$

Set a circle out to this radius, and for the bottom part add the length of neutral line E F on to each end of pattern. The pattern for the top part will be obtained by measuring the length of neutral line C D, and setting along on the pattern. It will be noticed that the pattern for the top part is just twice the diameter of the pipe greater in length than that for the bottom portion. This, of course, follows from what was said in connection with the quarter-bend.

To make proper allowance for the thickness of metal, all the above patterns should be set out to dimensions taken from the centre line of the plate sections in elevation.

CHAPTER XXX.

KETTLE AND JUG SPOUTS, HANDLES, ETC.

How to Make a Kettle Spout.

THE making of a kettle spout, to the novice, is just one of those jobs for which it is somewhat difficult to find a beginning or ending without previous instruction. Spouts are usually made up from one piece of sheet metal, the marking out of its shape presenting no great difficulty. In Fig. 259 the necessary lines required for the development of the pattern are shown laid out. The spout is first straightened out, as it were, in imagination, by making line *a c* equal in length to the curve *a b*; the diameter at *c* being made the same as the spout end. The centre line A C on the pattern is cut off the same length as *a c*, the lines D D and E E being drawn square across, and their lengths fixed by marking off A D and C E respectively equal to three and a quarter times *a d* and three and a quarter times *c e*. The lines C E and A D are each bisected in G and F respectively, and the line G F drawn and produced to H, F H being made equal to *a h*. The angle F H K is next constructed by drawing an arc of the same radius and length as *m l*; H K then being measured off the same length as *h k*. The compasses are next set to a radius of a little more than one and a half times *l p*, and the arcs drawn as shown at P. In the same manner the arcs are drawn at points marked N. The lines N N are put into position by constructing them to make the same angle with K H as *n n* makes with *k h*. The small lug shown at R is the usual shape for sheet iron, and assists in forming the heel of the spout; the dotted lug shows a suitable shape for a copper spout. Instead of the ends N N being made straight, it will be an advantage to

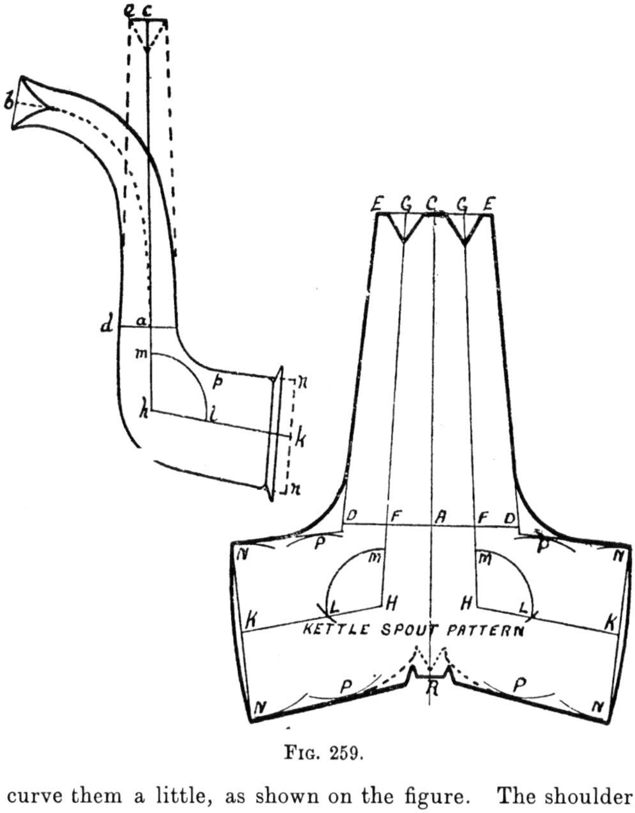

FIG. 259.

curve them a little, as shown on the figure. The shoulder curves are drawn to touch the lines N P and E D, and should be to a radius of about one and a half times that for the inside of the spout.

In working up, the plate is first bent a little, and then

stretched on the shoulders, as shown in Fig. 260. This stretching enables the two edges of metal to come together

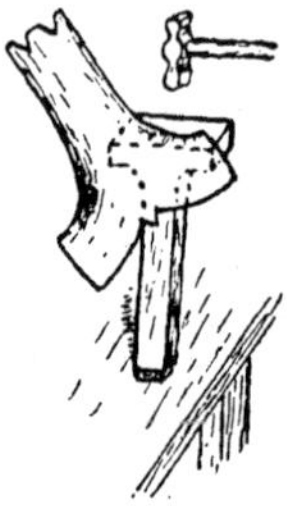

FIG. 260.

FIG. 261.

when the tapered pipe portion is formed, as seen in Fig. 261. When the edges are lapped over a little and carefully laid down, the seam can then be brazed, as shown in Fig. 262.

FIG. 262.

FIG 263.

The usual method of brazing is to bend a strip of sheet brass to the form of the joint, and when on the fire, start the brass to run at one end and gradually work along the seam to the other. A good sound brazed joint can quickly be done in this way. After the seam is hammered, the heel is formed, as shown in Fig. 263, by turning the small lug down first, and then lapping the sides upon it. The back seam is brazed, and the heel of spout carefully shaped on a block tool, as shown in Fig. 264. The edge of the spout mouth is now trimmed, and a groove and collar formed around it

FIG. 264.

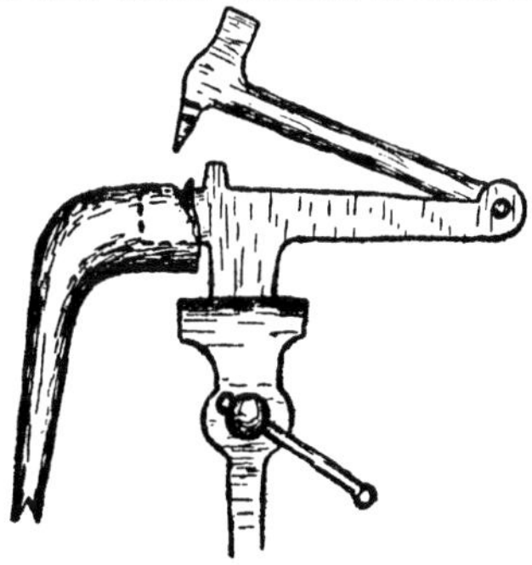

FIG. 265

—to fasten to kettle body—by a hand bumping-swage, as seen in Fig. 265. The spout is next filled with lead for the

FIG. 266.

purpose of bending, the end being first stopped by twisting a piece of stout brown paper around the outside for a distance of about 2 in., and over the end. The bending is carried out as shown in Fig. 266, the tool being shaped to the diameter and bend of the spout end. The bending must be done gradually, to avoid buckles on the underside or cracking on the top. If any buckles appear, these must be hammered out before the lead is melted from the spout. A small crack on the top of the spout can be repaired by hammering a piece of wire flat for a short distance, wrapping it around the pipe, brazing, and cleaning. After the lead is run out, the end should be rounded up and the lips opened somewhat and trimmed with a V-file. In a kettle

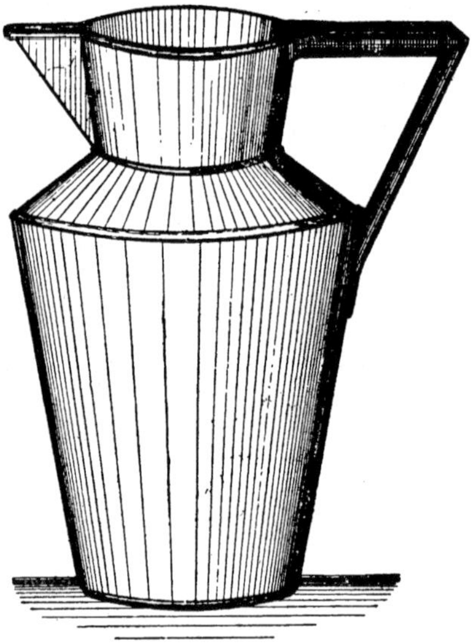

FIG. 267.

factory, it might be mentioned, all the above operations of cutting and shaping are carried out by the aid of presses fitted with suitable dies.

Square Spout for Conical Jug.

A square spout for a jug, as seen in Fig. 267, represents a good example of flat sheet surfaces fitting on to a conical surface. It may be applied in a variety of ways other than in the case shown. The setting out of the pattern is illustrated by Fig. 268, in which an elevation of the jug neck and spout is also shown. Before the pattern can be laid out, the line *n l* must first be obtained, this being done as follows: From the centre, *o*, describe the arc *k t*, and draw *n m* equal to half the width of the spout. Produce *a b* to *c*, and bisect *b c* in *d*. Draw *f e* square to *o e*, pass-

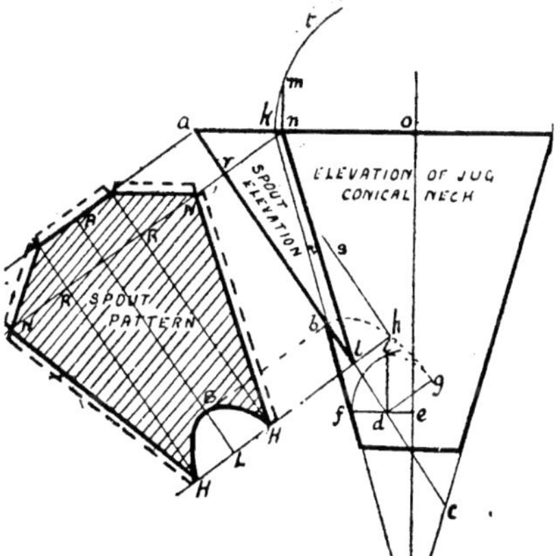

FIG. 268.

ing through *d*, and on this describe an arc of circle to meet the line *d i* in *i*. Now draw *d g* perpendicular to *b c*, and equal in length to *d i*. A quarter-ellipse should now be described on *d b* and *d g*, as shown. The line *s h* is next drawn parallel to *a c*, and at a distance from it equal to *m n*, to cut the ellipse in *h*. The perpendicular *h l* is then dropped on to *a c* to fix the point *l*, and thus determine the line *n l*.

The pattern is projected as shown, L H being equal to

l h, and R N equal to *r n*, the curve H B H being, of course, twice the part of ellipse represented by *b h* on the elevation. To fit exactly on to the conical surface, the edge H N should be slightly hollow; but this, if found necessary, can be put

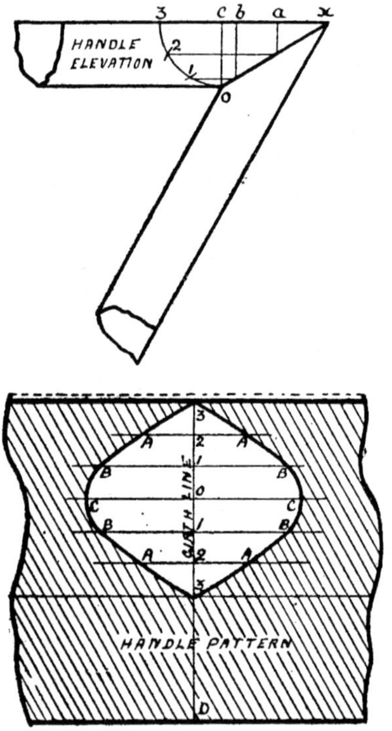

FIG. 269.

right when bending the flange over. If a not very particular job, there is really no need to cut away the part H B H, as the edge H H can be curved around the neck at the part where it fits.

Curved Spout for Conical Jug.

The method for setting this out will be exactly the same as that shown for the sponge bath lip in Chapter XVI.

Half-Round Jug Handle.

The jug in Fig. 267 is fitted with a half-round hollow handle, and if it is desired to make this in one piece, the pattern for same can be struck out as shown in Fig. 269. A quarter-circle is described on 0 c and divided into three equal parts, lines then being run along to the joint line 0 x, and up, to give the points a and b. The girth line of the pattern is laid out by taking six divisions, each equal in length to one of the arcs on the quarter-circle added to 3 D, which is the width of the handle, and equal to twice 3 c. The points A, B, and C on the pattern are found by marking off 2 A, 1 B, and 0 C respectively equal to $x\,a$, $x\,b$, and $x\,c$ from the elevation.

In forming this handle section, the part 3 3 will, of course, be shaped into a semicircle, whilst the portion 3 D will turn over to give the flat. The joint will run along one edge, and after this is formed, the mitre can be made by simply bending along the line 3 D until the two curved edges, 3 C 3, come together.

CHAPTER XXXI.

VASES, BRACKETS, DUSTPANS, ETC.

THERE are a great many different things that can be constructed in sheet metal which are particularly suitable for making by the amateur. It is true that some of them can be bought for a few pence, but the amateur with the true workman's instinct will find an immense amount of joy in the feeling that he has constructed something for his own use or pleasure.

FIG. 270

We shall now give one or two examples of sheet metal work, which, on account of their simplicity of construction and the few tools required in their manufacture, can readily be made up. Many neat-looking ornaments, such as vases, candlesticks, flowerpots, jugs, wall-brackets, pedestals, and such like things, can be made up out of strips of metal mitred together, with very little trouble.

Fig. 270 gives a view of a candlestick that may be constructed out of either sheet zinc, copper, or brass. For those who have not attempted work of this kind before, it will, perhaps, be the best plan to commence with thin zinc, say, No. 10 (zinc gauge) To simplify the work as much as possible, a square form of candlestick has been chosen, which is made up out of four strips of metal jointed at the corners. To mark out the shape of a strip a half-elevation of the candlestick is first drawn, as in Fig. 271. Each point is numbered as shown, and it will thus be seen that the total length of a strip must be equal to the sum of these numbered lines. Set the lengths 0 1, 1 2, 2 3, etc., down a line which will form the centre line of the pattern, and draw lines through these points square to the centre line. The width of the pattern at the different parts is obtained by setting on each side of the centre line of pattern the length of the line with the same number which is drawn from the point to the centre line in elevation. The points found are now joined with straight lines, and the pattern for one strip is complete. The greatest accuracy must be aimed at in setting out a strip, as any inaccuracy in the pattern will cause endless trouble in jointing the

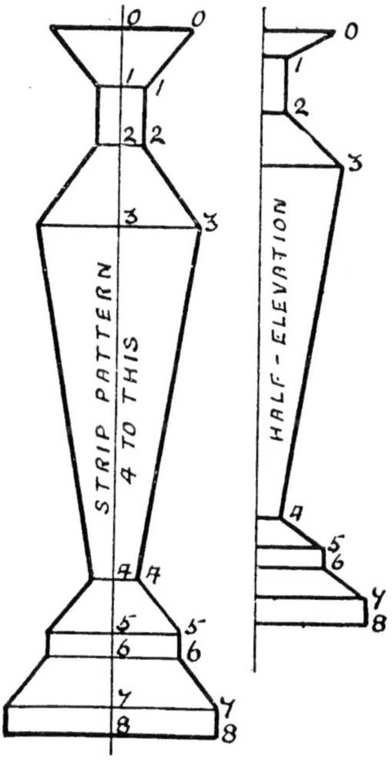

FIG. 271.

strips together. Four pieces are cut out to the pattern, and the lines for bending carefully marked. The strips can be bent to the required shape over a sharp edge of any kind, either on a bar of iron or a piece of timber. In bending, care must be taken that the centre line of the strip be of exactly the same form as the outline of candlestick in elevation. It will be as well to cut out the half-elevation in cardboard, and use it as a template to try on the bent metal strip, and thus by continual bending get them to the exact shape. In thin zinc the strips can be bent by the hand without using either mallet or hammer; but in any job on which the hammer is used, it should be seen that no hammer marks are left on the sheet, or else the finished appearance of the article will be anything but pleasing. If the strips are marked and cut correctly, and bent to the exact shape, there should be no trouble in making them fit together. None of the operations should be hurried, but great pains taken over the work, and this will save a lot of time and bother in the afterwork. In fitting the strips together it will be the best plan to tack them all in position before proceeding to completely solder along one corner joint. The best way to fix a corner will be to bring together two strips, and tack with solder the two points 0 0, then do the same with 1 1, 2 and 2, etc. After the four strips are tacked together at all the corners, the candlestick should be examined and tested as to being properly square. It should also be placed upon a level table to see if there is any twist in it. When in good shape the joints should be soldered down, as much of this being done from the inside as possible. In this case both the foot and top can be soldered from the inside, the joints of the body being done from the outside. In soldering, care must be taken that the iron does not get too hot, or else the flat parts of the strips will buckle, and cause the surfaces to have an ugly appearance. This is especially so with light sheet zinc. A square bottom is now cut out, allowing about $\frac{1}{8}$ in. all round for bending over an edge. The bottom is tacked at each corner inside the foot, and then soldered along each edge. If required the foot of the candlestick can be weighted by first of all stopping up the stick at 4 by soldering in a small square of zinc, then filling up the foot with sand before soldering in the bottom. A stopper might also with advantage be soldered at the bottom of the neck; this would be best done before the last strip is tacked in. The superfluous solder must be scraped off the joints and the corners carefully filed up, and if the stick be cleaned and polished, the job is finished. A small quantity of killed spirits can be used to clean the zinc, and oil and whiting to polish, or finish with a good polishing paste.

If a candlestick is made out of copper, the solder at joints can be coated with copper by applying a solution of sulphate of copper. It will then be an advantage if, after well polishing, the surface is lacquered.

Hexagonal Vase.

Fig. 272 shows a sketch of a simple kind of hexagonal vase that can be made up either of tinplate, zinc, galvanised iron, brass, or copper. A half-elevation, Fig. 273, shows the exact shape or section of one side of the vase. From the point where the centre line meets the base line a joint line making an angle of 30° with the base line is drawn. The required angle to set out will, of course, depend on the number of sides the vase has. The general rule for obtaining the number of degrees is a simple one: "Divide 360 by twice the number of sides." In the present case the vase has six sides; hence the angle to set out is 360 divided by $12 = 30^{\circ}$. The profile of vase in elevation is now divided up, giving points 0, 1, 2, 3, etc., up to 16. Dotted lines perpendicular to the base are drawn through each of these points, and continued across the base line to meet the joint line To set out the shape of one of the strips a centre line is drawn, and along this the lengths 0 to 1, 1 to 2, 2 to 3, etc., taken from the elevation, are marked. Lines square to the centre line are drawn through each point, and the lengths of these cut off equal to the length of the line with the same number between base and joint lines in elevation. Thus, to mark off line 0 0′, turn to point 0 in the elevation, follow the dotted line down to base line, and measure along the continued line between the base and joint lines; this will give the length 0 0′. In the same way obtain and set along the lengths 1 1′, 2 2′, etc., each side of the centre line in pattern. Carefully join these points up with curves or straight lines as required. It will be noticed that where lines are straight in the elevation the corresponding lines on the pattern will also be straight. For instance, lines 0 to 1, 7 to 8, and 13 to 14 are straight in the elevation, hence 0′ to 1′, 7′ to 8′, and 13′ to 14′ will be straight in the pattern. Remembering this, it will always act as a guide in joining up the points in the pattern for a strip.

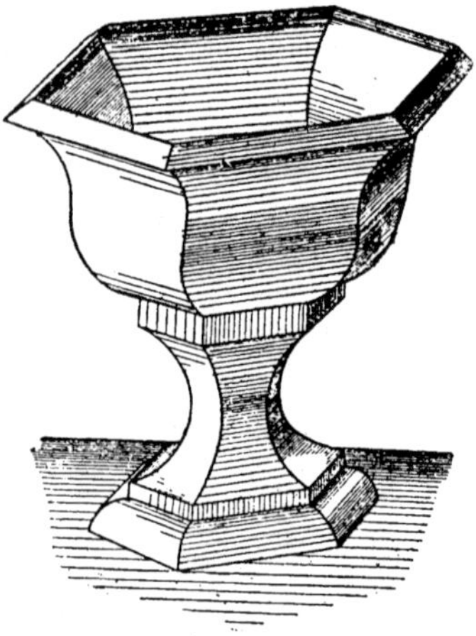

FIG 272

The bending of the parts can be carried out in the same manner as in the case of the candlestick, the curved portions being bent over a wooden roller. Before tacking the strips together a template (Fig. 273) for 120° should be cut out of

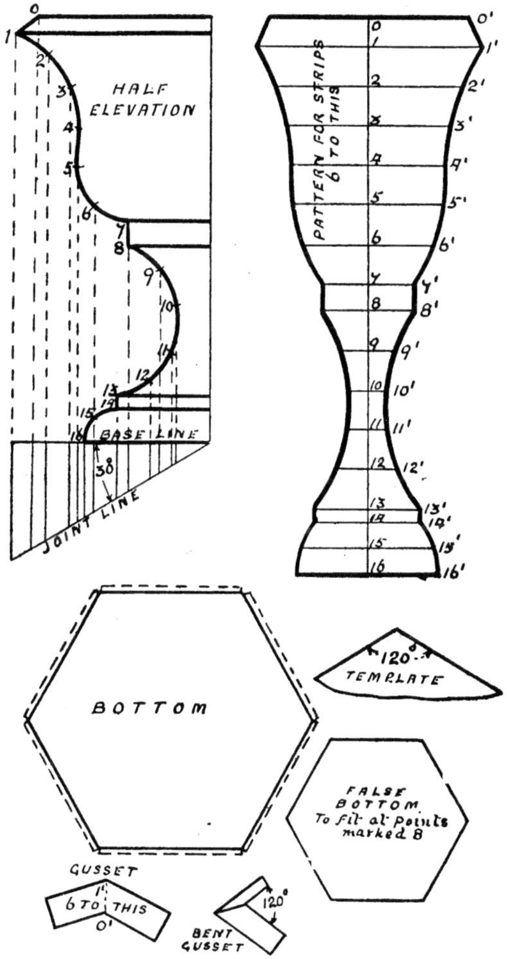
HALF ELEVATION
BASE LINE
JOINT LINE
PATTERN FOR STRIPS 6 TO THIS
BOTTOM
TEMPLATE
120°
FALSE BOTTOM To fit at points marked 8
GUSSET
6 TO THIS
BENT GUSSET

FIG. 273.

a bit of sheet metal. This can be used for trying in between each pair of strips as they are being tacked together. After all the strips are tacked in position, and before completely soldering, the diameters should be measured to see if they are all the same. The vase should also be examined to observe if it has any twist.

In this shape of vase it will be found that all the joints can be soldered down the inside. A bottom should be cut out, as in Fig. 273, allowing a small margin all round for turning an edge up. This edge is slipped inside the foot, and will facilitate the soldering, besides strengthening the edge of foot.

The top corners of vase will be much stronger if a small gusset (Fig. 273) is soldered over each joint. This gusset can be marked out from the top part of strip pattern, its centre line being equal in length to $0' 1'$.

Any size vase can, of course, be made. It may, however, act as a guide to know that the drawings have been made to scale for a vase $10\frac{1}{2}$ in. diameter (across the flats) at top.

If the vase be a large one, and made out of tinplate, zinc, or galvanised iron, its appearance will not by any means be inartistic if painted a dead chocolate, green, or any other colour in harmony with its surroundings.

Tobacco or Biscuit Box.

A sketch of a square box is shown in Fig. 274, the body being made up in four pieces, and jointed at the corners. The lid is in form a square pyramid, and is worked up from one piece, as will be further explained.

To set out the patterns for the different parts, it will be necessary to draw the shape to which the sides of the box must be bent. This is shown on the half-sectional elevation in Fig. 275.

The pattern for one side of the body is obtained by marking down a girth line, and setting along it the lengths 4 to 5, 5 to 6, etc., up to 15, as taken from the sectional elevation. It should be noticed that the lengths from 4 up to 7 are obtained by measuring around the small circle on the section, which represents the bead on the top edge of box. After the total girth is set out, then lines square to the

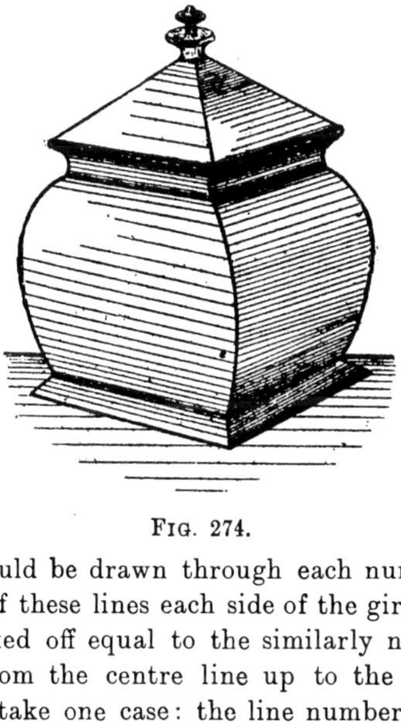

FIG. 274.

girth line should be drawn through each numbered point. The lengths of these lines each side of the girth line should then be marked off equal to the similarly numbered lines which run from the centre line up to the curve on the section. To take one case: the line numbered 9 9 on the pattern should be the same length as the line numbered 9 9 on the half-sectional elevation. After all the lengths have been carefully cut off, then the points should be joined up with an even curve. On account of the foot having straight sides it will be noticed that the cut on the pattern which forms the foot will be made up of straight lines.

The lid of the box is pyramidal in shape, and therefore

the making of the pattern is a simple matter. With radius equal to 0 1 on the elevation describe the pattern circle (Fig. 275). Draw a line touching this circle, and on each side of the point of contact cut off distances 1 A equal to the length of 1 A on the elevation. After one line A A is drawn, then the other three lines with the same letter can be drawn around the circle as shown. Strips now require to be added to form the rim of the lid. Take off the lengths 1 to 2 and 2 to 3 from the elevation, and transfer to the pattern, as seen by the same numbered lines. Now draw lines through the points parallel to A A, and cut these off to the corresponding lengths on the elevation. That is, make 2 *a* on the pattern the same length as 2 A on the elevation, and 3 *b* on the pattern the same as 3 B on the elevation. A small lap, as shown by the dotted lines, should be allowed on one edge of the lid pattern for jointing.

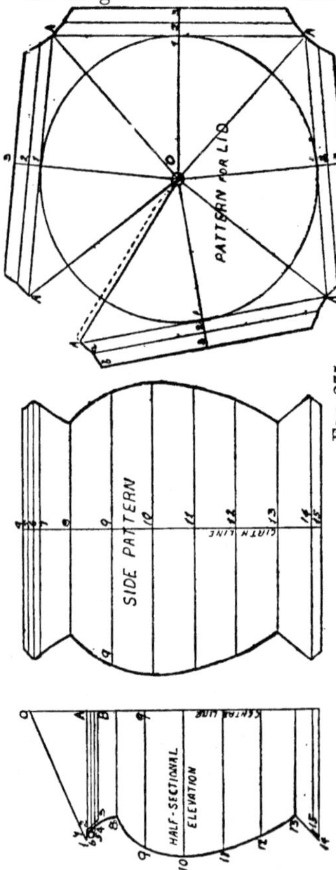

FIG 275.

Now to make up the article. If made of copper or brass, it will be essential to coat the inside of the sheet metal with tin. This can be done in the ordinary way by first cleaning the surface of the metal, sprinkling over with a flux, such as salammoniac, and then putting a few bits of tin on the sheet and heating over a gas or clean coke fire, and wiping off with a piece of dry cloth or tow. To avoid the tin running on to the side of sheet that is not required to be coated, it is a good plan to first brush its surface over with some whitening paste.

In shaping the four side-pieces that go to form the body, the small bead at the top should first be put on. This can be done by bending the edge along, doubling it over a piece of wire of the right size, carefully tucking the edge in, and then withdrawing the wire. The edge at the bottom of the foot should next be folded over and lightly flattened down. Each of the four pieces can then be formed into shape, and it should be remembered in connection with this that the centre line on the pattern must conform to the exact shape of the half-sectional elevation.

Before proceeding to completely solder down any one corner, all the pieces should be tacked together, and the body tested as to shape, and also if level across the top and bottom. The edges of the sheet down the corners should be brought into contact as far as possible, so as to avoid any appearance of solder on the outside of joint. The soldering should, of course, be done down the inside of corners, a fair body of solder being left on so as to strengthen the joint. The bead around the top may also be made stronger at the corners by bending small pieces of wire at right angles, and inserting in the bead before tacking.

The pattern for the bottom is not shown, as it will be simply a square piece of sheet metal the size of which will be equal to the length of line drawn through the point 13 on the side pattern. The bottom plate will, of course, be tinned on one side, and fastened to the body by soldering all round.

The sheet metal for the lid can be brought into shape by

bending along each of the corner lines marked O A until the end lines of the pattern come together. The joint should then be formed by fixing the lap on the inside of the lid and soldering down. The double edge to form the lid can then be bent, as shown in the sectional elevation. A hole is made in the centre of the lid, and a knob to suit the individual taste soldered in.

After cleaning away all superfluous solder, the outside of the box should be polished and coloured, lacquered, or treated in any other way suitable to the likes of the individual.

Whilst the making of a square box has been described, the above remarks will apply to a box of any number of sides, the only difference being that the pattern for the body will have to be marked out as explained in connection with the next example.

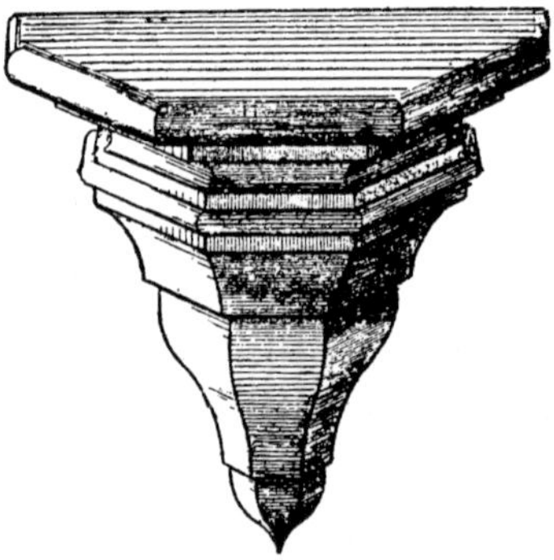

FIG. 276.

Wall Bracket.

Another piece of work that can be made by the amateur, who exercises carefulness and patience, is the wall bracket, as shown in Fig. 276.

The shape of a wall bracket can be made up by any number of pieces; but that in the figure is partly octagonal, the three front and two side pieces together forming five sides of an octagon. The whole number of parts in the bracket will be seven, three front, two side, and the top and back pieces.

Any convenient section for the moulding can be chosen, either simple or complex, to suit the skill of the operator in working up, and for the bracket to give the best effect when hanging from a wall.

The setting out of the various patterns is shown in Fig. 277. The section of the moulding is first set out. A base line is drawn square to the centre line; and as the article is octagonal in shape, a joint line will be set off, making an angle of—

$$\frac{360}{\text{twice number of sides}} = \frac{360}{16} = 22\tfrac{1}{2}^\circ.$$

This angle can be set on either side of the base line, whichever is most suitable. The section line of moulding is then divided up into any convenient number of parts, and figured as shown by the numbers 0 to 23. Perpendiculars to the base line are then drawn through each point and along to the joint line, as seen by the dotted lines.

The pattern for one of the three front strips will be marked out by first laying down the girth line, the lengths being taken step by step between the numbers from the section line. Through these points lines square to the girth line are drawn, and their lengths on each side cut off equal to the corresponding line between the base and joint lines. Thus, to give an illustration, lines 0 0 and 3 3 on the pattern will be respectively equal to lines 0 0 and 3 3 as indicated between base and joint lines.

It will be seen that the cut on a side strip is exactly the same as that for a front piece, and marked out in precisely the same way. The width of the strip is obtained by making the top line equal in length to the top line of a front strip or twice the length of line 0 0 between base and joint lines, and then drawing a line parallel to the girth line. Perhaps the most convenient and accurate way of marking out the strips would be to set out the shape of a side piece first, and then use this for a pattern from which to obtain the shapes of the other four pieces.

The pattern for the back can be easily drawn out, for the exact shape of half of it is as shown by the figure which is

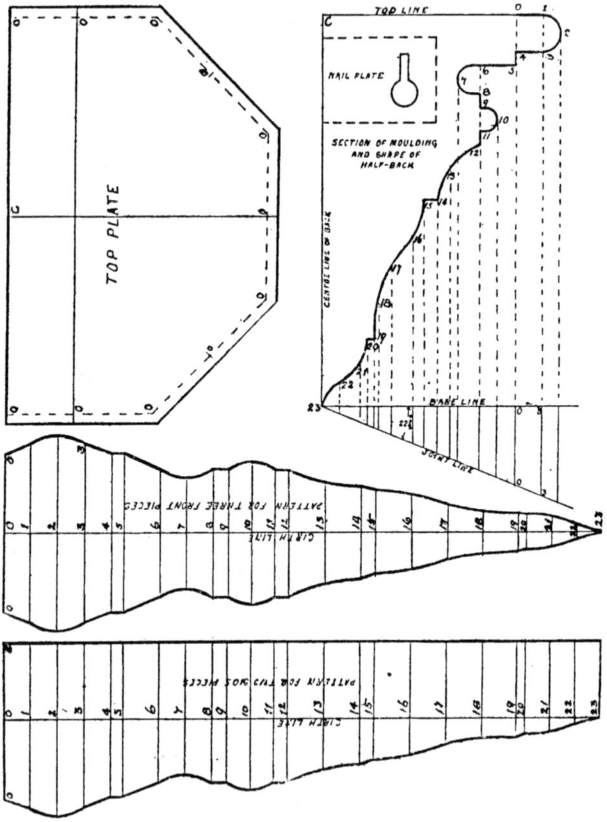

Fig. 277.

bounded by the top line, centre line of back, and moulding section on Fig. 277.

The shape of the top plate is also shown on the same figure, the dotted lines representing the exact shape around the inside of top of bracket. The lengths as marked being obtained from the lines with the same number on the section of bracket.

The top and back can be made in one piece; but this will cause some inconvenience in soldering, as all the joints should be soldered down the inside, the top plate being soldered on last of all.

To hang the bracket from the wall, a good plan will be to solder or rivet a plate on the inside of the back and to put two key-shaped holes right through the two thicknesses of metal, as shown in Fig. 277.

The bracket can be made out of sheet zinc or other suitable material, and after all the joints are carefully scraped, painted some colour that will harmonise with its surroundings.

Phonograph Horn.

The making of a phonograph horn in segments, as illustrated by Fig. 278, is particularly suitable for amateur's work, as it can be readily constructed with few tools and at little cost of material. It may be made out of tinplate, zinc, brass, aluminium, or hard rolled copper. The horn, as shown, is made up in twelve strips jointed together and fitting into a thimble.

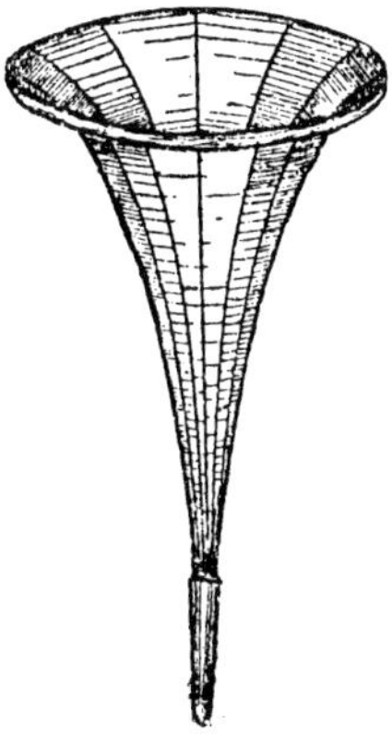

FIG. 278.

To obtain the pattern for a strip or segment, the profile or section of one strip is set out, as shown in Fig. 279. A joint line is drawn, making an angle of—

$$\frac{360}{\text{twice number of strips}} = \frac{360}{24} = 15°$$

with the base line. The section curve is divided into any number of parts, four being chosen in this case. The length of this curve is carefully set out to form the girth line of the pattern for a segment. This is done by making the lengths 0 to 1, 1 to 2, etc., on the pattern equal in length

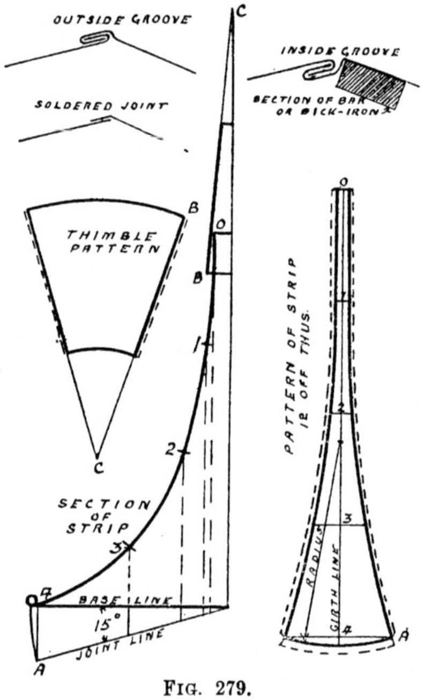

FIG. 279.

to the parts of the curved line 0 to 1, 1 to 2, etc., on the section. Lines are drawn across through each point on the girth line, and these cut off on each side equal in length to the corresponding line between the base and joint lines. Thus line 4 A on the pattern is the same length as the line 4 A on the section, and so with the other lines through points 3, 2, 1, and 0, all measured between base and joint lines. A piece of hoop iron or a lath should now be bent to pass through each point, and the pattern curve drawn in. If it is desired that the bell-mouth of the horn should come an exact circle when the strips are joined together, then the compasses must be set to a radius equal in length to the joint line, and the curve at the bottom of the strip pattern marked along as shown. The allowance for wiring around the mouth must be added on as seen in the pattern by the dotted curve.

The strips can be jointed together either by soldering or grooving. If soldered, the allowance for the lap will be as the dotted line on the right-hand side of the pattern. If grooved, an allowance on both sides will have to be made, the lap on one side being twice the width of that on the other. The double lap is shown on the left-hand side of the pattern.

The pattern for the thimble will develop out quite easily, the surface being that of a frustum of cone, and being marked out as explained in Chapter XII.

To make up the horn, the strips will first be bent so that the girth line will have the same shape as the section or profile. If to be soldered, the small lap will then be slightly bent over with a mallet, so as to lie on the adjoining strip. The strips should all be tacked together before any joint is completely soldered down. The laps and soldering should be on the outside of the horn, the joint being made as neatly and cleanly as possible. The wire edge on the bell-mouth should now be turned over, the ring of wire inserted, and the edge hammered down with the mallet, and carefully tucked in with the pane end of hammer.

If the joints of the horn are to be grooved, then the single edge must be edged up and half of the double edge turned down, this, of course, taking place after the strips are shaped. The strips can now be hooked together and grooved (see sketch of outside groove, Fig. 279) by placing on a square bar or bick-iron and hammering the groover gently along the joint. The grooving of the narrow part of the horn will present some little difficulty; but this can be overcome by fixing on the small end of bick-iron, or by the amateur on a piece of round bar-iron held fast in a vice, or by other means.

If it is desired to have the outside of the horn plain, and consequently the groove formed on the inside, this can be accomplished by placing the joint on the edge of bick-iron or bar, hammering down with mallet to form groove, and then flattening the groove with the mallet or hammer in the usual way. A sketch of this method of forming an inside groove is shown on Fig. 279.

Dustpans.

Of all household utensils, perhaps the most difficult to obtain is a strong, serviceable dustpan. After having put up with broken handles, cracked corners, and other defects of the modern dustpan, the writer some years back devised and made a pan out of aluminium (Fig. 280), which seems to be making a fair bid towards old age without showing any signs of collapse. The dustpan is simple in construction, and can be quite easily made by an amateur.

FIG. 280.

A sheet of aluminium 14 in. by 11 in. by about $\frac{3}{64}$ in. thick is required for the body, and for the handle and washer a piece about 9 in. by 5 in. The sheet for the body is marked out as shown in Fig. 281, and cut down the corner lines as indicated. The back is bent up square, and the corner flaps turned inside. The sides are now bent up square, and the corner flaps of them turned on to the back. The $\frac{3}{8}$ in. edge on the back is turned over and hammered down on to the two corner flaps, as seen in Fig. 280. The $\frac{1}{8}$ in. edges on the sides are turned over, and the edges of the inside corner flaps turned over the sides. Thus the two corner flaps are firmly held without the use of rivets, and the corner cannot be pulled or knocked apart. This method of forming the corner also gives the additional advantage of two thicknesses of metal at the corner—the part of the dustpan that is usually the most strained. A

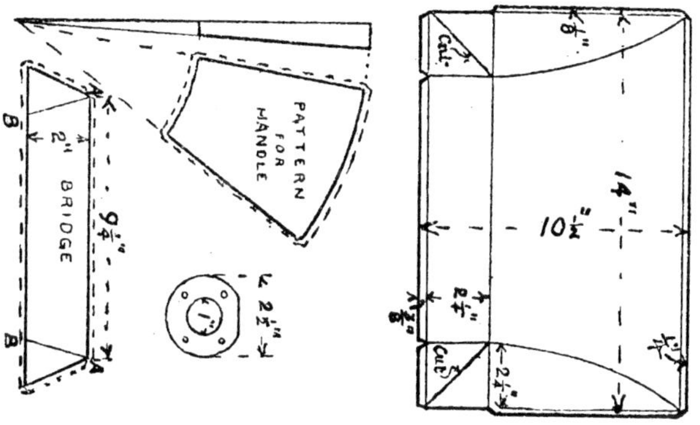

FIG. 281.

lap of $\frac{1}{4}$ in. is turned over on the front edge of the pan, thus stiffening and keeping straight this part.

The handle is $1\frac{1}{4}$ in. diameter at one end, 1 in. at the other; and 6 in. long. This is shown set out in the usual manner. An edge is turned over on the end of handle to protect the hand from the raw edge of the sheet. The handle is jointed down with a small groove, after which the washer is slipped on, and a small flange thrown over on to it. The washer is now riveted on to the back of the pan, and there is no danger of the handle coming loose. A hole should be put into the handle by which the pan can be hung up.

There is very little necessity to put a bridge on the pan; but, if required, the pattern can be cut out as shown in Fig. 281. To fix to the pan the bridge should be bent along the lines A B, and the outside edges doubled over and slipped under the edges on top of pan before these are hammered down, thus forming a kind of groove or knocked-up joint.

Whilst aluminium is somewhat costly for an article of the above description, it is the cheapest in the long run, being relatively strong, and of little weight when made up into the pan.

Fire-Shovels.

A fire-shovel is another common article that can quite readily be worked up by the amateur. A simple design is that of Fig. 282, the pattern for the body of which being shown in Fig. 283. The sides and back can be bent up and

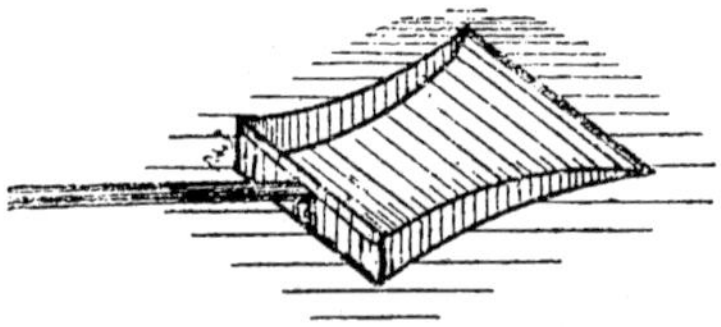

FIG 282.

jointed in the same manner as the dustpan, and will make a very good job in that form, especially if the four flaps are riveted down in the corners. The usual plan, however, and the simplest to follow, is to cut out the plate (in, say,

16 or 18 S.W.G. iron or steel), as shown in Fig. 283. Holes are punched in the two flaps and back, and these bent up and riveted. After the flaps are riveted, the top edge of the

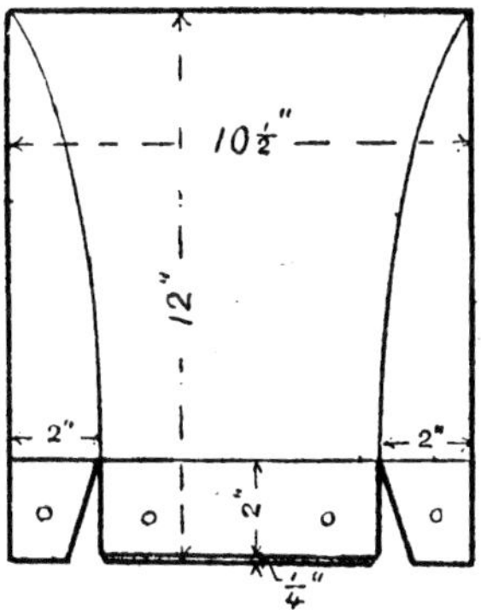

FIG. 283.

back is then turned over. A handle can be formed by bending a piece of 1 in. by ¼ in. flat iron, shaping it according to fancy or skill. It should be firmly riveted to the back, and also to the bottom of the shovel body.

Hand Scoops.

The cone surface, as we have seen, plays a most important part in building up the shapes of a multitude of articles. A simple application, and one that can be readily understood by the amateur, is in the construction of a hand scoop, as shown in Fig. 284. It will be seen that the handle, thimble, and back of scoop are formed by parts of cones of different dimensions. The patterns for the handle, thimble, and back are shown set out in Fig. 284 in the usual way; the letters on the lines of the different patterns being the same as the lines in the elevation to which the compasses have been set for the various radii.

The pattern for the front part of the scoop can be obtained by treating it as a portion of a straight pipe. A semicircle is drawn on the line E 4, and the bottom half divided into three equal parts. Lines square to E 4 are run through the points until they cut the front edge of scoop. It will be observed that the top edge of scoop body cuts the semicircle at the point O. A girth line is set down for the pattern and lengths 0 to 1, 1 to 2, etc., marked along as shown. Lines are drawn through these points perpendicular to the girth line, and cut off the same length as the corresponding line in the elevation. These are shown cut off by the dotted lines, which are projected from the ends of the lines in the elevation. In practice, however, lines should be measured directly from the figure, and lengths marked off without projection. The projection method is used in the illustration to better explain how the various lines are obtained. An unbroken curve is now drawn through the points, and laps added on to the net pattern as required.

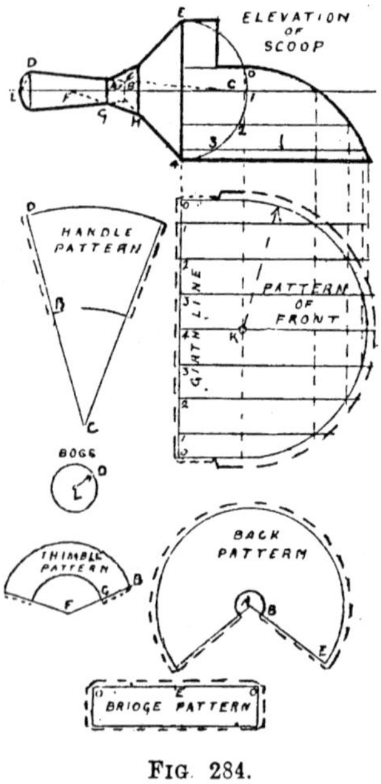

FIG. 284.

When the front curve of the scoop is not required to be of any specific shape it is usually marked out on the pattern as a semicircle. Thus, in this case if centre K be chosen and half the length of the girth line taken as radius, the pattern curve will come out practically the same as that which has been obtained by the method previously explained. But in all cases where the shape is definite the first method must be pursued.

The bridge pattern will be a straight strip, its length being equal to twice the length of the arc 0 E on the semicircle. On the patterns for the front, bridge, and back, allowances have been made for wiring, for grooving bridge to front, and for knocking up along the joint E 4. If there is to be no wiring, or if the joints are to be soldered instead of grooved, then the allowances must, of course, be somewhat different.

The radius for marking out the boss blank will be equal to the line L D in the elevation.

The handle, thimble, and back will be fastened together by firmly soldering; also the boss will be just let into the end of handle, soldered, and cleaned off. A small disc should be soldered in the back at B to block up the hole, or this can be accomplished by cutting the back pattern out as for that of a complete cone.

CHAPTER XXXII.

PLATER'S WORK, TANKS, SHELLS, ETC.

Allowance for Metal Thickness.

IT is absolutely essential in the making of patterns or templates to cover for the necessary allowance for the thickness of sheet or plate if the different parts that form the article are to fit together correctly. In general sheet metal work the allowance to be made for thickness is not so important as in plate work; but, in any case, if a good-fitting job is required some thought must be exercised, so as to make the requisite modification of pattern to cover for the different thicknesses of metals. In plate metal work it is of the greatest importance that patterns should be so marked out that the thickness of plate is properly allowed for, as in this class of work a job is completely botched if rivet-holes are half-blind, and have to be gouged, reamered, or drifted.

To illustrate the method adopted in allowing for thickness, suppose the following experiment to be carried out: A straight bar of metal is taken and a line N N (Fig. 285) marked along the centre of one side; also two parallel lines are drawn across the bar, such as E C and F D. Now if the bar be bent as in the lower figure, the lines will fall into the positions shown. If the line E F be measured both before and after bending, it will be found to have lengthened in bending, and in the same manner if C D be measured, it will be found to have shortened. The line A B, however, will be the same length as before bending. From this it is evident that the whole line N N will remain of constant length as the bar is bent. This line is called the "neutral axis," and in every bent bar or plate it will be possible to find the position of some line that has been unaltered in length by the bending.

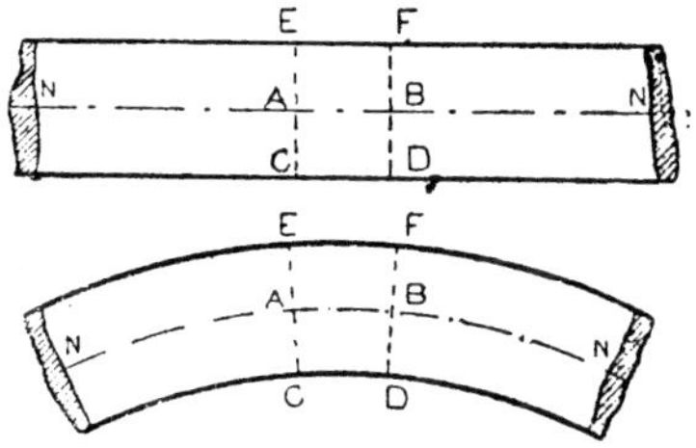

FIG. 285.

Every plate metal worker who is interested in the principles underlying his trade should make several experiments on bars and plates similar to the one abovementioned. If the plates or bars are bent hot, care must be taken that they are uniformly heated, or else the elongation and contraction will be unequal. For instance, if the outside of the bar is hotter than the inside, most of the draw will take place on the outside, on account of the bar being softer; but if the inside be the hottest, then nearly all the draw will be on the inside.

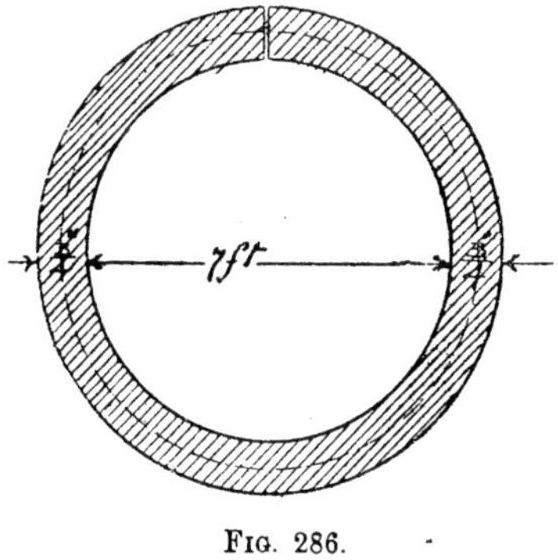

FIG. 286.

In square, flat, and round bars the neutral line will always pass through the centre of the section; similarly, if sheets and plates are bent, the neutral line will be at the middle of the thickness of metal. Angle-iron, tee-iron, and other sections will be dealt with later.

If it is required to obtain the length of a plate to bend into a complete circle, as in Fig. 286, this can be done in two ways—either by setting out and measuring along the centre line of the plate, or by calculation. Suppose the inside diameter to be 7 ft., and the thickness of metal $\frac{3}{4}$ in., then the diameter of the circle formed by the neutral line will be 7 ft. $0\frac{3}{4}$ in. Multiply this by $3\frac{1}{7}$ we have—

$$84\tfrac{3}{4} \times \tfrac{22}{7} = 266\tfrac{5}{14} \text{ in.} = 22\text{ft. } 2\tfrac{5}{14} \text{ in.}$$

If the number 3·1416 be used to represent the ratio between the circumference and diameter of a circle, then the above will run out—

$$84{\cdot}75 \times 3{\cdot}1416 = 266{\cdot}2506 = 266\tfrac{1}{4} \text{ in.}$$

In all work where accuracy is required, the number 3·1416 should be used.

It will be noticed that before proceeding to calculate, the thickness of metal was added to the inside diameter, and it will thus be seen that the girth of plate to form a circle will always be $3\frac{1}{7}$ times the thickness of the metal greater in circumference than the circumference of the inside of pipe.

If a plate is to be bent in any form, such as Fig. 287, its length in the flat can be obtained by first setting out the required shape, and then measuring along the centre line of the section. To mark the plate for bending, the distance along the centre line up to C (the centre of the bend) must be measured, and this set out from the edge of the plate. In bending the mark must be kept right in the centre of the bend.

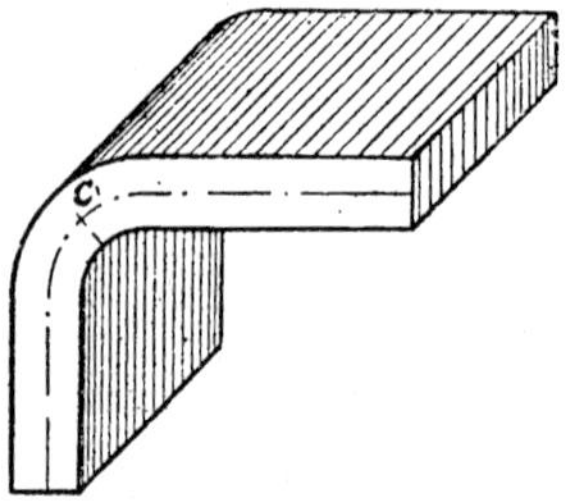

FIG. 287.

To bend a plate with rounded corners, as in Fig. 288, the required length in the flat can be found as in the last case, or it can be calculated as follows: Suppose the inside diameter to be 2 ft. and the inner radius of corners 3 in., and the plate 1 in. thick; then the radius at the corners to the centre line of plate will be $3\frac{1}{2}$ in. And if the four quarter-circles which form the corners be added together, they will make up a complete circle of 7 in. diameter. The length of plate, therefore, to cover for the four corners will be $7 \times 3\frac{1}{7} = 22$ in. If 3 in. be taken from each end of the inside diameter, this will leave 18 in. of flat on each side. And if $4 \times 18 = 72$ in. be added to the 22 in., the total

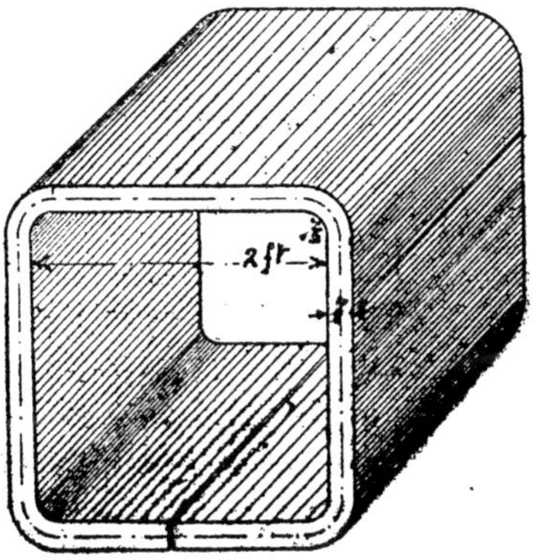

FIG. 288.

length of plate will be 94 in. To mark the plate for bending, it should be remembered that the distance apart on plate of corner lines will be $94 \div 4 = 23\frac{1}{2}$ in. If the joint is at the centre of a flat side, as shown in the figure, the marks for bending the bottom corners will be $23\frac{1}{2} \div 2 = 11\frac{3}{4}$ from the butt edges of the plate.

In bending the plates care must be taken so that they are bent to the proper radius, or else the diameter will not come out correctly. In the workshop all kinds of methods are in vogue to make the allowance for a rounded corner; but none are correct without they are based on the above calculations.

To centre-punch mark along the edge of a plate for flanging, the width of the flange should be set out as in Fig. 289, and the line G H measured. This will give the distance of the centre-punch marks from the edge of the plate. After being flanged, the marks should be in the position K. If a section of the flange is set out in this manner, the proper position of the rivet-hole centres can be determined for both plates.

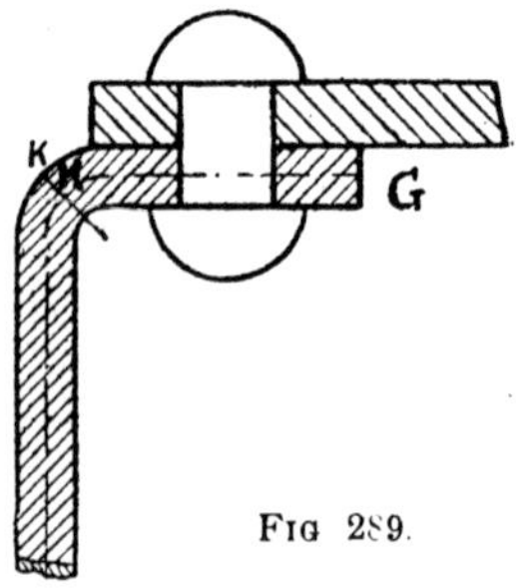

FIG 289.

Fig. 290 shows the plan that can be adopted to obtain the lengths of plates and pitch of rivets, where two corner plates

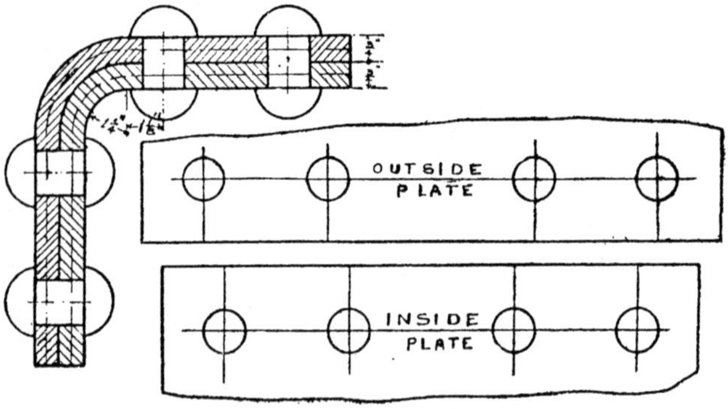

FIG. 290.

or bilge plates are jointed together. The joint is set out as shown, and the length of each plate found by measuring along the centre line of the section. To find the pitch of rivet holes to mark on plates, the neutral lines on each plate are measured between the centre lines of the two innermost rivets. Or this pitch can be determined by calculation thus:—

Suppose the plates are $\frac{3}{4}$ in. thick, and the inside radius of inner plate $1\frac{1}{4}$ in., and the distance from centre of inner rivet to beginning of curve $1\frac{1}{8}$ in. Then the length of neutral line on outside plate between the centres of inner rivets will be —

$$\frac{2{\cdot}375 \times 3{\cdot}1416}{2} + 2{\cdot}25 = 5{\cdot}98 \text{ in.}$$

And the corresponding length on inner plate will be:—

$$\frac{1{\cdot}625 \times 3{\cdot}1416}{2} + 2{\cdot}25 = 4{\cdot}8 \text{ in.}$$

The difference of the two thus being 1·18 in.

Where plates are bent into quarter circles, as in this case, the difference of pitch between the innermost pair of rivets can readily be worked out by the use of the following rule:—

Difference of Pitch =

$$\frac{\text{Twice the thickness of plate} \times 3{\cdot}1416}{4}$$

The pitch of rivets on the flat part of the plates will, of course, be the same on both plates.

A useful application of this method of obtaining the lengths of plates or bars can be made by blacksmiths and whitesmiths. If a round, square, or flat bar is to be bent into any shape, then all that is necessary to do is to set out the required design, as in Fig. 291, mark in the centre line, and measure for length of bar in the straight.

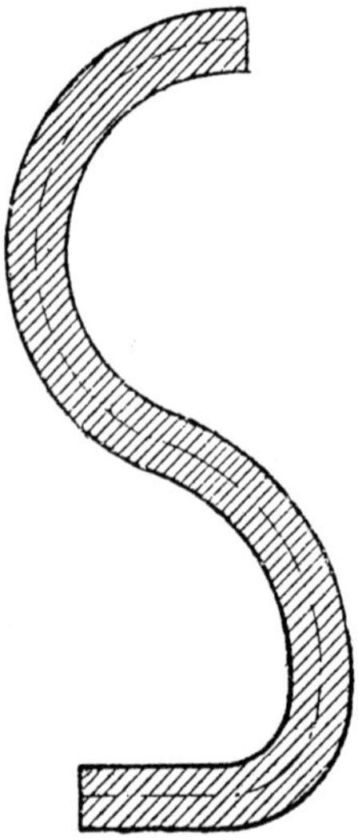

FIG. 291.

Cylindrical Shell Plates.

In setting out plater's work for boilers or other similar class of work, a high degree of accuracy is required if joints are to be properly constructed, and the various parts made to fit together as they ought to do. The settings out for the inside and outside plates of a cylindrical shell are shown in Fig. 292. The thickness of metal is purposely drawn out of proportion to the diameter, so as to better exhibit the construction lines. The lengths of the plates can be obtained, as previously stated, by measuring the lengths of the centre lines of each ring in section and setting out for inside and outside plates respectively. A much better plan, however, and one that will give more accurate results, is to calculate the lengths of the plates. Thus, suppose the inside diameter of inner tier of plates is 12 in. and the plates 1 in. thick, then the girth of outside plates will be :—

$$15 \times 3{\cdot}1416 - 47{\cdot}1240 \text{ in.}$$
$$\text{And inside plates } 13 \times 3{\cdot}1416 - 40{\cdot}8408 \text{ in.}$$
$$\text{Difference in lengths} = 6{\cdot}2832 \text{ in}$$

It should be observed that the difference in length between the inner and outer plates is $2 \times 3{\cdot}1416$, and this gives us a rule by which we can always determine the difference between their lengths:—

Difference = twice thickness of plate × 3·1416
Or thickness of plate × 6·2832

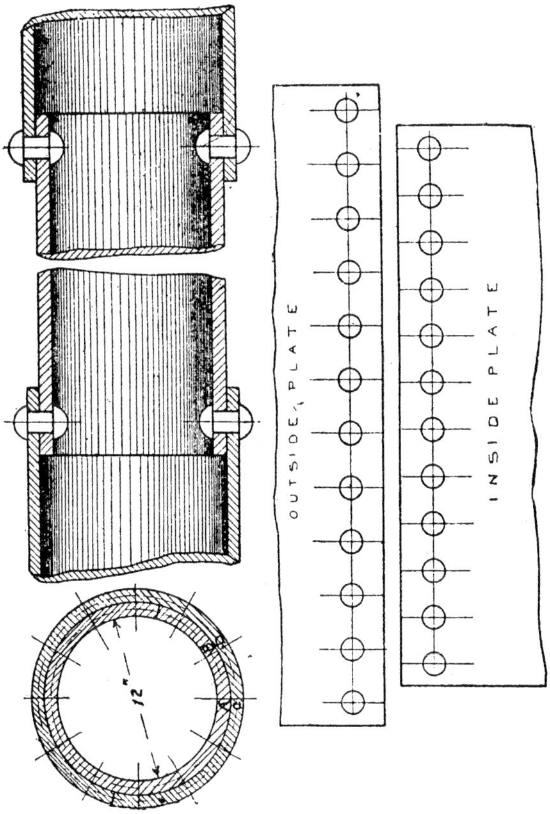

Fig. 292.

If $3\frac{1}{7}$ be used instead of 3·1416, then this difference will always be—

Thickness of plate × $6\frac{2}{7}$

For an accurate-fitting joint, the calculation of this difference is really of more importance than the exact girths. It should be borne in mind that before proceeding to calculate, the thickness of the plate should be carefully gauged. A plate may be called a certain thickness; but as plates are usually rolled to a given weight per square foot, the thickness may be a little more or less than that stated. Consequently, if the calculations are based on a given thickness, and the plate happens to be a shade thinner, the joint will be slack, and if the plate is thicker than that allowed for, the joint will be too tight.

The pitch of the rivet-holes in the two plates can be measured directly from the centre line circles on the section of the two rings. Thus the length along the arc from A to B will be the pitch of the holes on inner plate, and the length measured along the curve from C to D will equal the pitch of holes in outer plate.

Whilst the above method is accurate enough for rough work, or for jobs bringing in only a small part of a circle, it is not of much use where very particular work is wanted. The pitch can be determined by arithmetic from the following rule—

$$\text{Pitch of holes} = \frac{\text{diameter of neutral circle} \times 3{\cdot}1416}{\text{number of holes in circle}}$$

Thus in the present case:—

$$\text{Pitch of holes in outer plate} = \frac{15 \times 3{\cdot}146}{12} = 3{\cdot}927 \text{ in.}$$

$$\text{Pitch of holes in inner plate} = \frac{13 \times 3{\cdot}1416}{12} = 3{\cdot}4034 \text{ in.}$$

When the distances between the hole centres run out to such awkward figures as those above, we are confronted with a fresh difficulty in not being able to set the compasses, with exactness, to this length. So that, in practice, it is a good plan to mark the two end holes and then carefully subdivide the distance; the calculations above giving considerable aid. Usually, the centre of end holes would come on the end lines of net template; but in the present case no lap has been allowed for so as to simplify the problem. As the holes are arranged in Fig. 292, it will be observed that the distance from edge of plate to first hole will be equal to half the pitch on each plate.

The way to calculate the required pitch for any given thickness of plate, and the proper formation of the various riveted joints, will be dealt with later.

In most of the better-class boiler work the plates are rolled and the joints tacked together before drilling, the bulk of the holes being drilled in position. In this way holes with irregular walls are obviated, the joint left stronger than with punched holes, and no stresses set up in the joint. The calculations for lengths of plates and pitch of rivets will, of course, have to be carried out whether the plates have punched or drilled holes, or holes drilled after the plates are shaped and fixed.

Tanks.

A most interesting example of a particularly simple method of jointing is that used in the construction of tanks, when the plates are flanged and lapped and no angle-iron used. Fig. 293 shows the outside view of such a tank. It will be noticed that each face of the tank has two lines of rivets upon it: hence it can be seen that the tank will be constructed of six plates, each plate having two flanges. Fig. 294 is a view of an inside corner which should readily explain the arrangement of the laps on the three plates. An outside view of the same corner is shown in Fig. 295. It will be observed from both of these views that the laps are so formed as to leave a hole right in the corner of the tank. After the tank is riveted up along the laps, the holes at corners are either drifted or reamered out, and a special corner rivet put in as shown on Fig. 293.

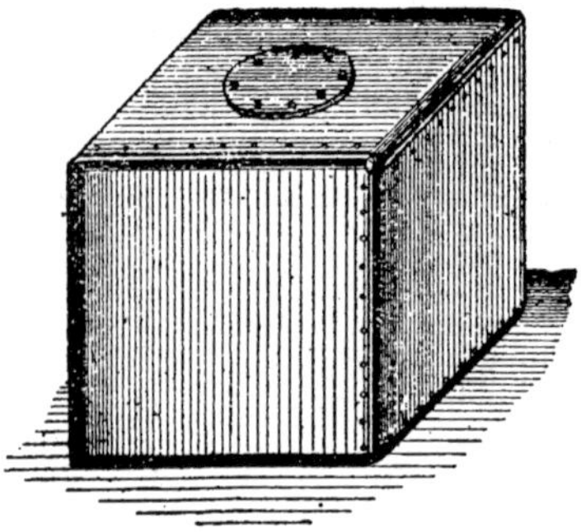

FIG. 293.

The setting out of a plate is explained by the aid of Fig. 296. A cubical-shaped tank has been chosen for the sake of simplicity. The length, breadth, and depth of the *inside* being the same, by inspection of the arrangements of the plates and joints, it will be seen that each plate is exactly the same: hence if a template be made of one plate, the other five can be marked from it.

A section of one plate covering two joints is first set out as shown (the thickness of plate is enlarged somewhat to better show the construction lines), and from this both the length and width of plate can readily be obtained. The length of plate will be found by measuring the distance along the centre line of section, and the width will be equal to the length of flat part of plate—in this case A B. The length and breadth of plate can, of course, be calculated as in the cylindrical shell. Thus, suppose the inside dimen-

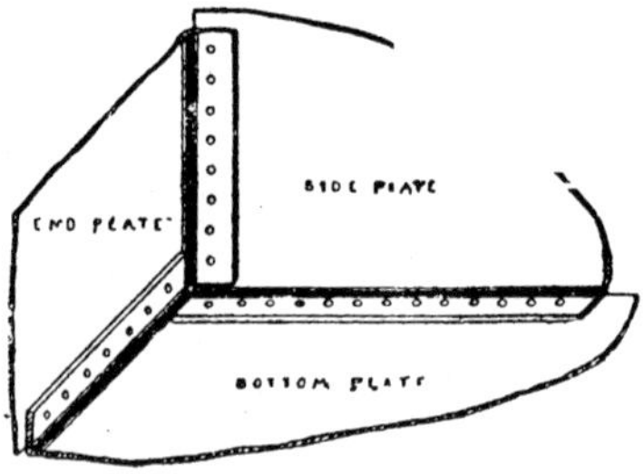

FIG. 294.

sion of tank to be 4 ft., and inside radius of plate to which the flange is bent $\frac{1}{4}$ in., thickness of plate $\frac{1}{8}$ in., and lap $1\frac{1}{8}$ in.

Rule for width of plate:—

"Deduct twice the thickness of plate and twice the inside radius of flange from the inside dimension of tank."

$$\text{Width} = 48 - 2\left(\tfrac{1}{4} + \tfrac{1}{8}\right) = 48 - \tfrac{3}{4} = 47\tfrac{1}{4} \text{ in.}$$

Rule for length of plate:—

"To the width of plate add $3\frac{1}{7}$ times the radius of centre line and twice the lap."

$$\text{Length} = 47\tfrac{1}{4} + \tfrac{5}{16} \times 3\tfrac{1}{7} + 1\tfrac{1}{8} \times 2 = 50\tfrac{27}{56} = 50\tfrac{1}{2} \text{ in. (nearly).}$$

There should be no difficulty in marking out the holes on plates, as each line of holes is exactly the same. A good

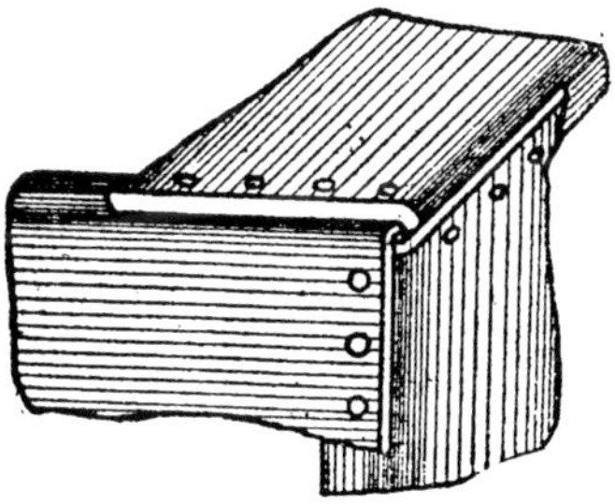

FIG. 295.

method to pursue is to make a template of hoop iron or a batten of timber, with a line of holes carefully set along at the required pitch. From this, all the holes can be marked, and it will avoid the lifting about of a heavy template, which would happen if one of the plates were used for this purpose.

All rivet-holes can of course be punched before the plates are flanged. The gauge for marking the width of flange from edge of plate will be the distance measured along the centre line of section from the end of plate to middle of bend. It can also be calculated by the following rule: "To the lap add one-quarter of $3\frac{1}{7}$ times the radius of centre line of section."

Width of flange—

$$= 1\tfrac{1}{8} + \tfrac{5}{16} \times 3\tfrac{1}{7} \times \tfrac{1}{4} = 1\tfrac{83}{224} = 1\tfrac{3}{8} \text{ in. (nearly).}$$

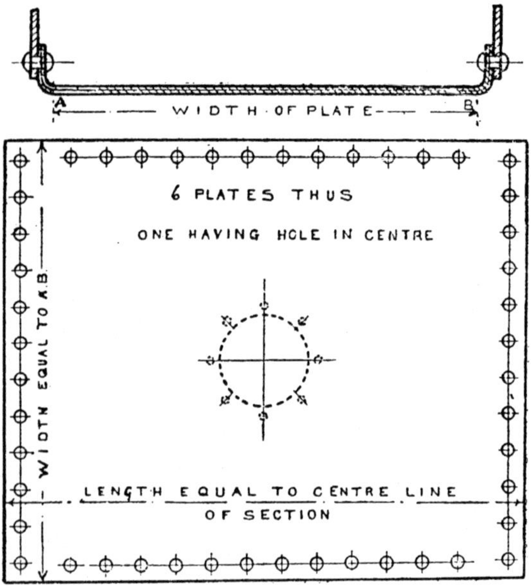

FIG. 296.

The cover is bolted on with set screws, and the holes for these are drilled and tapped to suit the screws.

Whilst a cubical tank has been described, the above remarks and calculations can of course be easily modified and applied to suit the case of any sized tank with either open or closed top.

CHAPTER XXXIII.

PLATER'S DOUBLE-CURVATURE WORK.

DOUBLE-CURVED work in wrought iron or steel plates is, of course, much more difficult to manipulate than in the softer metals, and, on account of the greater resistance that iron or steel offers to being drawn or stretched, greater accuracy is, in consequence, required in the marking out of the plate shapes. At the best, it is only possible to approximate to the real shape of plate wanted, and, in any case, theory is not of much use in this class of work, without it is tempered with experience. Another point to remember is that the amount of stretch or contraction in any particular plate depends very much upon its treatment in working into shape. In all cases it should be aimed to hollow or raise a plate in a natural manner—that is, to work it up as near as possible to the conditions that would obtain if it were stamped or drawn in a pair of dies. We purpose giving one or two typical cases of this class of work, beginning with a

Curved Pipe-Bend.

A sketch of the bend is shown in Fig. 297, on which it will be seen that the back and throat of the curved portion is made up in two pieces, and the cheeks in three; the joints being broken as shown.

The construction lines for the templates are obtained as shown in Fig. 298. An elevation of a segment is first set out; this really showing the elevation of the pieces combined. A semicircle is described on 0 8 and divided into eight equal parts, lines square to 0 8 being run down from each division point. Then, with C as centre, arcs are run around from the foot of each perpendicular, as shown. The complete circumference of the pipe is divided into four equal

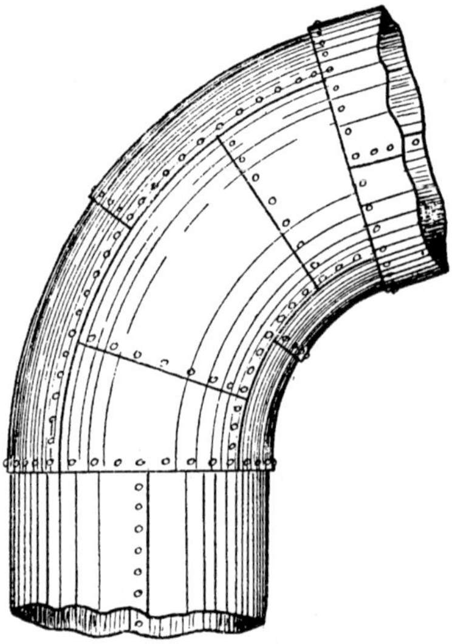

FIG. 297.

parts by the longitudinal joints: hence the length of the girth line for each pattern will be equal to four divisions from the semicircle. To deal with the back pattern first. The girth line of four equal divisions is laid down, and cross lines drawn as shown, these latter being cut off equal in length to the correspondingly-numbered arc in the elevation. Thus 0 0^0 on the pattern equals 0 $0''$ in the elevation, 1 1^0 equals $1''$ $1'$, and 2 2^0 equals $2''$ $2'$. In working the plate into shape it will be found that the line 0^0 0^0 lengthens slightly, and that the edge 2^0 2^0 will shorten a little; hence, as it will manifestly be an advantage to be on the safe side, the best plan will be to draw the arc 2^0 0^0 2^0 to pass through the point 0^0, as first found, and make it somewhat flatter

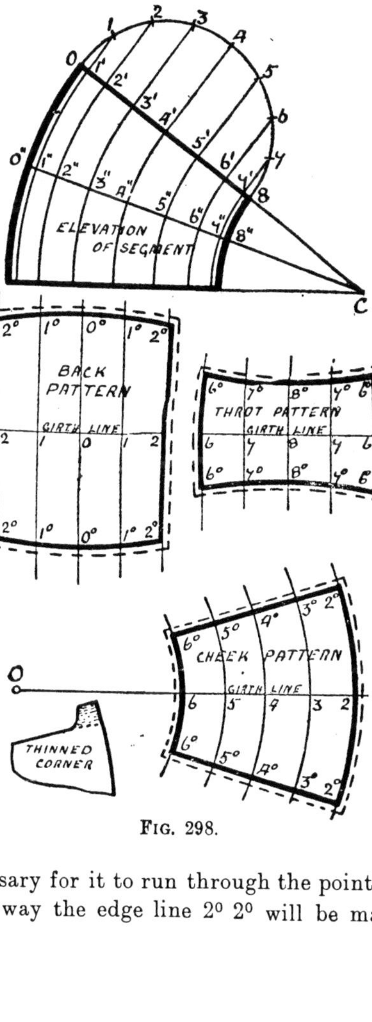

FIG. 298.

than is necessary for it to run through the points numbered 2^0. In this way the edge line 2^0 2^0 will be made slightly longer. As the joints are lapped, the bottom end of the pattern will require to be made just a shade narrower than the top; this can be allowed for by deducting one and a half times the plate thickness from the width.

The throat pattern can be laid out in a similar manner to the back, and here it may be seen that the centre line will shorten slightly in working the plate into shape, and the side lines lengthen somewhat. This difference can be allowed for by making the end arcs 6^0 8^0 6^0, to pass through the points 6^0 6^0, as first found, and slightly flatter than required to pass through the original position of the point 8^0. Here again the pattern must be one and a half times the plate thickness narrower at one end than the other. It will also be an advantage to slightly curve the side edges, as shown on the pattern.

The shape of the plate for the cheek can be laid out by first making the radius O 4 on the pattern equal to C $4''$ on the elevation; then on each side of the point 4 setting two lengths from the semicircle, to make up the girth line, 6 2. Now, using O as centre, arcs are drawn through each point on the girth line as shown, these being cut off respectively equal in length to the corresponding arc in the elevation. Thus 4 4^0 equals $4''$ $4'$, 3 3^0 equals $3''$ $3'$, and so on for the others. It will be found that the points 2^0 to 6^0 lie practically on a straight line, hence this can be drawn in as seen. For a bend of a very sharp curvature it will be an advantage to lengthen the arc 2^0 2^0 just a little, as this will contract somewhat in bringing the plate into shape. Allowance for joints will be added, as shown by the dotted lines.

If instead of making the throat portion in two plates, as shown in Fig. 297, it is desired to make it out of one plate, then the curves at the plate-ends will be somewhat flatter than shown on the throat pattern, as the draw will be less, on account of the longer plate.

In work of this description it is not advisable, only under exceptional circumstances, to put any holes in the plates until after they are shaped. If there are many bends a "cradle"

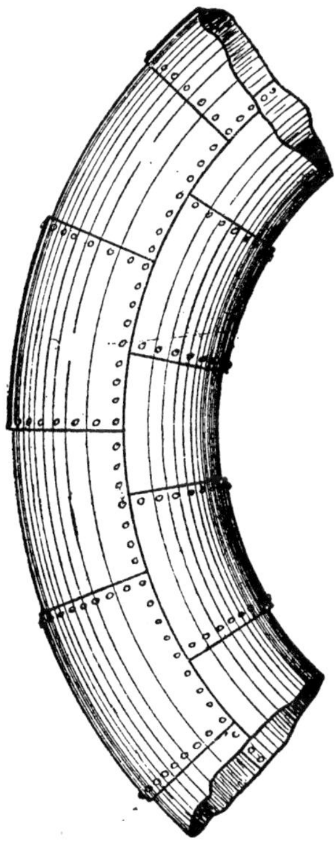

Fig. 299.

punching template for each plate might then very conveniently be made for marking the joint holes.

Instead of having four plates to make up the complete

girth of the pipe, as in Fig. 297, two plates may be used, as in the

Furnace Blast Pipe,

part of which is shown in Fig. 299. Here it will be seen that the longitudinal joints run around the middle of each

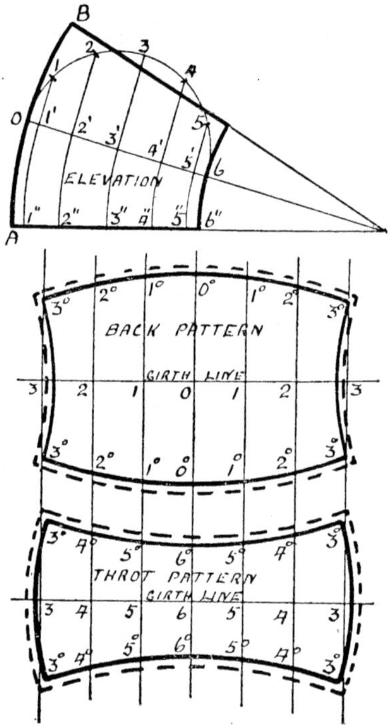

FIG. 300.

side of the pipe, while the transverse joints come to the middle of opposite plates.

The patterns are shown set out in Fig. 300. An elevation of a segment is first drawn, the arc A B being made equal to the width of a segment between the centre lines of the rivets on the back of the pipe-bend. The radius O C, of course, represents the radius of curvature of the back of the bend. A semicircle is constructed upon 0 6, and divided into six equal parts, perpendiculars from each division point then being run down as shown. The girth line, 3 3, of the back pattern will be laid out equal in

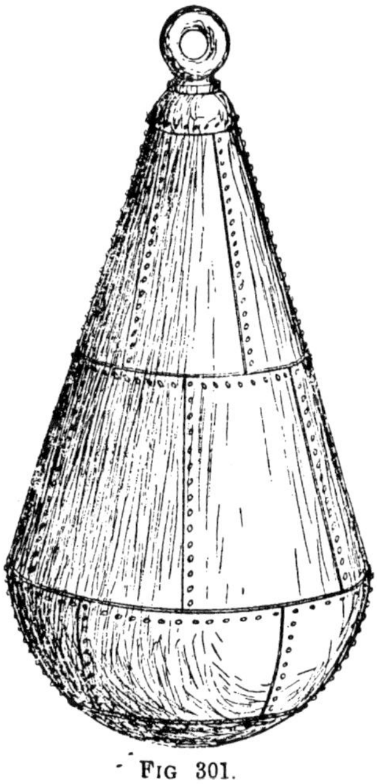

FIG 301.

length to the semicircle and divided into six equal parts, cross lines being run through each division point. These construction lines are then cut off the same length as the similarly-numbered arcs in the elevation. Thus 3 3° equals 3′ 3″, 2 2° equals 2′ 2″, and so on for the remaining lines.

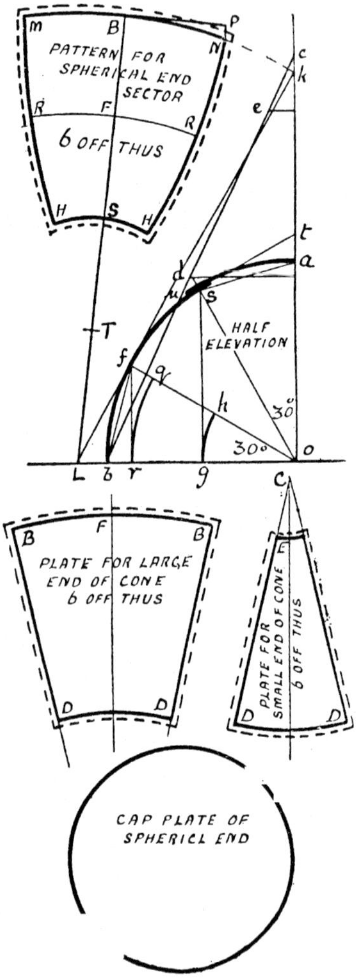

FIG. 302.

In working the plate up it will be found that the line 0° 0° will lengthen somewhat: hence it will be an advantage to make the arc 3° 0° 3° slightly flatter than would be required to pass through the position of 3° as obtained by using the length of the arc 3′ 3″. Instead of being left straight, as in Fig. 298, it will be an advantage to curve the side edges as shown, the radius used being slightly longer than C 3′. As the girth line 3 3 on the back pattern will lengthen in hollowing, the side curves should be drawn to pass through the points 3° 3°, in this manner shortening the line 3 3 somewhat.

The pattern for the throat segment can be struck out in a similar manner to the back, the lines used being those having corresponding numbers on pattern and elevation. Here, again, it will be an advantage to curve the side edges to a radius slightly greater than the centre radius, C 3′, in the elevation.

As the curves at the ends of the patterns come out practically as arcs of circles, there is really no necessity to use all the lines as shown. All that is wanted being the points 3^0, 0^0, 3^0 on the back pattern, and 3^0, 6^0, 3^0 on the throat pattern. The lengths of lines to obtain these can also be calculated if required, and thus the necessity of drawing any kind of elevation avoided. In ordinary practice, however, it is generally the safer plan to use an elevation for obtaining lengths of construction lines.

In working the plates hot, care must be taken that they are drawn or hollowed as uniformly as possible, as the plates, if worked too much in any one particular part, will be pulled out of the shape that the pattern has been designed to produce.

As the plate segments are arranged to fit alternately outside and inside, it is evident from what has been said previously that the girth of the outside plate must be $3\frac{1}{7}$ times the thickness longer than that of the inside plate. It will also be necessary to thin the four corners of the inner plates on the back and the four corners of the outer plates of the throat. A sketch of the method of thinning is shown at the bottom of Fig. 298.

Patterns for Buoy-Plates.

We may consider the buoy, shown in Fig. 301, as being constructed of a cone and a hemisphere. It will be seen from the position of the joints that the girth of the buoy is divided up into six plates.

Four patterns will be required, and these are all shown set out in Fig. 302. The patterns for the conical part will be laid out as explained in the earlier chapters; the radii for the bottom tier of plates being *c b* and *c d*, and for the top tier *c d* and *c e*. The arc B F on the large plate will equal in length the arc *b f*, this being one-twelfth the circumference of the cone base. The length of the arcs, marked D D, at the small end of the large pattern, and the large end of the small pattern, should be the same length; or, if the thickness be taken into account (which it always should),

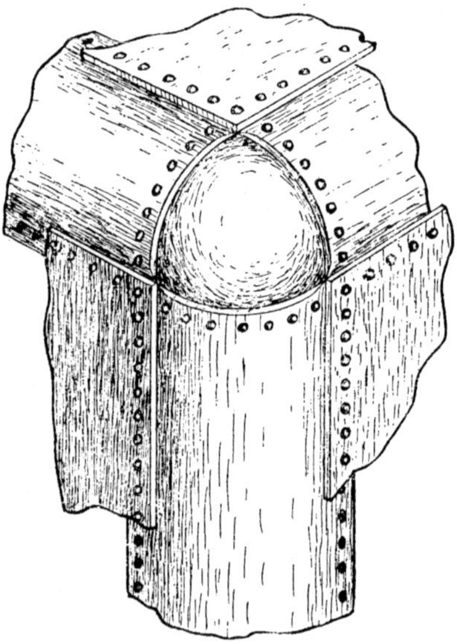

FIG. 303.

the arc D D on the small pattern—seeing this plate is one of the inner tier—will be the thickness of the plate less in length than on the large pattern. Two corners on each of these plates will need thinning, and, of course, the rivet-holes can be put in before the plates are rolled.

The hemisphere is made up in seven pieces: six sectors and a circular centre plate. First let us mark out the shape of plate for one of the gores, or sectors. A quarter-circle, *b o a*, is drawn, and the lines *o f*, *o s* set along to make angles of 30° with *o b* and *o a* respectively; the arc *b s* will then give one-sixth the circumference of the hemisphere. Next draw L *k* square to *o f*, and, using L as centre, run around the arc *k* M. Now lay down the line L B in any convenient position, and draw B P square to it, and equal in length to the arc *f b*; then run down P N parallel to B L, cutting the arc *k* M in N. Next mark B M equal to B N. The points F, S, and T are then determined by making B F equal to the chord *b f*, F S equal in length to the arc *f s*, and S T equal to *s t*; the latter line being drawn square to *o s*. Using T as centre, and T S as radius, an arc is now drawn through S and cut off equal in length, on each side of S, to the arc *g h*; that is, S H equals *g h*. The point *g*, it will be noticed, is determined by dropping a perpendicular from *s* on to the line *o b*. The line *f r* is next drawn square to *o b*, and the arc *r q* run around. Now, using L as centre and L F as radius, the arc R R is drawn, the lengths F R on each side being measured off equal in length to the arc, *r q*. Choosing a suitable radius (one that will give an arc to pass through the points H, R, M) the side curves are now drawn. Allowances for laps are afterwards added as shown.

In work of this character, where the operator has had little experience, it is always best to experiment on a model pattern, this being marked out for a similar article drawn to a small scale. Such a pattern is shown marked out at the bottom of Fig. 304. The model pattern can be cut out of sheet iron or other metal, and before working into shape, its surface should be firmly marked with crossed lines, as shown, these being $\frac{1}{8}$ in. or $\frac{1}{4}$ in. apart. When this pattern has been worked up to the proper curvature, it can then be examined, and by measuring between the lines on its surface it can be ascertained which part has been extended or contracted. If it does not work up to the exact size required, then, by careful examination and measurement, the deficiency can be determined, and this allowed for when striking out the pattern for the full-size articles.

The dimension of disc for the spherical segment will be found by using *a u* as the radius to mark it out.

By the exercise of some thought and careful experiment the pattern for any other kind of gore for an egg-ended boiler, still, or other vessel can be struck out with a good degree of approximation.

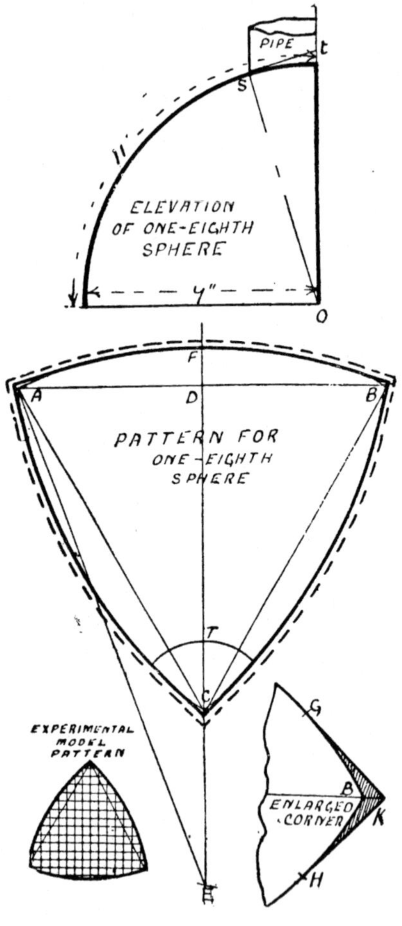

FIG. 304.

Rounded Corner for Tank.

A rounded corner-plate for a tank, motor-car hood, or

other object can be set out in the flat very much the same as explained in connection with Fig. 302. A sketch of the corner-plate is shown in Fig. 303, and, on consideration, it will be seen that its surface can be imagined to be that of one-eighth of a complete sphere. In practical plate work it would, of course, be an advantage to break the joints somewhat differently to that shown in the sketch; but this example, as the joints are arranged, will serve to illustrate the setting out of patterns for objects that come out as a part of a spherical surface.

In Fig. 304 the necessary setting-out required for the pattern for one-eighth of a sphere of 14 in. diameter is shown. The circumference of a 14 in. sphere will be 44 in., one quarter of this, of course, being 11 in. First construct an equilateral triangle of 11 in. side, lettered A B C, in the figure. It is found from experiment that the radius giving the best curve for the sides of the pattern is $2\frac{1}{4}$ times the radius of the sphere, which in this case will be—

$$2\frac{1}{4} \text{ in.} \times 7 \text{ in.} = 15\frac{3}{4} \text{ in.}$$

Bisect A B in D, and join to C, producing the line D C outwards. Now, using A as centre and $15\frac{3}{4}$ in. as radius, mark the point E. This will give the centre from which the arc, A F B, can be described. In the same way the other two arcs can be constructed. A set-square should now be put upon each corner, and two tangential lines, mutually perpendicular, drawn from the arcs. This is best shown by the enlarged corner at the bottom of Fig. 304. Here B G and B H represent the side arcs, and K G, K H the pair of mutually-square lines. It will thus be seen that the small shaded area is added on to the pattern to make it work up correctly. Allowance for laps will be added as shown.

As mentioned in the last example, it is always advisable for the inexperienced in this kind of work to make up a model sector before proceeding to the larger job. A pattern for this is shown at the bottom of Fig. 304.

If the pattern A B C is to be one of the four gores to make up into a hemisphere, with a pipe fitting centrally, as shown in the elevation, Fig. 304, then the part to be cut away on the pattern can be determined by drawing *s t* square to *o s*, and using the former line as radius for the pattern cut; that is, the radius C T on the pattern will be equal in length to the line *s t* in the elevation.

CHAPTER XXXIV.

PATTERNS FOR IRREGULAR ARTICLES.

Rectangular Pipe Fitting on Conical Hood.

In several cases of ornamental and other work it may be necessary to joint together square, or rectangular, and conical pipes, as shown in Fig. 305. The pipes may fit concentrically (which means having a common centre line) or their centre lines may not coincide, but be parallel. In either case, the method of laying out the pattern-shapes will be the same; but as the former will require a less number of lines than the latter, we will show the setting out in connection with that one.

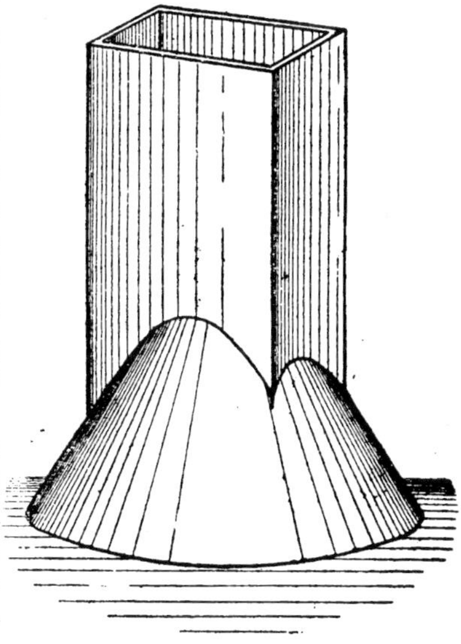

Fig. 305.

The drawing of the patterns, both for the rectangular and conical pipes, is shown in Fig. 306. A quarter-plan and half-elevation is first drawn.

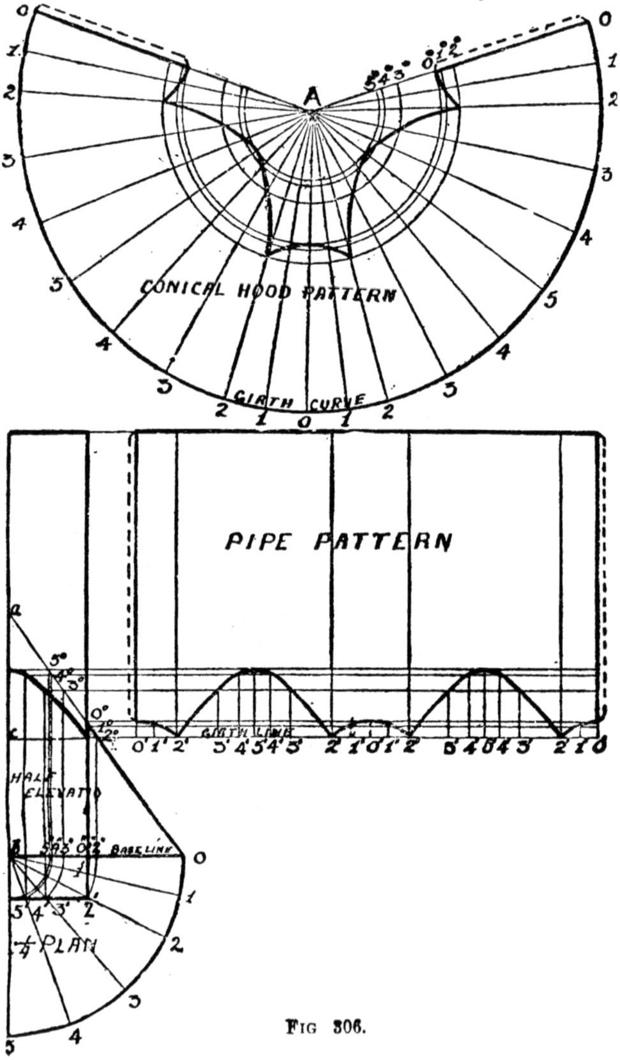

CONICAL HOOD PATTERN
GIRTH CURVE
PIPE PATTERN
GIRTH LINE
HALF ELEVATION
BASE LINE
¼ PLAN

FIG 306.

The point *b* is joined to 2′, and produced outwards to 2; the arc 2 0 then being divided into two equal parts and the arc 2 5 into three equal parts. The division points, 1, 2, 3, etc., are then joined to *b*, thus fixing the points 1′, 2′, 3′, etc. Taking *b* as centre, and *b* 5′, *b* 4′, etc., respectively, as radii, the points 5′, 4′, etc., are turned on to the base line, thus giving the points 5″, 4″, etc. Perpendiculars are then run up from the latter points to meet the outside lines of the curve in 0^0, 1^0, 2^0, etc. A girth line is then laid out for the rectangular pipe pattern by marking the distances 0′ 1′, 1′ 2′, 2′ 3′, etc., as taken from the same numbered line in the plan. Perpendiculars are next run up from the girth line division points, and these cut off by projecting along the required heights from the elevation. That is, a line is drawn along through 5^0 to cut off all the perpendiculars drawn up through the points marked 5′ on the girth line; the same for 4^0, and so on for each number. (In practice it will be better to mark out the pipe pattern away from the elevation, the heights for these lines then being measured from the line *c* 2^0 up to the respective points.) The points as found are then joined up, and the rectangular pipe pattern is complete.

The pattern for the conical part can be struck out by first setting the compasses to *a* 0 as radius, and describing the pattern circle. The length of the girth curve can be measured out by making the respective lengths the same as those with the similar number on the quarter-plan. The division points of the girth curve are then all joined up to the centre, A. To mark off the points on the radial lines, to form the cut, the compasses are opened out to *a* 5^0, *a* 4^0, etc., on the elevation, and these distances used to mark around the arcs on the pattern. Thus, A 5^0 on the pattern equals *a* 5^0 on the elevation, and so on for the other lengths. Where these arcs cut the correspondingly numbered radial lines will give points on the pattern cut as shown. These being joined up with an even curve will complete the pattern. Any allowance for laps can be added according to the method of jointing adopted.

Tapered Square Pipe Fitting on Conical Dome.

A tapered square pipe fitting concentrically on to a conical cap, or dome, as shown in the half-elevation, Fig.

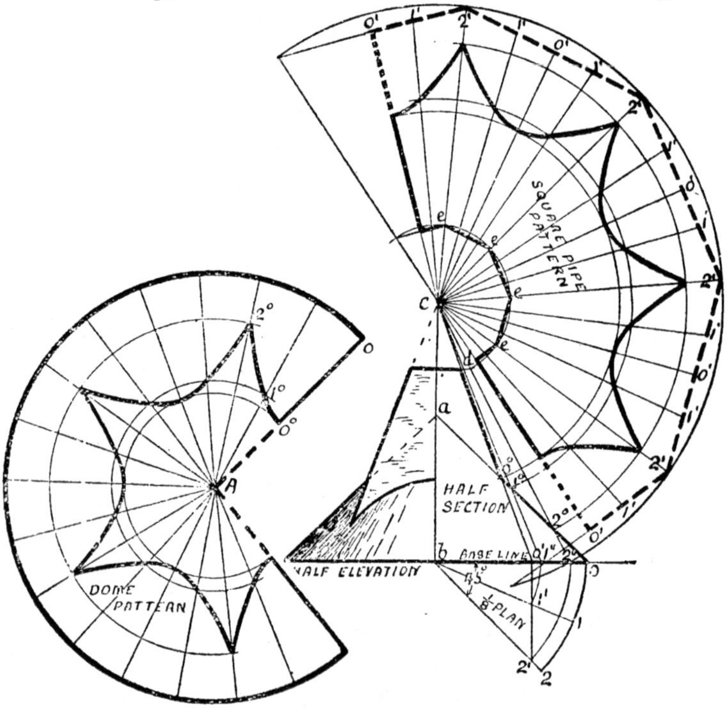

FIG. 307.

307, may for some kinds of ventilator, or other work, require to be made up in sheet metal.

All the setting out necessary to obtain the pattern construction lines can be done with a one-eighth plan and a half-section, as shown in Fig. 307. On the plan, *b* 0′ and 0′ 2′ each equal half the diameter of the base of the pyramid. The centre, *b*, is joined to 2′ and produced outwards to meet the cone base in 2, the arc, 2 0, then being bisected and the middle point, 1, joined to *b*. The points 2′ and 1′ are now swung up, around *b*, on to the base line, giving the points 1″ and 2″. These latter are then joined to *c*, and where the connecting lines cut the outside of cone in 0°, 1°, and 2° will determine the lengths required for the two patterns.

In marking the pattern out for the tapered square pipe, the compasses are put in centre *c* and opened to the point 2″, the outer arc then being described to this radius. The compasses are next set to the length of the side of the pyramid base, that is, twice the length of the line 0′ 2′ on the plan, and *five* lengths to this stepped around the arc on the pattern, thus marking the points 2′. The five chords are next drawn by joining the points marked 2′, the two end chords being bisected in the points 0′. It will thus be seen that there are three full sides and two halves to make up the complete pattern of four sides. The points 1′ and 0′ on the pattern are fixed by making the lengths of 2′ 1′ and 2′ 0′ the same as these lines on the plan. From each point radial lines are drawn to *c*, these being cut, to give points on the pattern curve, by drawing arcs around from the points 0°, 1°, and 2°. Thus, to give one instance, where the arc drawn from 1° intersects the radial lines 1′ *c*, will give points on the curve of the pattern cut. These are then all joined up with even curves, as shown. The cut to form the small end of the pipe is set out by producing the top line in the elevation outwards to meet *c* 2″ in *d*; then, with *c d* as radius, the arc is swept around, and where this intersects the lines *c* 2′, in *e*, on the pattern, will give the end of the top lines; these being then drawn in, as shown, by the lines marked *e e*.

The outer dotted pattern, it might be useful to remember, will, if bent into shape, give the portion of the tapered square pipe which fits *inside* the conical dome.

For the conical dome pattern the compasses are set to the radius *a* 0 in the elevation, and the circular arc marked out; sixteen distances being stepped around this, each equal in length to either of the arcs 0 1 or 1 2 in the plan. From each division point radial lines are drawn to the centre A; these are then cut by arcs drawn to the respective radii, *a* 0^0, *a* 1^0, *a* 2^0, from the half-section. Thus the length A 1^0 on the pattern equals *a* 1^0 on the section, and so on for the other lines. The points being joined with even curves, the pattern is now complete.

It is interesting to notice that the inner pattern, marked off by the dotted line A 0^0, will, when bent into shape, give the portion on the cone fitting *inside* the tapered square pipe.

If the centre lines of the pyramid and cone do not coincide, the patterns for the parts required can still be marked out by the method shown above, the only difference being in the plan, this, perhaps, requiring to be a half or a full-size plan, according to the position of the tapered square pipe on the dome.

Round Pipe on Conical Cap.

If a cylindrical pipe fits on to a conical pipe, both having the same centre line, it will be manifest that the cylindrical pipe will be cut square at the joint, and that the conical pipe will come out as a frustum of a cone. If, however, their centre lines do not coincide, but are some distance apart, and parallel, the two pipes will fit together as shown in Fig. 308.

Before attempting to mark out the pattern it will be necessary to first draw in the elevation of a joint curve. To do this, construct the semicircle, as seen in plan, Fig. 308, and divide it into six equal parts, running lines up through each division point square to the base line. Now, taking *b* as centre, and *b* 1, *b* 2, etc., as radii, swing on to the base line, thus determining the points 1′ 2′, etc. From these

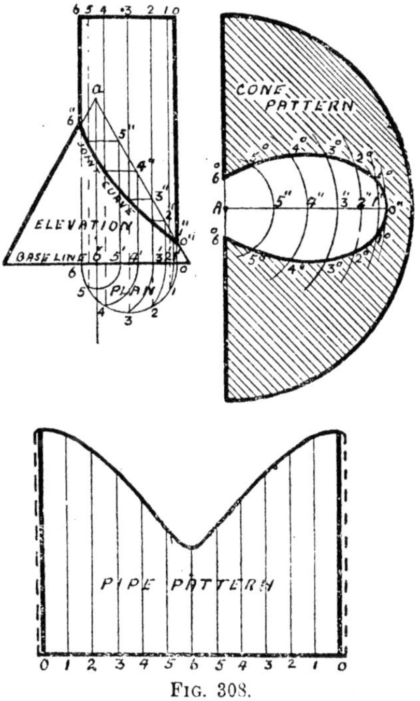

FIG. 308.

last points run perpendiculars up to intersect the outside lines of the cone in 0″, 1″, 2″, etc., and then from these draw lines across parallel to the base line to meet the perpendiculars already drawn from the points on the semicircle. The points of intersection of these two lines will lie on the joint curve.

The complete conical pattern is first marked out in the

usual way, and a centre line, A 0″, drawn. Along this the distances from *a* 0″ in the elevation are set; that is, A 5″ equals *a* 5″, A 4″ equals *a* 4″, and so on. With A as centre, arcs are then drawn through the points 5″, 4″, etc. The lengths of these are carefully measured off equal to that of the corresponding arc on the semicircle in plan. Thus 1″ 1° equals 1 1′, the arc 2″ 2° is the same length as 2 2′, and so on for the others. The points so found when joined up with an even curve will give the cut required.

For the cylindrical pipe the pattern will be marked out in the usual way, the lengths of the construction lines being measured from the top end down to the joint curve.

Circular Tapered Pipe Fitting on Conical Dome.

If the centre lines of these two conical surfaces coincide, it is evident that each part will come out as a frustum of a cone; but if the centre lines are not common to each, but parallel, then the two parts will fit together as shown in Fig. 309.

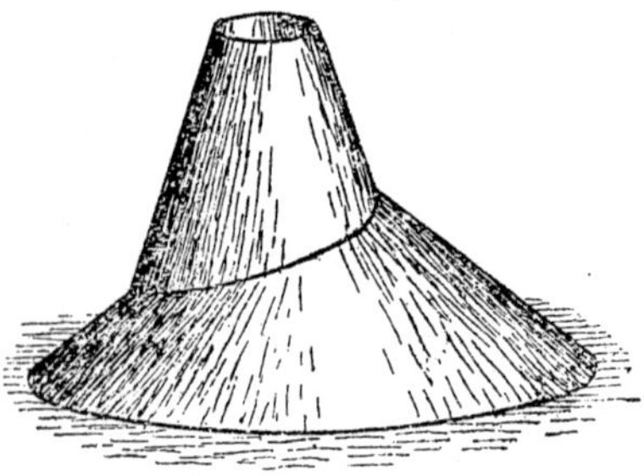

FIG 309.

In the latter case the problem becomes somewhat more difficult.

The setting out of the patterns is shown in Fig. 310. Just as in most other cases that we have dealt with, the first thing to be done is to obtain an elevation of the joint curve. For this purpose an elevation and half-plan of the dome and pipe is constructed. Through the points 0″ and 6″ lines parallel to the base line are drawn, cutting the centre line of the large cone in the points *e*

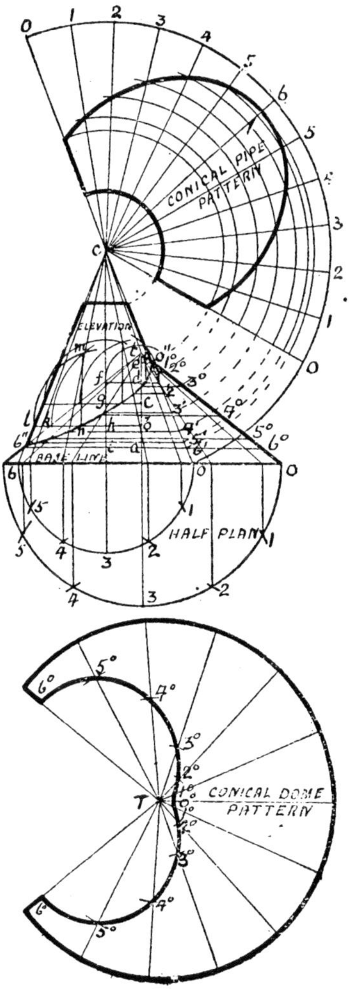

Fig. 310.

and *a*. Then line *e a* is divided into, say, four equal parts, giving points *b c* and *d*. Cross lines are then drawn through each of these points to cut the centre line of the small cone in *f*, *g*, *h*, and *i*. Now taking *b* as centre, and *b k* as radius, draw an arc of a circle; then, with *h* as centre, and *h l* as radius, construct another arc to cut the former in *m*. From *m* drop a perpendicular on to *b l*, thus obtaining *n*, which will be a point on the joint curve. In the same manner, circular arcs can be described on the lines passing through *c* and *d*, and further points on the joint curve obtained. These being carefully joined up will give the elevation of the joint curve as required for getting the lengths of the pattern construction lines.

For the conical pipe pattern it will be most convenient to produce the sides down to the base line, and on this describe a semicircle, dividing it into six equal parts, as shown. Perpendiculars are then run up from each division point to the base and joined to the apex, *c*, of the cone. Where these radial lines cross the joint curve, lines square to the centre line are run to the outside of the cone, giving the points 5′, 4′, 3′, etc. The lengths *c* 5′, *c* 4′, etc., are then transferred by running the arcs around on to the correspondingly numbered radial line of the complete cone pattern, thus giving points for the pattern cut. These being connected with a fair curve will give the required pattern cut for the conical pipe.

In the same manner the pattern for the conical dome can be struck out. The semicircle in plan is divided as before, perpendiculars run up, and radial lines drawn. From the points where these latter intersect the joint curve, cross lines are drawn to meet the outside line of the conical dome in the points 5^0, 4^0, 3^0, etc. Then for the pattern, these lengths are marked off on the construction lines. That is T 5^0, T 4^0, etc., on the pattern will respectively equal *t* 5^0, *t* 4^0, etc., on the elevation. Any laps for jointing can, of course, be added as required for either of the patterns.

Round Pipe with Spiral Joint.

If a pipe is required to have a twisted seam as shown in Fig. 311, the rake of the pattern strip can be quite easily determined by the method of construction as seen in the figure. The pipe girth is laid out first and the line A B drawn perpendicular and made equal in length to the height of half a twist. Line B C is then marked off equal to half the pipe circumference, and the pattern completed as shown.

If A B had been made equal to the pitch of the spiral seam—that is, the vertical height of one complete twist—

then B C would have to be made equal to the complete pipe girth.

Sheet Metal Worm.

Sheet metal screws for moving grain along a trough, as shown fitted to a shaft in Fig. 312, are usually made up out of rings which are shaped by hammering and riveted together. Being a twisted surface it is not strictly developable, but a very good approximation can be obtained by the diagram shown in Fig. 312. A C is set out equal to the shaft circumference, and A B made the same height as the pitch; the

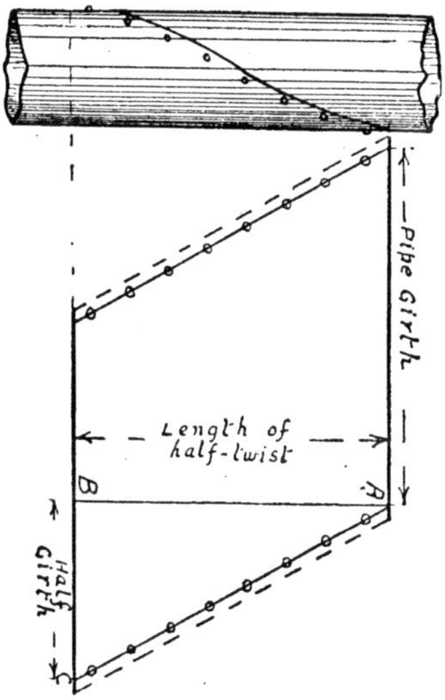

FIG. 311.

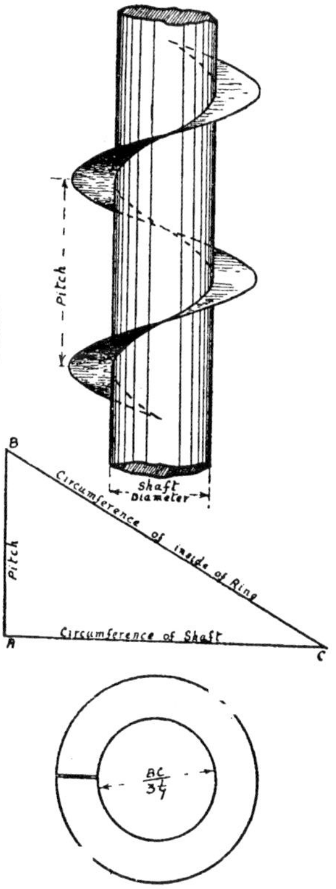

FIG. 312.

line B C will then give the length of the inner circumference of the pattern ring; the corresponding diameter being determined by dividing B C by $3\frac{1}{7}$.

By calculation it can be shown that the outside of the ring is a little longer than that required for the outside of the spiral, but when the pitch is small and the ring narrow the difference is very slight. To give the twist without buckle the rings will require to be carefully hammered, the blows falling heaviest on the inner part of the ring.

Twisted Rectangular Pipe Bend.

A peculiar application of the last case can be made to that of an oblong pipe bend, as shown in Fig. 313. In this the top and bottom pieces will be formed by a quarter of a ring as explained in connection with Fig. 312. The side

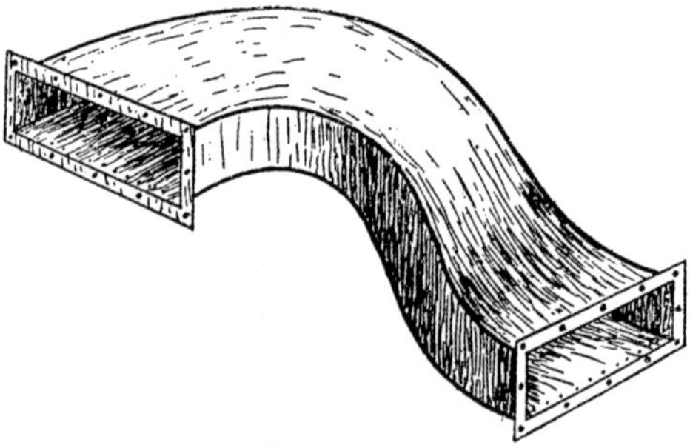

Fig. 313.

strips will be straight, with the ends just curved to cant the bend to meet the straight lines of pipes.

The patterns are shown marked out in Fig. 314. For the inside and outside patterns the heights B E and D F will respectively equal the difference in level of the two lines of piping—that is, it will correspond to a quarter of the pitch, as shown in Fig. 312. A B will equal the girth *a b*, and C D that of *c d*. The distance F C will give the length of the quarter-circle S S, and from this the radius O S can be calculated, and the pattern thus set out.

Junction of Straight and Bent Round Pipes.

If a straight pipe is to fit on to a curved bend, as shown in Fig. 315, it will be necessary to obtain the shape of the joint line before the pattern can be set out. This is very simply done by describing the semicircles, as seen on the figure; dividing up in the same manner, then running lines down and arcs around from centre C to intersect; thus giving points on the joint curve. The pattern (not shown) will then of course be set out in the usual way.

Any other junction of a straight with a curved pipe, either of the same or unequal diameters, can be marked out in a similar manner.

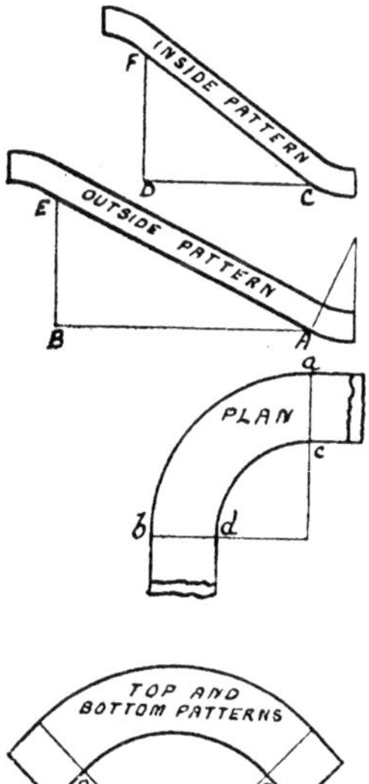

FIG. 314.

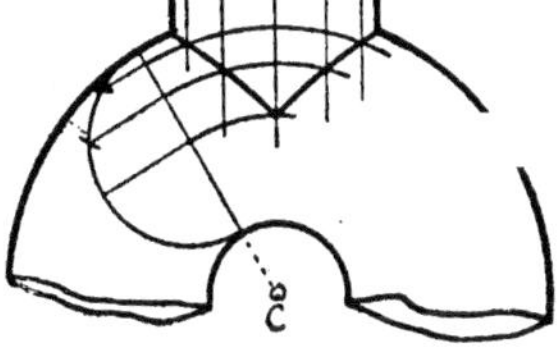

FIG. 315.

Oblique Square Connecting Pipe.

When two square pipes, having their ends cut level, need connecting, this may be accomplished by joining them with an oblique square pipe, as shown in Fig. 316.

The setting out of the pattern is obviously so simple that there is no need for further description.

In bending up it should be remembered that the *ends*

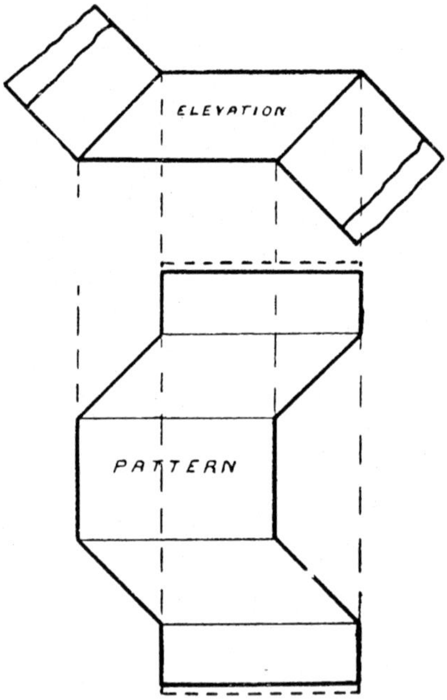

FIG. 316.

of the connecting pipe will be square, its cross section, of course, being rectangular.

Tapered Oblique Square Connecting Pipe.

In a similar manner to the previous case, when two square pipes of unequal sizes require connecting, the intermediate pipe will come out as a frustum of an oblique square pyramid, as shown in Fig. 317.

For the pattern B D will be drawn perpendicular to C B

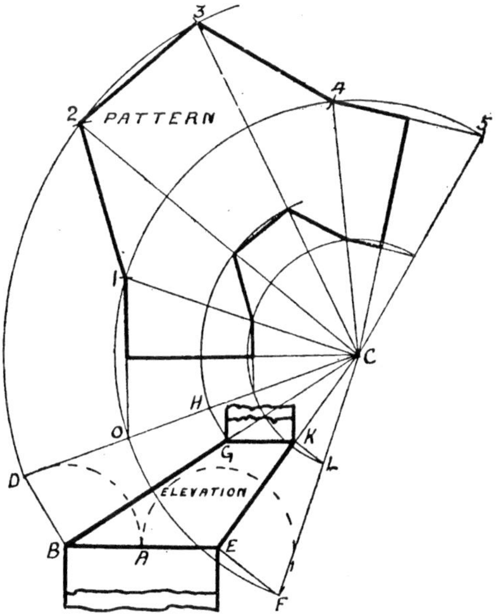

Fig. 317.

and equal in length to B A; D then being joined to C and H G drawn square to B G. In the same way points F and L can be determined. Now using C as centre, arcs are swept around from points D, F, H, and L. Then commencing, say, at 0 the distances 0 1, 1 2, etc., each equal to B E, are marked off. The remaining part of the work should require no further description.

Square Hopper or Outlet on Round Pipe.

The setting out shown in Fig. 318 is for an outlet fitting on the underside of a pipe, but the method of laying out the pattern will be exactly the same if the case is that of a hopper resting on top of the pipe.

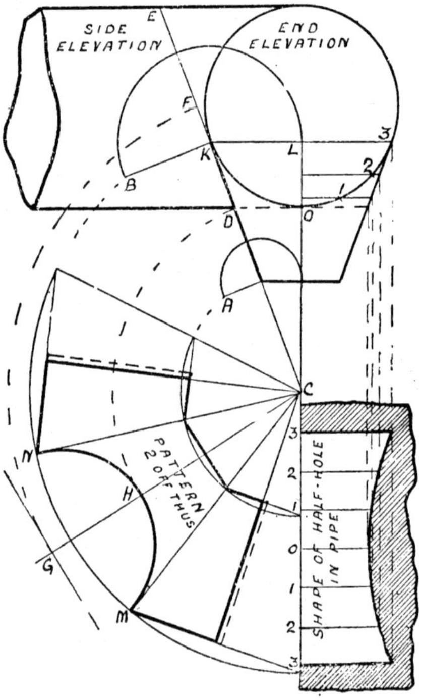

FIG. 318.

Geometrically this is the intersection of a pyramid and a cylinder.

In striking out the patterns the point B is obtained by

drawing the line B K square to K C, and making it equal to K L, the point A being fixed in a similar manner. Now, using C as centre and C B as radius an arc is swung around, and on this three full widths each equal to K 3 cut off. The two outside lengths are each bisected, as it will be seen that the pattern is for making the article up in two pieces. The curve passing through H on the pattern can be set out by first getting the middle point F of D E and then cutting C G off equal to C F; fixing H by making C H equal to C D; drawing a line through G at right angles to G C, and by the trammel method (Chapter XXI.) constructing the part ellipse M H N. The major axis of the ellipse will, of course, be E D, and the minor axis the diameter of the pipe.

The hole is shown marked out by projection; the girth being measured from the end elevation along the arc 0 to 3. It could manifestly be set out in any position, the lengths of construction lines being taken by measurement across the circle in the end elevation.

Spherical Surface Dome in Sectors.

A flat dome like a gasholder top (Fig. 319) can have the patterns for its sectors struck out by assuming that each ring or tier of plates forms part of a cone surface. Thus a pattern for the outer ring in Fig. 319 will be laid out by

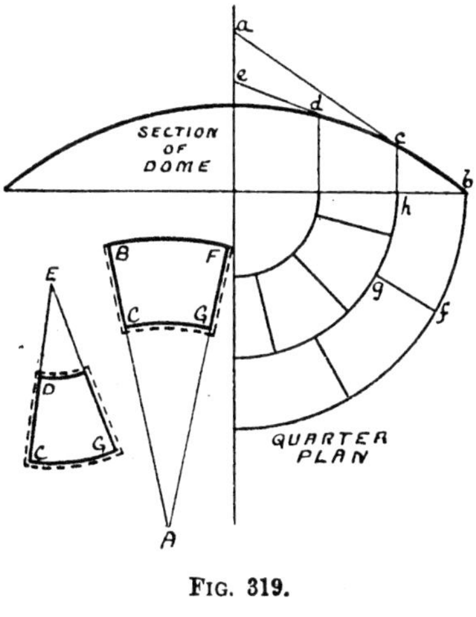

FIG. 319.

making radius A B equal to *a b*; A C equal to *a c*, and the length B F equal to *b f*. In the same way the next tier of plates can be dealt with.

A dome pattern of this character can also be made up by the strip method, as shown in the chapters on roofing work.

Cylindrical Pipe on Spherical Dome.

A pipe fitting as above is shown in Fig. 320. Before its pattern can be developed a series of points on the elevation of the joint line will have to be determined. As the method of finding each point is the same the construction for one

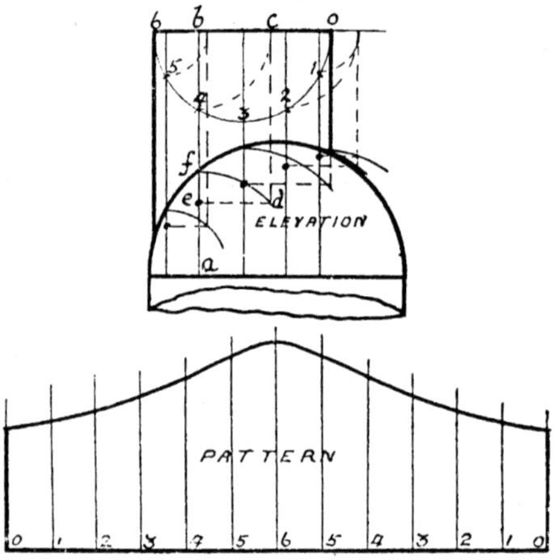

FIG. 320.

point only (4) is shown. With centre *a* and radius *a f* describe the arc *f d* of indefinite length, then with centre *b* and radius *b* 4 describe the arc 4 *c*. From *c* run down a perpendicular to meet the arc in *d*; and to fix *e* run a line square across from *d*.

Having determined all the joint line points the pattern can be set out in the usual way.

Conical Spout Fitting on Conical Vessel.

Perhaps the most complicated patterns to mark out are those for objects where the two parts fitting together are both conical. Such a case is shown in Fig. 321.

As usual the first thing to do is to locate points on the joint or curve of intersection, and when this is done the

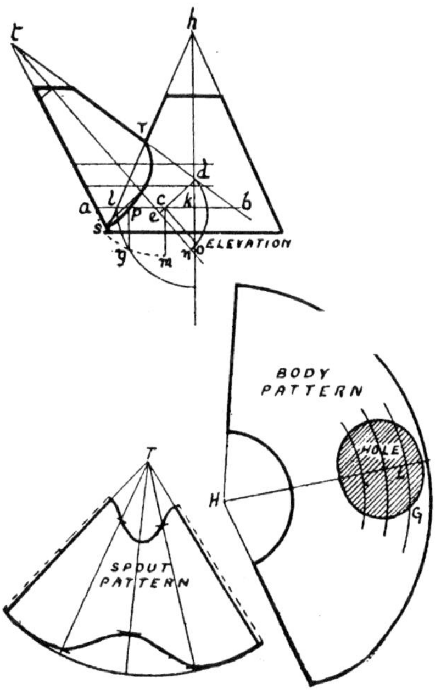

FIG. 321.

ordinary method of getting out a pattern for a part cone (Chapter XIV.) can be applied.

The obtaining of one point (p) only is shown, as all the others will be found in exactly the same manner. The centre line *t n* of the spout cone is first drawn and *r s* divided into four equal parts, and lines drawn across as shown by *a b* and the others. The middle point, *c*, of *a b* is next determined and the line *e d* drawn through it square to *t n*. A quarter-circle is constructed on *e d*, and *c o* drawn parallel to *e n*. The line *c m* is next drawn perpendicular to *a b* and cut off equal to *c o*. The quarter ellipse, *a g m*, is now constructed by the trammel method (Chapter XXI.), and the point *g* determined by describing the quarter-circle on *l k* to intersect the ellipse. A perpendicular is now run up from *g* to cut *a b* in *p*; which will be a point on the elevation of joint curve. In the same way points can be found on the other two lines.

There is no need to describe the marking out of the spout pattern, as this is done in former chapters, but the method of obtaining the shape of the hole on the body pattern is worth considering. Mark off H L equal to *h l*, and draw around the arc, cutting off L G equal in length to the arc *l g* on the elevation. In the same way other points can be found which, when joined up, will give the shape of the hole.

Oblique Circular Hood Fitting on Round Pipe.

The intersection of an oblique cone with a cylinder, as shown in Fig. 322, presents a way by which a circular-mouthed hood can be run into a vertical pipe.

The determination of the joint line and the method of getting the lengths of the pattern lines will be shown for one point only. The plan and elevation of the oblique cone and the setting out of the full pattern are just the same as shown in Chapter XVIII. A line is first run up from *b* to *b'*, joined to *t'* and cut through at the point *a'* by a line run up from *a*. Now using *t* as centre draw the arcs *a c* and *b* B; joining B to *t'* and drawing a line up from *c* to cut B *t'* in A (or A can be found by drawing a line across from *a'* to cut B t'). The line T A on the pattern is then marked off equal to t' A in the elevation. In the same way the other points for the pattern can be obtained.

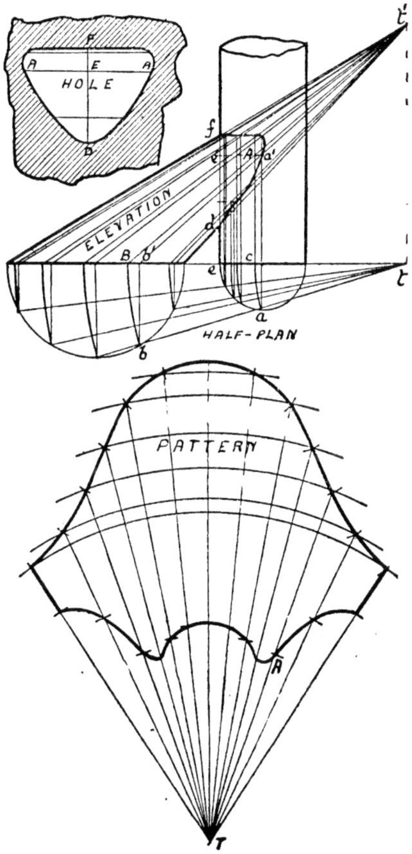

FIG. 322.

The shape of hole for the cylindrical pipe can be drawn by making D E and D F respectively equal to *d e* and *d f*, and the distance E A equal in length to the arc *e a*. In a similar manner the other widths of the hole can be determined.

Gusset Plate for Round Pipe Elbow.

The exact shape of the pattern for a gusset may be found as set out in Fig. 323.

A quarter-circle is described, divided into three equal parts, and lines run up to meet the joint line of gusset in $0'$, $1'$, $2'$, and $3'$. The middle line $0'\,3^0$ is next drawn, and lines run across and cut off equal in length to the quarter-circle lines; that is, $0'\,0'' = a\,o$, $1^0\,1'' = b\,1$, and $2^0\,2'' = c\,2$. From the dotted curve thus obtained the girth line of

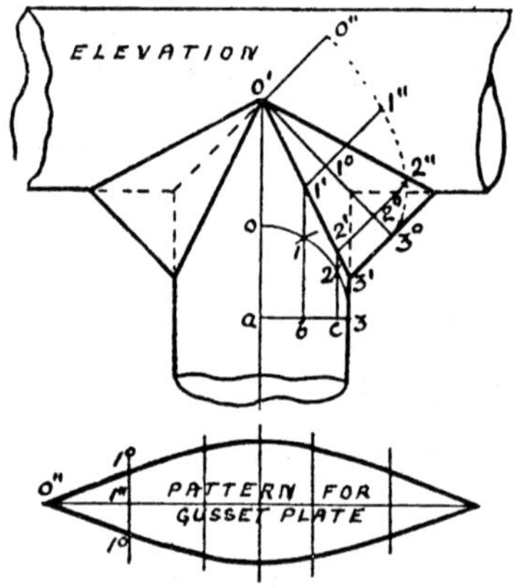

FIG. 323.

pattern is measured, set down, and construction lines drawn across; these latter being cut off equal in length to the lines on gusset elevation. Thus, to give one example, $0''\,1''$ on the pattern will equal $0''\,1''$ on the elevation, whilst $1^0\,1''$ will be the same length as $1^0\,1'$.

Round Pipe Elbow with Twisted Arms.

A peculiar case of an elbow is that shown in Fig. 324, where one arm is twisted so that the elbow would not lie flat on a plane surface, or geometrically, when the centre lines of the arms are not in the same plane.

This is usually made up with the middle piece telescopic, so that the elbow can be twisted into its proper position. In any case the correct angles that the arms make with the

FIG. 324.

middle pipe must be determined, and it will perhaps be an advantage, also, to show how the pattern for the middle part may be set out in one piece.

In Fig. 325 a plan (a b) and elevation (a' b') of the centre line of the middle pipe is shown. The angle for the bottom elbow can be found by drawing b B square to a b and making it equal to c' b'. If a line then be drawn through a perpendicular to a b the bottom angle will be determined as indicated. The top angle can be set out by making a B′

equal to *a* B, and drawing a line through B′ square to B′ *b*. Having found the angles the patterns can be struck out as shown in Chapter II.

To make the pattern for the middle pipe in one piece it will be necessary to find the true length of one line on the

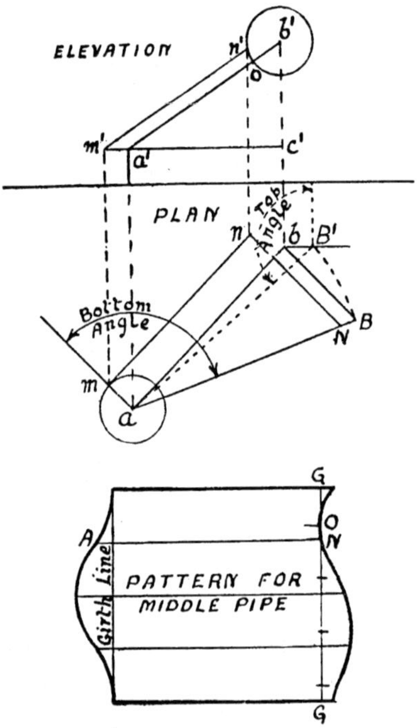

FIG. 325.

pipe and use this to set out the length of the pattern. Draw *m n* parallel to *a b*, and project down from *n*′ to determine the point *n*. Now draw *n* N perpendicular to *n m*, and so fixing the point N on *a* B. The length *a* N will be the true length of the side line of the pipe. Having set out the pattern cut for the bottom elbow (shown passing through A) the girth line should be divided into four equal parts, and lines run along as shown. The side line A N is made equal to *a* N from the plan. Next draw the line G G to pass through N and mark the distance N O equal to the length of the arc *n′ o* in the elevation. The point O will be the throat part of the top elbow, hence the curve must be drawn as shown passing through the point N.

The patterns for the arms and the construction for the curves are not shown, as these will come out as explained in the early chapters.

CHAPTER XXXV.

SHEET METAL JOINTS.

There are really only five ways in which the edges of sheet and plate metals can be fastened together—viz., by soldered, brazed, welded, grooved, and riveted joints. But whilst we are limited to the use of one or other of these forms of jointing, there are numerous modifications of them in practice.

The sketches of joints shown are enlarged somewhat, to better exhibit the layers of metal. (1) shows the ordinary lap-joint, as used in soldering together the edges of tin-plate, zinc, or galvanised iron, the width of lap running from about $\frac{1}{8}$ in. in thin tinplate up to $\frac{1}{4}$ in. in galvanised iron.

To make a soldered joint is not a very difficult matter; but there are a few things that want to be taken notice of if the job is to be carried out successfully. The fluxes (anything that is used to assist the flow of metals) used are various; but those commonly in use are "killed spirits," and ready-prepared soldering fluids. "Killed spirits," or "spirits of salts," as it called, takes a good deal of beating for all-round work, as by its use almost any metal can be soldered, with the exception, perhaps, of aluminium. It is prepared by dissolving as much scrap zinc as possible in hydrochloric acid, the resulting liquid being known chemically as a solution of chloride of zinc. If the edges of the metal are clean, they can be lapped over without any preparation, and the spirits applied along the joint with a brush about $\frac{1}{8}$ in. wide—a good brush can be made with a few bristles fixed into a strip of double-over tinplate. Before using the soldering-bit it should be seen that it is

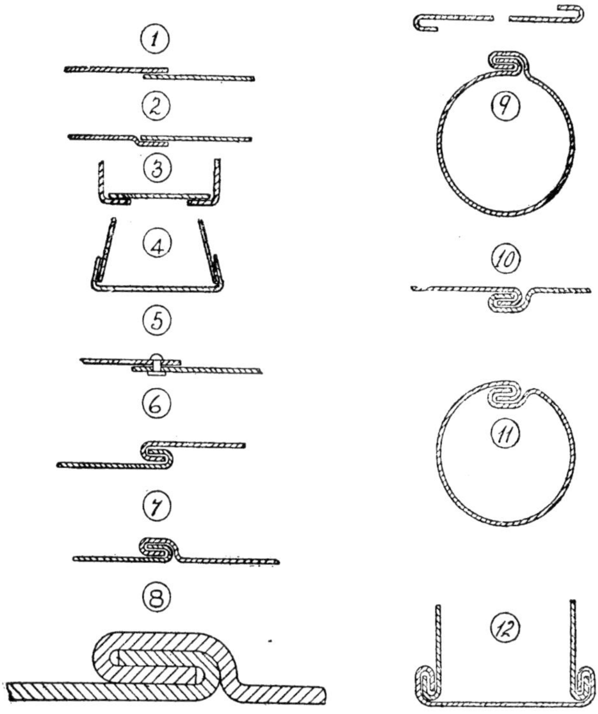

properly tinned, and if not, get to dark-red heat, file the point about $\frac{3}{4}$ in. along, dip in spirits, and then apply solder.

The mistake that the novice usually makes in soldering

a joint is to stick the metal on like glue or putty, instead of holding the soldering-iron long enough against the joint for the solder to be properly melted and the joint to get sufficiently hot for the solder and sheet metal to firmly adhere together. Instead of using the extreme point of

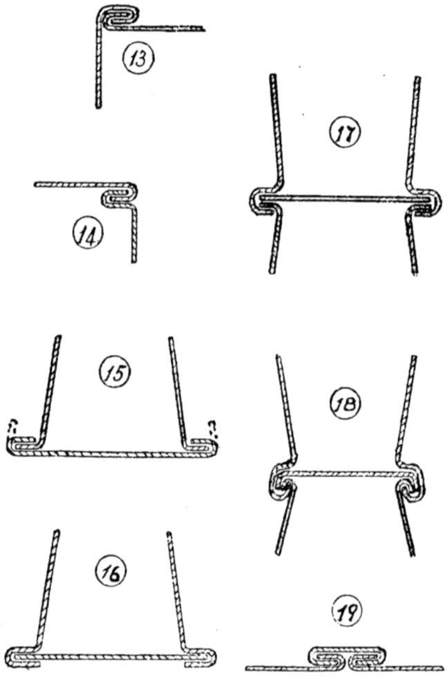

the soldering-iron to run the solder on joint, an edge of the square point should be used to draw along the solder. In this way a greater quantity of heat will be transmitted to the joint, and thus a better and quicker job made.

The soldering-iron must be watched, so that it does not get red-hot, or else the tinning on its point will be burnt off, or, what is worse, form a hard skin of bronze, which is somewhat difficult to file away. When the soldering-iron is drawn from the fire it can be cleaned by quickly dipping the point into the spirits, and also in this way one can judge as to its proper temperature. If when dipping the iron into the spirits much smoke is given off, or the liquid

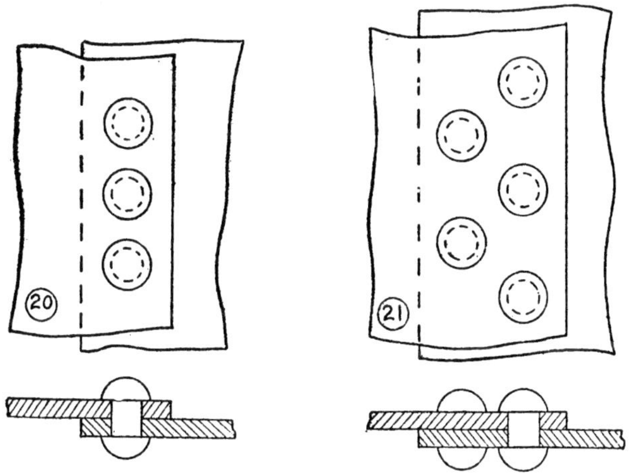

spurts about, the iron is too hot; or, on the other hand, if small bubbles of spirit adhere to the soldering-iron it is not hot enough. An hour's practice should teach one the proper temperature at which to use the bit.

In soldering zinc or galvanised iron, if the soldering-bit is too hot the joint will be very rough on account of some of the zinc being melted from the surface of sheet and mixing in the solder. For tarnished zinc and galvanised iron, the spirits should not be quite "dead," that is, the scrap zinc should be withdrawn from the acid before the boiling action has quite ceased. It is, perhaps, a better plan, though, to freshen up the "killed spirits" by adding a small quantity of neat acid.

In soldering copper, brass, and black iron, the edges of metal should be carefully cleaned before the lap is formed. One of the tests of a good soldered joint is that the solder shall have run right through the joint, and if this be done, and the joint properly cleaned with soda and water, there is little danger of corrosion from the use of chloride of zinc. The great drawback to the use of this flux is in the corroding action that takes place if any be left about the joint; perhaps the chief evil being when it is not properly driven out from between the laps with the running solder.

The next joint (2) is known as a countersunk or flush joint, and is used either for soldering or riveting where one face of the article is required to be level or flush. The crease also adds stiffness to the joint, and assists to keep the edges of the metal on the lap close down to the surface of the sheet or plate; (3) is a joint that is sometimes used for fixing a bottom in an article, either by soldering or riveting; the edge of the body of vessel is turned or flanged inside; (4) is an edge-over joint, generally used for readily attaching bottoms to articles by soldering and occasionally by riveting; (5) shows a riveted joint for sheet metal, the width of lap usually being about six to eight times the diameter of rivet. It is not the general custom to punch holes in the laps of thin metal before lapping over: the rivets are, as a rule, drawn through the two thicknesses of metal with a rivet-set or fetcher-up (Fig. 326), hammered down, and then snapped with the cup on upset. In making this kind of a joint the plan followed is to place the rivet on a stake or bar, bring the joint over it, and tap with the hammer, the position of the rivet being at once seen

FIG. 326.

by the slight mark or bright spot on the sheet; the article is then moved until it is found that the rivet is in its right position on the lap, it is then drawn through as before mentioned. Workmen—such as bucket makers and others of this class—become remarkably skilful in this blind kind of riveting.

The grooved joint (6) and (7) is perhaps the most universally used of all forms of jointing, and whoever invented it certainly conferred an enormous benefit on all classes of sheet metal workers. In making the joint, the edges of the metal are bent over, either with a mallet on a hatchet-stake or in a folding machine, and hooked together as in (6), and the seam placed on a bar or other tool, and grooved by hammering a groover (Fig. 327) whilst it is being moved along the seam. Care must be taken that the groover does not cut or mark the metal on either side of the groove. The joint is now flattened down with, in the case of thin metal, a mallet; and thick metal, a flat-faced hammer. An enlarged view of the finished seam is shown in (8). In a shop where much pipe or other grooved work is done, it is worth while having a grooving machine. The same kind of joint is shown in (9) as a longitudinal seam for a pipe. It should be observed that one edge of sheet is folded down and the other up.

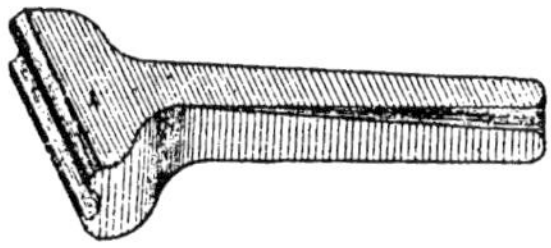

FIG 327.

In (10) and (11) a countersink or inside groove is shown. The use of this is to avoid having projections on the outer surface of an article. It is also used in jointing the zinc lining in coal-buckets, scoops, and similar things. For work of large diameter, where a groover can be used inside, this joint will be made in the ordinary manner; but for small work the edges are hooked together as in (6), the article slipped over a bar with a square edge, and the groove sunk with mallet or hammer.

(15) is known as a "paned down" joint, and is a ready means of edging a bottom on to an article. The body is stretched or flanged first, and the edge of bottom is turned up all round, as shown by dotted lines, the bottom slipped on to body and then paned down. In (12) the same joint is shown bent over again. This is known as a "knocked-

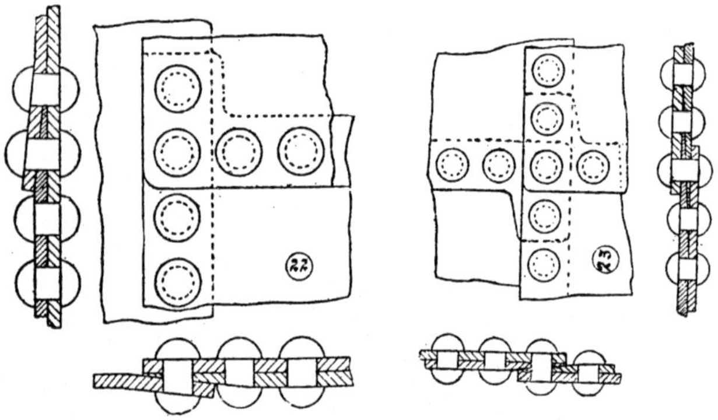

up" joint, and is commonly used for fastening bottoms on to all kinds of sheet metal vessels. If made properly, both this and the ordinary grooved seam should be water-tight without being soldered; but, of course, a better job is made if soldered as well. If an article is made up in black iron and then galvanised, no soldering should be needed. A similar joint to above is shown in (13), and in this form is used in seaming the corners of boxes or trunks. Another modification of this joint is shown in (29) and (30), and is used for joining together two pipes—end on, and also for attaching a neck or collar to the body of an article.

A cash-box joint is illustrated by (14); the object of this is to get the knock-up inside the box, and also to have the outside of corner flush.

Two further methods of jointing are explained by (16) and (31): they are sometimes used in fixing bottoms to articles.

The joint in (32) and the bottom seam in (33) are ways

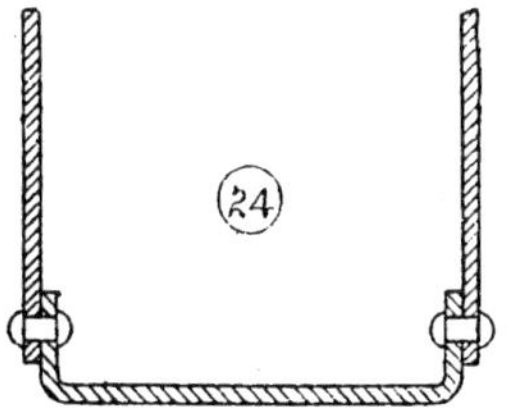

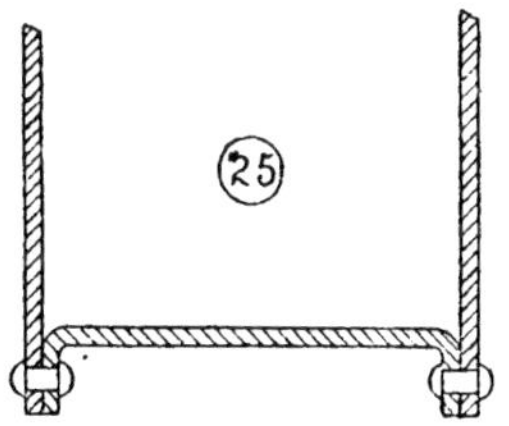

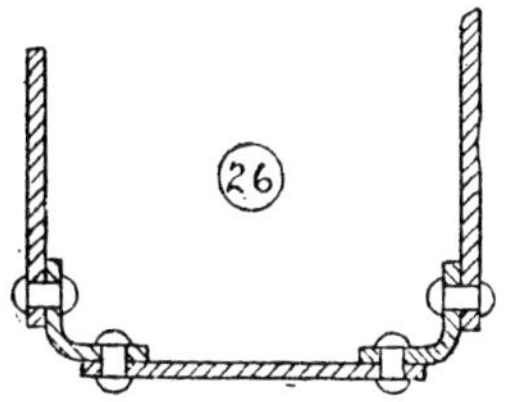

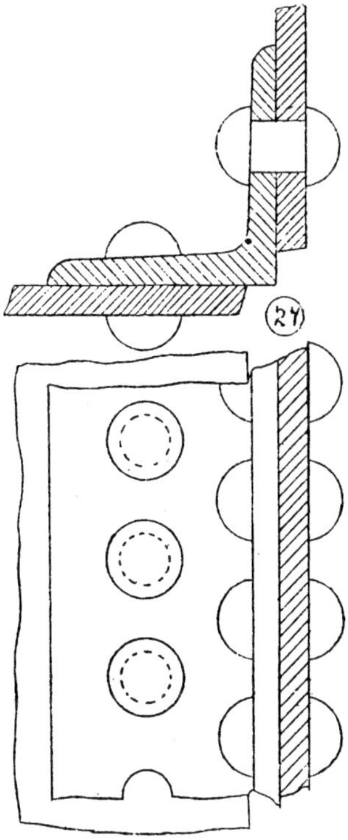

that are employed to attach bottoms and tops on articles by the use of the spinning lathe or other machine. The method of jointing in (33) can also be used with advantage in fixing the bottom and top on to a closed vessel such as a cylindrical hot-water tank or other similar vessel.

To fasten a bottom and foot in an article with one joint, as in the case of a coal-bucket, the plan of joining shown in (17) and (18) is followed. The latter joint being, of course (17), knocked up.

In (19) a sketch is given of what is known as a double-grooved joint. This is an excellent plan for firmly holding

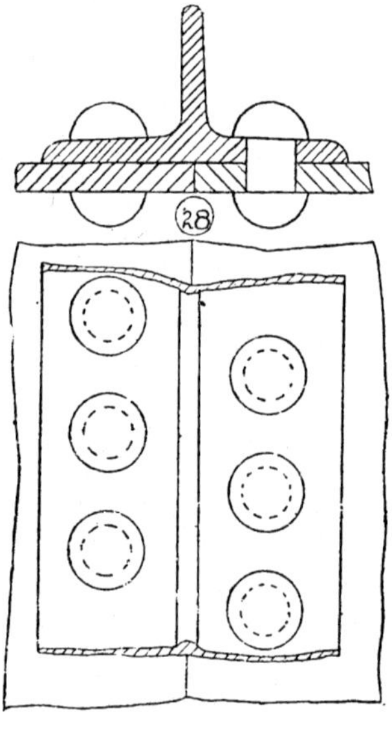

together the edges of round or straight-sided articles made out of heavy metal which is too strong to be grooved in the ordinary way. It will be seen that the strap is a separate strip of metal, which, after being bent, is slipped over the two edges, and then hammered down.

Soldering and Brazing.

Every mechanic who is a worker in any kind of metals should at least be able to make a simple soldered or a brazed joint. To acquire a knowledge of the operations is not at

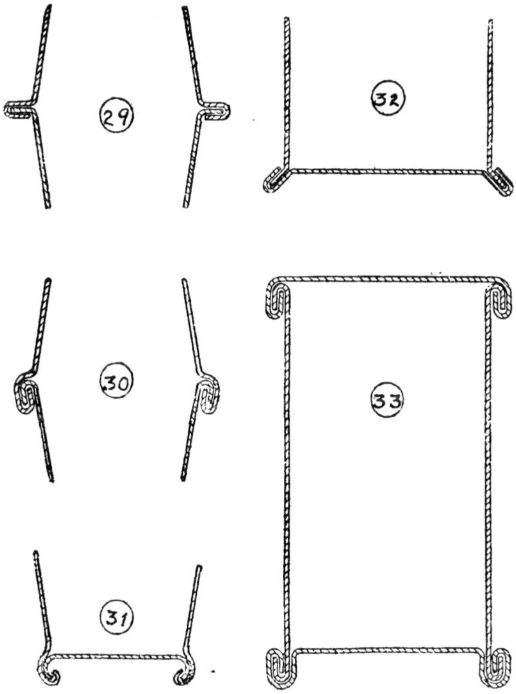

all difficult, a working acquaintance being readily obtained after a few hours' practice.

The operations of soldering and brazing are not analogous to those of gluing, gumming, or cementing, as it is not simply a question of inserting some adhesive substance in between the two surfaces of the joint, and thus sticking the metals together. When two edges or surfaces of metal are soldered or brazed together, the solder or spelter actually alloys with the metal to be soldered for some small distance

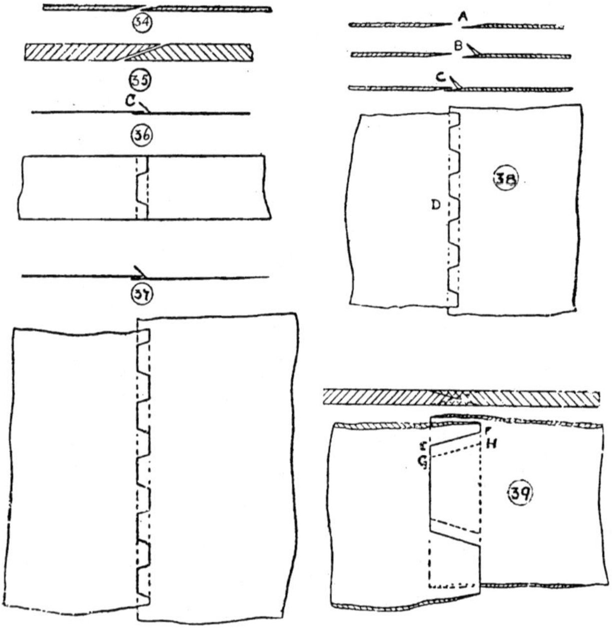

beneath the surface; hence the solder or spelter penetrates into the pores of the metal, and thus obtains a firm grip. If a joint be cut through and the section examined under the microscope, no clear line of demarcation between the solder and the metal can be observed. For instance, if the metal soldered is copper, it will be noticed that the bottom layers are yellow, the solder having combined with the copper and formed a bronze. In a brazed joint the spelter will have alloyed with the copper and thus formed a brass.

Considerations such as above will lead to the conclusion that for a joint to be properly made the temperature of the melting solder or spelter and of the joint to be soldered is of some importance.

Before proceeding to describe the methods of making the above kind of joints, it will be as well to first consider the subject of solders and fluxes.

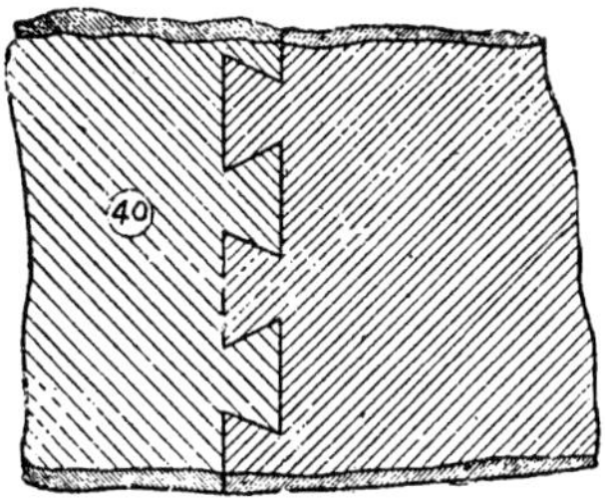

In making or choosing a solder the requirements of a good solder should be kept in mind. They are as follows:—

1. The melting-point must be below the melting-point of the metals to be soldered.
2. The solder must flow readily.
3. The solder must firmly unite with the metals to be soldered.
4. The solder must be strong.

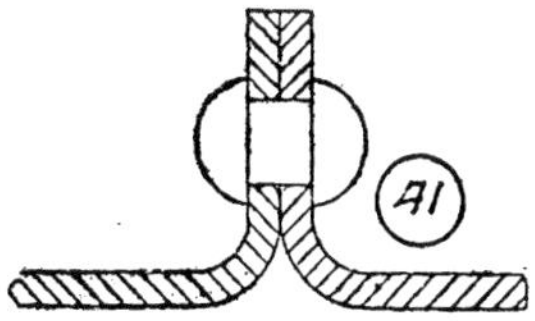

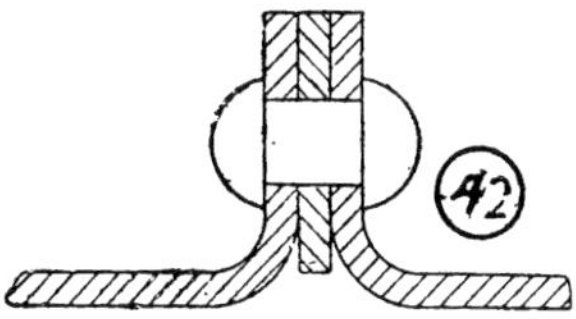

Let us consider the above requisite properties of a good solder or spelter. In the first place, it would manifestly be foolish to attempt to solder a metal with a solder whose melting temperature was higher than that of the metal to be soldered, as before the solder commenced to run the sheet itself would have a hole melted in it. So that, in soldering the softer metals, such as block tin and pewter, care must be taken to choose the proper solder.

For the solder to properly permeate every part of the joint it is, of course, necessary that it should become liquid or thin, so as to flow readily. To obtain this property all foreign substances must be kept out of the solder. Thus, to give an illustration, if a small quantity of zinc gets into a soft solder composed of lead and tin, it makes it become thick or pasty in use.

From what has been said at the commencement it will be readily understood that the solder must be of such a nature as to alloy with the metals to be soldered, or else it will be impossible to make a firm joint.

For iron, copper, or brass work that is to be subjected to pressure, it is essential that the joint shall be as strong as possible. Hence, in making joints for this kind of work a brazing spelter must be chosen that will give the best results.

The following is a table of a few of the soft solders in ordinary use:—

	Lead.	Tin.	Melting-point.
Coarse plumber's solder ...	3	1	480° Fahr.
Ordinary " " ...	2	1	440 "
Blowpipe " ...	1	2	340 "
Fine tinman's " ...	2	3	335 "
Ordinary " " ...	1	1	320 "
Pewterer's solder ...	1 Bismuth	1 2	201 "

It is interesting to notice the change in the melting-points of the solders from that of the metals which form them.

Thus, lead melts at 620° Fahr., and tin at 440° Fahr., yet when these are alloyed together in equal proportions to form ordinary tinman's solder, the melting-point drops to 320° Fahr. This is one of the advantages that is derived from the alloying of metals.

It is generally the best plan to make one's own solder, as much of that which can be bought is unreliable. Besides which, without some guarantee that the solder contains the required proportions of lead and tin, there is no knowing whether or not there is more lead in the solder than has been bargained for. Tin being about ten times the price of lead, a small reduction in the quantity of tin makes a considerable difference in the value of the solder. A rough test of the quantity of tin in a solder is by listening to the characteristic "cry" of the tin when the solder is bent.

In making solders, the lead and tin are melted together, the metals properly mixed, and the scum or oxide skimmed off the surface. And before pouring into the mould, it is a good plan to dust a little resin on the surface of the solder, and let it burn away. In lieu of a cast-iron mould, a bar of small angle-iron can conveniently be used for running the sticks of solder.

It will be noticed that pewterer's solder melts some degrees below the boiling-point of water; but it does not of necessity follow that boiling-water will melt away the solder from the joint on a pewter vessel, as the solder, by virtue of alloying with the pewter, will, in this case, have its melting-point raised.

The following table gives the composition of the ordinary hard solders or spelters:—

	Copper.	Zinc.	Silver.
Iron work...	2	1	0
Copper and thin iron work ...	3	2	0
Brass work	1	1	0
Thin brass work...	8	8	1

The term "spelter" should not be confused with the same name that is applied to ingot zinc, as a hard solder is essentially a brass, whilst, of course, ingot zinc is almost pure zinc, and is principally used in galvanising. It will be seen that the first spelter has the same composition as ordinary brass, and it might be here said that sheet brass is often used, instead of brazing spelter, as it is sometimes found to be more convenient to put along the joints. In bent joints, such as that in a kettle spout, a strip of brass can be cut that will lie along the whole length of the joint.

In practice there is really very little need to trouble about the composition of brazing solders, as they are usually sold in a graded form, numbers 1, 2, 3, etc., the coarse being used for iron and the finest for thin brass work. Silver solders, mostly composed of copper and silver, are used principally in jeweller's work, with which we are not here concerned.

The fluxes used in soft-soldering are "killed spirits," resin, resin and oil, tallow, and for pewter Gallipoli oil. Soldering fluids are sold ready made-up, and these are probably composed of crude chloride of zinc, with some salammoniac in solution. A lump of salammoniac is sometimes used for cleaning the point of the soldering-bit, and powdered salammoniac is used as a flux in various tinning operations.

Borax is almost generally used as the flux for brazing. There are, however, several advertised substitutes; but the principal ingredient in these is probably borax in some form or other.

It may be noticed in passing that the object of using a flux is to assist the solder to flow, and to keep the part of the joint which is being soldered from contact with the atmosphere. The air being kept from contact with the surface of the joint, no oxides can form, consequently the melted solder is free to unite with the heated metal. In many cases, too, the flux has a cleaning action, removing any thin film of oxide that may have formed on the surface of the sheet previous to soldering.

Brazing Joints.

Brazing joints are important, as they present to us the somewhat peculiar instance in which it is possible to make a joint as strong as the solid plate. In ordinary riveted joints it is never possible in workshop practice to make a joint as strong as the rest of the plate, the strength of the joints varying from about 55 per cent. in single-riveted joints up to about 80 per cent. in treble-riveted. In a properly made brazed joint, however, either in iron, copper, or brass, the joint will be found to be as strong as the rest of the plate. The present writer has made many experiments on the strengths of brazed joints, and has invariably found that when properly made the joint is as strong as the sheet or plate.

The sketches numbered (34) and (35) show sections of the ordinary wedge or scarf joints, which are used in thick or heavy work, such as steam-pipes. The edges of the plates are thinned down to form the scarf as shown. There is some difference of opinion as to the proper length of the scarf; but the writer has found in all his tests that if the length of the scarf be made equal to three times the thickness of the plate, it gives a joint which is stronger than the rest of the plate. Perhaps in the thinner metals that are brazed together with this form of joint, it will be found convenient to make the scarf a little longer than above. To ensure the spelter properly running into the joint, the surfaces of the metal should be carefully cleaned, and borax water allowed to run through between the metals before attempting to put the work on the fire. One can generally assure that if the metals are clean and carefully fluxed, the spelter will follow the flux. Another point to remember is that the joint must not be too tightly clamped, or else the spelter will not be able to work its way into the joint. When the job is being brazed, if the spelter is not running properly through the joint, a good plan is to gently tap the plate, which will set up a slight vibration at the joint, and thus assist the spelter to percolate through the joint. At the same time, the melting spelter should be kept dusted with borax powder.

After brazing, the surplus spelter is removed, spent borax and oxide scaled off, and the joint hammered. Excessive hammering should be avoided, as the metal at the joint becomes hard and brittle, and at the best the joint is never as ductile as the rest of the plate. Where there is danger of the joint cracking under pressure it should always be annealed, so that the metal will be soft, and thus stretch somewhat before coming to the breaking point.

This kind of joint is also used in jointing or repairing band-saws. The saw is usually thinned down over a length of two teeth by filing or grinding. To hold the saw in position the writer has generally found it convenient to make a couple of plates, as shown in Fig. 328, fixed the band in between, and bolted together. To braze, the borax and spelter are put in between the joint, and the joint gripped with a heavy pair of blacksmith's tongs, previously made red-hot.

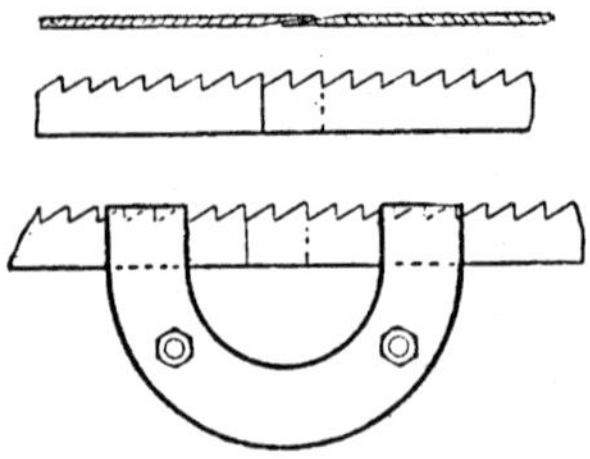

FIG. 328

Sketch number (36) explains the way in which the two edges of a band or hoop can be brazed together by first cramping the edge of one end.

Number (37) illustrates the general method in use for making a brazed joint in thin sheet iron. The cramps are first cut as shown on the right-hand piece, and every alternate cramp lifted, as seen in the section. The left-hand plate is then slipped in and the cramps hammered down. In ordinary sheet-iron work there is no need to clean the edges, as the fused borax sufficiently removes the scale on surface to allow the spelter to come into contact with the iron. After brazing, the joint is usually hammered to remove any inequalities of surface and to chip away the remaining borax and oxide of iron.

The method followed in making a brazing joint in copper or brass is shown in number (38). The edges are first thinned by hammering, as in A, and then properly cleaned. Cramps are nicked and lifted, as at B, and the edges brought together, as at C, and then hammered down, as at D. In copper work it is usual to cut the cramps by holding a strong knife on the slant, and drive it into the metal with a hammer. This plan of cutting ensures that when the cramps are hammered down there will be no thick edges at the side of the cramps, and that the joint can be made perfectly flush and the same thickness as the rest of the plate. An enlarged view and section of one cramp is shown in number (39). The line E F and the dotted line G H show how the plate is cut on the slant, so that the cramps may be thin on the sides as well as at the ends.

Joint number (40) shows a dovetailed method of jointing that is occasionally used in brazing together two plates of iron or steel where the surface is required perfectly flat. It is a difficult joint to make on account of the accurate work required in fitting, and when done, is not any better job than the ordinary scarfed joint.

In brazing a side seam on a circular vessel, the operator should be careful not to let the spelter run away from the joint. To avoid this, the brazing-mixture should be placed just along the joint. A plan often adopted is to bend the body of the article into some such shape as Fig. 329, and fasten with a pair of dogs or clips. The sharp curve about the joint will ensure the spelter running only along the joint. After brazing, the article can, of course, be readily shaped to its proper form.

The joints in circular work are often held together by passing binding-wire around the article and twisting up tightly. To keep the joint from springing open in sheet-iron work, a good method is to bend or roll the sheet to a much smaller radius than required; pull out and let the joint spring together. And again, if the cramps are carefully knocked down, beginning at the points first, these should materially help to keep the joint from opening on the fire, and thus do away with the necessity of binding with wire.

The three sketches of Fig. 330 show the way in which a bottom can be fixed and brazed

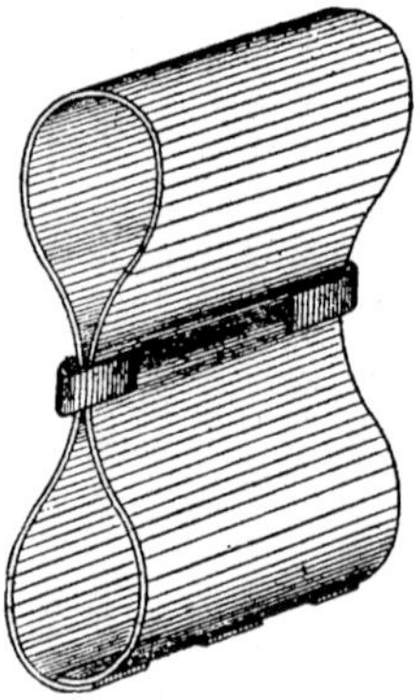

FIG 329.

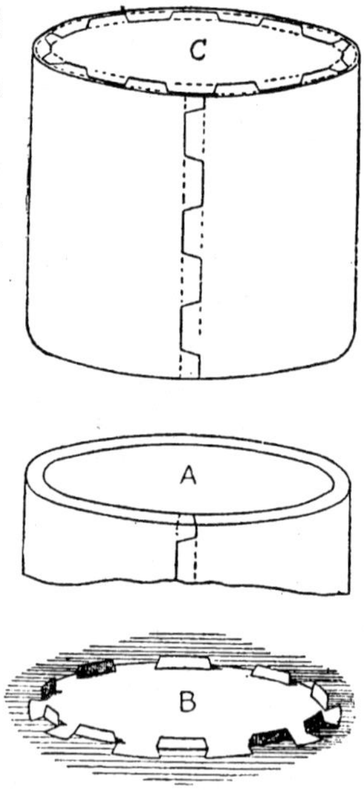

FIG. 330.

into a round article made of sheet iron. The bottom edge of body is turned or flanged in with a mallet, as shown in A. A circular bottom is cut out a shade less in diameter than the inside of vessel, and the cramps snipped and turned up as in sketch B. The bottom is now slipped into the body, and the cramps hammered down over the edge, as shown in C. In brazing, the article should be tilted on the fire so as to ensure the spelter being concentrated on the joint.

In all brazed joints it should be observed if the spelter has run through the joint and fastened the cramps on the outside, as this is the test of a good solid braze.

Brazed Outlet or Tee-Pipe.

Figs. 331 and 332 show outside and sectional views of the way in which an outlet may be brazed on to a pipe. The

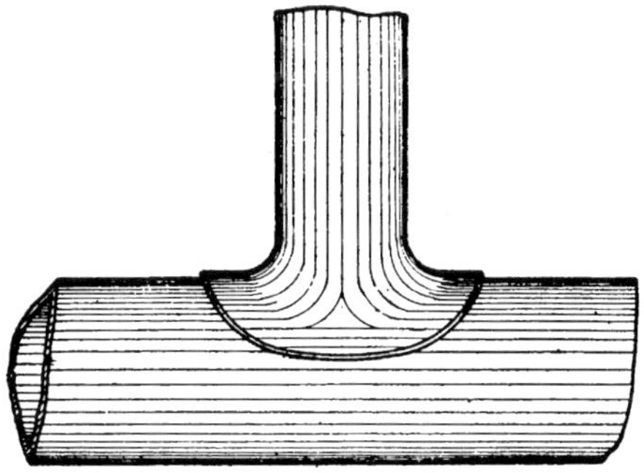

FIG. 331.

outlet is flanged to fit on the main pipe as shown, whilst the hole in the latter is made small to begin with and gradually worked out to the required size. After being thoroughly cleaned the two are wired together and carefully brazed.

Pipe Flange Brazing.

In brazing on pipe flanges (see sketch of four-way piece in Chapter XXIX.) great care must be taken that both they and the pipe ends are properly cleaned. The flange should be slipped a little over the edge of the pipe, and the latter turned over on to the flanges to prevent the spelter from running through. Fireclay should be rubbed around the collar, and if the pipe is a brazed one, along the seam to

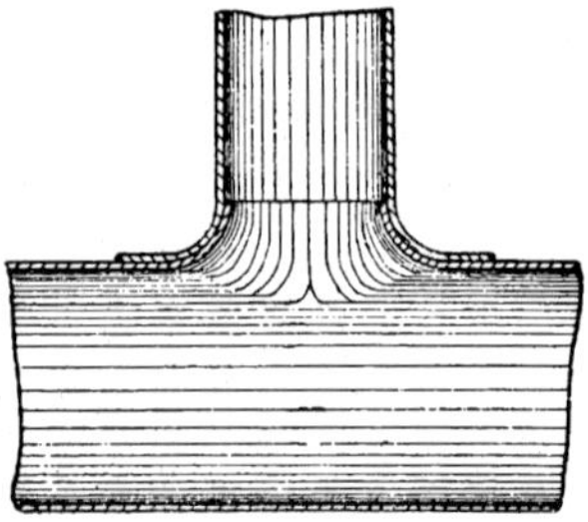

FIG. 332.

protect from fire. To keep the heat on the joint a sheet-iron stopper should be placed in the pipe just above the flange, with fireclay rubbed around its edges. In brazing, great care must be taken to ensure the spelter running through the joint. While hot the brazed flange should not be treated too roughly, as the brass is very brittle in that state. When cold, the face of the flange should be cleaned up, and all superfluous spelter removed.

CHAPTER XXXVI.

RIVETED JOINTS.

THE making of a good sound riveted joint is one of the most important operations in plate metal work; hence in this chapter we intend to consider a few of the main points that should be taken into account in the designing of a properly constructed joint. To design a riveted joint to give the best possible results with any given material for some particular purpose is not by any means a simple matter, and in the more complicated cases is somewhat outside the scope of a plater or boiler maker's work. We shall therefore deal only with the common forms of joints.

In the first place, it should be remembered that in ordinary practice it is never possible to make a riveted joint equal in strength to the solid plate, the relative strength of joint to plate varying from 50 to 90 per cent., according as to whether it is single, double, or treble-riveted, lapped or butted, punched or drilled, or iron or steel plates and rivets.

To increase the strength of the joint, it has been proposed to thicken up that part of the plate which forms the joint. Whilst theoretically there is no doubt but what this plan would give a joint equal in strength to the rest of the plate, practically it would not act on account of the cost and difficulty of rolling plates with thickened edges, and the awkwardness in their manipulation. In some cases welding is resorted to; but even in this, the uncertainty of the joint being welded right through makes it doubtful if a welded joint is, on the whole, any stronger than a riveted joint. For furnace plates there is not so much harm, as the joint here is in compression, whilst in the shell-plates the joints are, of course, in tension.

Diameter of Rivets.

With any given thickness of plate, the first thing to determine is the diameter of the rivet which is most suitable for the joint. And, in considering this, we shall see that there are several practical considerations which assist us in arriving at the best size. If the holes are to be punched in the plate we shall find that it is practically impossible to punch holes of less diameter than the thickness of the plate. Even with holes equal in diameter to the thickness of plate, it will be necessary to have a large clearance between the punch and die or else the punch will break, and this arrangement again will give a very much tapered hole. Therefore, in practice, it is not a good plan to punch holes of less diameter than $1\frac{1}{4}$ times the thickness of the plate.

On the other hand, to form a good joint the rivet must not be of too great a diameter, as this will entail a broad lap; and the rivets being set at too great a pitch, consequently it will be difficult to get a perfectly tight joint; and even if properly caulked, changes of temperature would soon cause the joint to open and leak. The difficulty of small holes is, of course, overcome when the holes are drilled; but here again we are met with a practical difficulty, and that is if too small a rivet is chosen a much larger cost is incurred in drilling and riveting on account of the greater number of holes required in the joint. Taking all things into consideration, the common rule of making the *diameter of rivet equal to one and a quarter times the square root of the thickness of plate* is a good guide to assist us at arriving at a suitable diameter. The rule is conveniently written in this form—

$$d = 1{\cdot}25\ \sqrt{t}$$

A very simple way of obtaining the required diameter (for those who cannot readily calculate) is shown in Fig. 333. A line is drawn, and along it a distance of 1·44 in. marked, and then on to this the thickness of plate is added. A semicircle is now described on the whole line, and a perpendicular run up as shown. The length of this line will give the required diameter of rivet. In Fig. 333 the construction is for $\frac{3}{4}$ in. plate, and it will be seen that the rivet diameter comes out nearly $1\frac{1}{16}$ in. The nearest ordinary size to the calculated or measured dimension will have to be chosen.

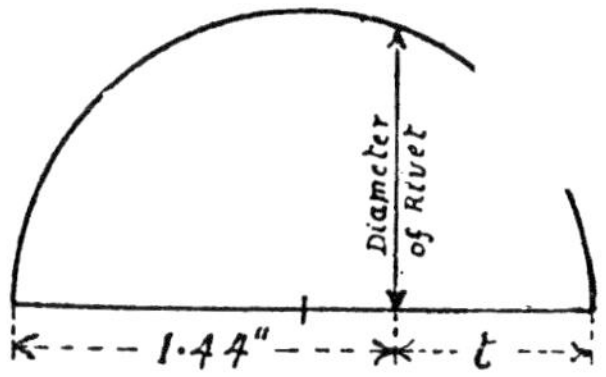

FIG. 333.

Thus for a $\frac{1}{2}$ in. plate—

$$d = 1{\cdot}2 \times \sqrt{\cdot 5} = \cdot 84 \text{ in.}$$

and the nearest stock size to this would be $\frac{7}{8}$ in.

Generally, the diameter of rivets to suit particular thicknesses of plates will be :—

Plate thickness ...	$\frac{1}{4}$	$\frac{3}{8}$	$\frac{1}{2}$	$\frac{5}{8}$	$\frac{3}{4}$	$\frac{7}{8}$	1
Diameter of rivet ...	$\frac{9}{16}$	$\frac{3}{4}$	$\frac{7}{8}$	$\frac{15}{16}$	$1\frac{1}{16}$	$1\frac{1}{8}$	$1\frac{1}{4}$

Rivets for general work, such as girders, roofing, and ship plating, are usually a little less in diameter than in the above table.

Pitch of Rivets.

The distance from centre to centre of the rivets can be calculated from the principle that the part of plate in between each pair of holes should be the same strength as one rivet. It may be put in the form of a rule as follows: "The area of plate between a pair of holes multiplied by the tensile strength of the plate material, is equal to the cross sectional area of rivet multiplied by the shearing strength of the rivet material," or—

$$(p - d)t \times T = d^2 \times \cdot 7854 \times S.$$

Where—
p = pitch of rivets.
d = diameter of hole.
t = thickness of plate.
T = tensile strength of plate.
S = shearing strength of rivet.

For iron T may be taken as 22 tons
" " S " " " " 19 "
" steel T " " " " 28 "
" " S " " " " 23 "

It should be noticed that d represents the diameter of the hole in plate, and this will for punched holes be about $\frac{1}{32}$ in. larger than the diameter of rivet for, say, $\frac{3}{8}$ in. rivets, varying up to $\frac{1}{16}$ in. for 1 in. rivets.

For the sake of clearness it will, perhaps, be as well to work out an example in the use of the above formula. Suppose we require to find the pitch of rivets for a single-riveted lap-joint, steel plates and rivets. If the plates are $\frac{1}{2}$ in. thick the diameter of rivet should be $\frac{7}{8}$ in. Adding $\frac{1}{40}$ in. on to this to allow for clearance, it gives a finished rivet diameter of ·9 in. So that we have—

$$(p - \cdot 9) \times \cdot 5 \times 28 = \cdot 9^2 \times \cdot 7854 \times 23,$$

from which we obtain—

$$p = 2 \text{ in.}$$

The above calculation is based upon the assumption that the holes have been drilled, and in cases where the plates are drilled in position, it will be an advantage to take the clearance as slightly less than that allowed.

For punched work it is important to remember that the operation of punching damages the plate for some small distance all around the walls of the holes. Investigation seems to show that the plate is fractured for a distance of about $\frac{1}{10}$ in. from the edge of hole. So that, in using the above rules for punched plates, $\frac{1}{5}$ in. must be deducted from the space between the holes before proceeding to use the equation to obtain the pitch. It thus becomes—

$$(p - d - \tfrac{1}{5})t \times T = d^2 \times \cdot 7854 \times S.$$

Suppose we want to find the pitch of rivets for a single-riveted lap-joint formed of iron plates and rivets. Plates $\frac{5}{8}$ in. thick, punched holes, and rivets 1 in. diameter. Adding $\frac{1}{16}$ in. on to rivet diameter for clearance, we have—

$$(p - 1\tfrac{1}{16} - \tfrac{1}{5}) \times \tfrac{5}{8} \times 22 = (1\tfrac{1}{16})^2 \times \cdot 7854 \times 19$$

or—$(p - 1{\cdot}06 - \cdot 2) \times \cdot 625 \times 22 = (1{\cdot}06)^2 \times \cdot 7854 \times 19$

from which— $p = 2{\cdot}5$ in.

For boiler work little attention need be given to the construction of joints with punched holes, as all good work is now drilled in position, one or two small tacking holes only being first put in the plates in the flat, the remainder being drilled after the plates are rolled and bolted together. Indeed, with a spacing arrangement attached to a drilling machine, there is no need to mark off the holes with the exception of those needed for tacking. It might be here remarked that after drilling, the plates are separated, and the burr or aris cleaned off, so that the plate surfaces may come into dead contact in riveting.

There is no need to calculate the pitch for every thickness

of plate, as the space in between a pair of holes is the same in each case. Thus, for a single-riveted lap-joint formed of iron plates and rivets, with punched holes, the

$$\text{pitch} = 1\tfrac{1}{2}\text{ in.} + \text{diameter of rivet},$$

and a similar joint of steel will have a

$$\text{pitch} = 1\tfrac{1}{8}\text{ in.} + \text{diameter of rivet}.$$

The space between the holes in a double and treble-riveted lap-joint will work out to about twice and three times, respectively, that of a single joint as above.

In general work, which has not to be subjected to much pressure, the pitch of rivets is usually taken greater than that shown in the above calculations.

Width of Lap.

The distance of the centre of rivet from the edge of plate is generally taken to equal one and a half times the diameter of the rivet, so that a single-riveted lap-joint would have an overlap of three times the diameter of rivet, and a double lap five times the diameter, and so on.

Caulking.

Where caulking is to be done for work which is to be subjected to pressure, it is important that the lap should not be greater than that named above, as the plates may spring in caulking or in use.

To caulk properly, the plate edge should be planed slightly on the bevel, as shown in joint No. 27. A too thin caulking tool should not be used, as this has a tendency to drive the metal under the edge, and thus spring the plate.

In arranging the position of joints on any kind of vessel, care should be taken that they are so placed that the riveting and caulking can be conveniently done.

Shapes of Rivets.

The heads and tails of rivets are of various forms, several of which are shown in Fig. 334. (*a*) shows a cup or snap-head and tail, the dotted lines also showing a nobbled head; (*b*) and (*c*) have pan or cheese tails, and (*d*) has a combined countersunk and pan tail, with a nobbled head; (*b*) and (*e*) show countersunk heads.

The required length of rivet to form any given shape of head can be calculated; but in practice, on account of the clearance varying and also the cup of the snaps not always being of the same size, it is the best plan to obtain the correct length by trial.

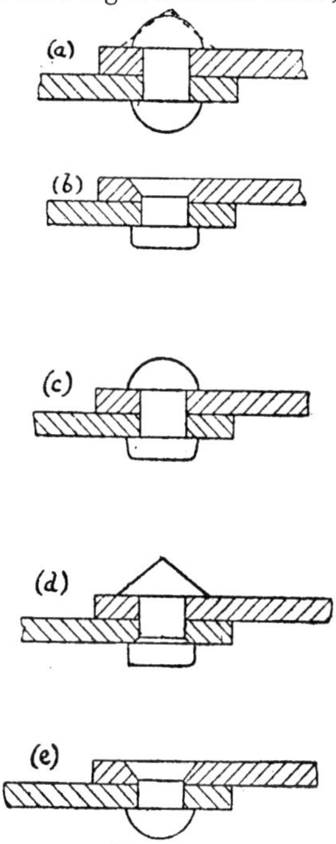

FIG. 334.

Riveting.

The bulk of riveting is now done either by hydraulic or pneumatic power; but where the work is done by hand, it should be observed that the rivet is made red-hot the whole length, so that in being hammered down it may swell and completely fill the hole. The difficulty of making rivets entirely fill the hole is one of the disadvantages that hand-riveting has as compared with machine-riveting.

Forms of Joints.

There are a multitude of different forms and combinations of riveted joints. A few kinds only, however, will be shown—just sufficient to explain the arrangement of the plates. No. (20) shows the ordinary single-riveted lap, and (21) the double-riveted lap with zigzag riveting; (22) explains how three plates can be joined by smithing or thinning the edge of the middle one, and (23) shows a similar arrangement for joining four plates by thinning the corners of the two middle plates. Sections across both joints in each of the last two figures are shown, and by reference to these the formation of the joints should readily be understood.

The method of fastening flanged ends in cylindrical and other shaped articles is shown in (24) and (25). And (26) shows how corner or bilge plates may be fixed in, these again being sometimes riveted on the outside instead of the inside, as shown. (27) is a method adopted when two plates need fixing square to each other, the plates being joined by an angle-iron. A form of butt-joint is exhibited in (28), the strap or stiffener being of tee-iron. Sometimes the rivets, instead of being zigzag, are placed opposite each other to form what is known as chain-riveting. Joint (42) explains the method adopted for joining the ends of tubes, a stiffening ring of flat-iron being placed in between the flanges, and (41) shows the same without the ring.

Strength of Joints.

The relative strength of joint to solid plate expressed in the form of a percentage will be equal to—

$$\frac{\text{pitch} - \text{diameter of hole}}{\text{pitch}} \times 100.$$

And using the example for $\frac{1}{2}$ in. steel plates, already calculated, the strength of the joint will be—

$$\frac{2 - \cdot 9}{2} \times 100 = 55 \text{ per cent.}$$

The strengths of all kinds of joints can be multiplied out in a similar manner. Generally, drilled joints, on calculation, show up about 5 per cent. stronger than punched plates; but, practically, this percentage does not represent the difference in value between the joints, as with drilling there is no need for drifting holes, and consequently no local stresses are set up in the plates.

Butt-joints, with double straps, are the strongest form of joint, the strength of a treble-riveted joint of this description being about 90 per cent. of the solid plate. For longitudinal seams in a boiler, this class of joint also has the advantage of the plates pulling directly on the rivets and not obliquely as with a lap-joint. On account of the uncertainty of the stresses that are set up in a lap-joint, it is questionable whether it ought ever to be used in the longitudinal seams of a boiler.

Bursting Strength of Cylindrical Shell or Pipe.

The bursting pressure of a solid shell or pipe can be determined from the following rule: "Multiply together the thickness of the metal and its strength in lbs., and divide by the shell radius in inches."

Thus, suppose a welded cylindrical boiler shell is 7 ft. diameter and $\frac{3}{8}$ in. thick steel plate. Assuming that the strength of the metal is 28 tons per square inch, we have—

$$\text{Bursting pressure} = \frac{\frac{3}{8} \times 28 \times 2240}{42} = 560\,\text{lb.}$$

If the shell is riveted the above would have to be multiplied by the percentage strength of the joint to obtain its correct bursting pressure.

The strengths of steel or copper steam pipes can be found in the same way.

The above calculations will also serve to get out the required thickness of metal for a shell of given diameter to stand a given pressure.

As a somewhat curious fact it is worth noting that a spherical vessel of same thickness and material will stand just twice the pressure of a cylindrical vessel of the same diameter.

Length of Angle, Tee-Bars, etc., for Rings.

The method of finding the lengths of flat-bars, etc., explained in Chapter XXXII., can also be applied to bars of irregular section. The important point is to find the position of the neutral axis. This will always pass through the centre of gravity of the section. The centres of gravity (see Chapter XXVIII.) can be found either geometrically,

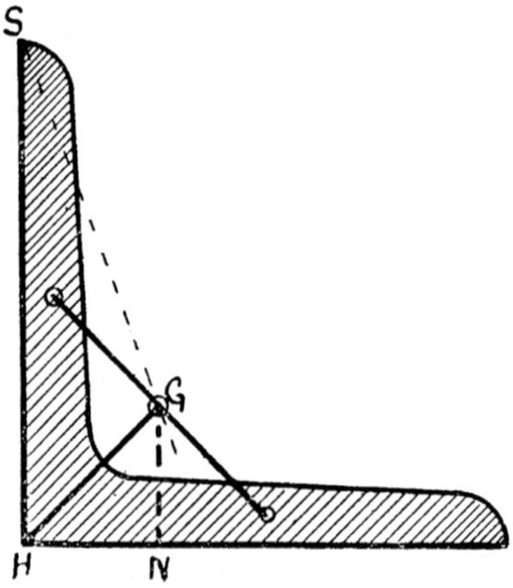

FIG. 335.

or better still for practical purposes, by the method of suspension. A section of the bar should be cut out of cardboard or sheet metal and suspended from a point (such as S in Fig. 335) and a vertical line drawn down. It should then be hung from another point (such as H) and another vertical line drawn. Where these lines intersect will give the centre of gravity of the section. In Fig. 335 this is marked by the letter G, and when the bar is being bent, as shown in Fig. 336, either to form a ring with an outside or inside flange the neutral line will pass through the point N. To get the required length of the bar in the straight, the ring will be set out and the neutral circle drawn, and its

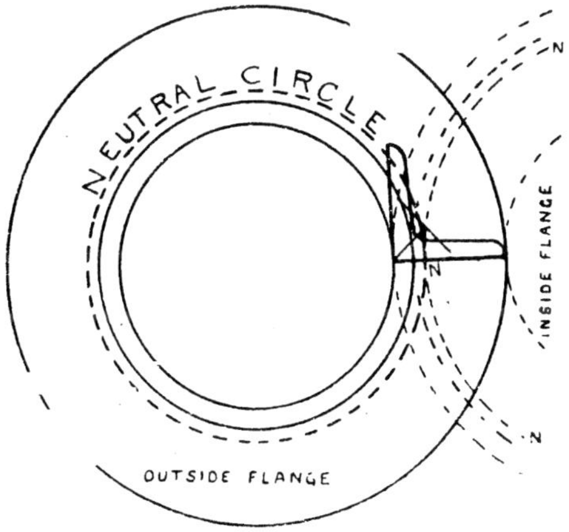

FIG. 336.

length measured or calculated. For an inside flange part of the neutral circle is shown as N N N.

For the above to be true it should be remembered that uniformity of heating and bending is demanded.

In punching angles it might be here mentioned that it is usual for the centre line of the holes to run down the middle of the inside of angle-iron.

To Calculate the Increased Length of a Bar when it is made Red-hot.

Problems on the expansion and contraction of metal bars are important, hence we give one example below.

Suppose a bar of iron is 10 ft. long and its temperature (that of the atmosphere), say, 30° C. It is placed in a furnace and got red-hot. How much will it lengthen?

A red heat is generally reckoned to be about 1,000° C. So that the increase in temperature would be 970°.

Now turning to the table of multipliers for linear expansion on page 451 we get that for iron, and our calculation works out as follows:—

$$\text{Increase in length} = 10 \times 12 \times 970 \times \cdot 000013 = 1 \cdot 5 \text{ in.}$$

So that the increase in length comes to about $1\frac{1}{2}$ in.

Calculations like above come in useful in making allowance when rings, bands, etc., have to be shrunk on.

Planishing or Flattening.

To the uninitiated the levelling of plates or sheets presents one of the most awkward jobs it is possible to have. Yet with the exercise of thought and some practice the difficulties soon disappear. Before commencing to hammer a plate the position of the buckle or looseness should be carefully noted, and the blows placed accordingly. There are only two ways in which a plate may be buckled. It may either be sagged in the middle, as shown on plate "A" in Fig. 337, or it may be tight in the centre and slack along the edges, as shown by plate "B." On one plate there may be a combination of these two ways of buckling; one half may be slack in the middle and the other half slack on the edges.

The cause of buckling is due to unequal contraction of the sheet or plate in cooling, so that one part becomes longer or shorter than the other. To bring the plate level, all strain must be removed, so that no one part of the surface shall be pulling against another. To do this all the short or tightened parts of the plate will require hammering. Thus in plate "A," where the buckle is in the centre, the hammer blows will need to be thickest at the outside of the plate, running away to nothing at the middle. In plate "B," where the edges are buckled or loose, the process will have to be reversed; the blows being placed about the middle and gradually dying away towards the edges.

The strength of the blows will, of course, depend upon the

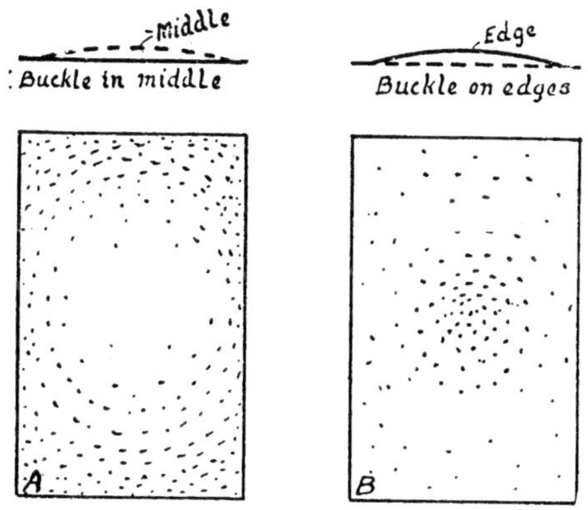

FIG. 337.

thickness of the plate. The inexperienced should always err on the side of light blows, as one heavy blow inadvertently given may require a hundred more to remove its ill-effects.

When particular work is required a flatter should be used to avoid the blows coming directly on to the surface of the plate. When the hammer alone is used great care must be taken so that its face edge shall not cut into the plate.

To obviate this, hammer faces for planishing purposes usually have a slightly outward curve.

Straightening rolls and other machines are now used for flattening plates and sheets, but where these are not available it is a good plan to run the sheet through the ordinary bending rolls a few times, reversing the sheet at each operation, as this tends to run the small buckles altogether and generally assists in determining how the sheet shall be hammered.

CHAPTER XXXVII.

SURFACE TREATMENT OF SHEET METALS.

ALL metals more or less oxidise or corrode when exposed to a damp atmosphere or corroding fumes. And if the oxide so formed is dissolvable in water or other liquid, or readily detaches itself from the metal, as in the case of iron, then rapid deterioration takes place. Although iron has many distinct advantages over other metals in the way of strength, working properties, and cheapness, yet it is the worst of the ordinary metals in offering resistance to the action of air and moisture when exposed to the atmosphere without some protective coating. Copper, lead, and zinc are all quickly acted upon by damp air, or if the atmosphere contains sulphur, carbonic acid, or other fumes, the metals very soon tarnish. The thin film of oxide or scale so formed, however, in the case of these metals holds tenaciously to the metal, and, consequently, acts in a very effective manner as a protecting skin for the metal underneath. Aluminium, it has been said, is not affected by the atmosphere, but this is not true. Probably what happens is the instant formation of a *transparent* film of oxide. Again, sheet aluminium will not stand continued exposure in a damp atmosphere, as a heavy oxide forms on its surface, and if the sheet be thin the metal becomes very brittle.

To protect the surfaces of metals from corroding influences, many methods are in vogue, such as galvanising, tinning, electro-plating, dipping, lacquering, enamelling, japanning, painting, oxidising, and, for special purposes, metals may be coloured by bronzing, steeling, gilding, etc.

Galvanising.

As galvanising is the commonest process adopted for applying a protective coating to sheet-iron work we shall explain the method followed with some fulness. Essentially the process consists in applying a thin film of zinc to the surface of the iron. We will first explain the plan followed for sheets and work on a large scale, and then give some hints of how best to deal with small articles. Before sheets can be galvanised all scale must be removed from their surface, and this is usually done in a pickling solution composed of equal parts of hydrochloric acid (or muriatic acid, as it is often called) and water. Lead-lined tanks are sometimes used for holding the acid; but the better plan is to have stone tanks, jointed with rubber packing, and held together with tie-rods. During the time the sheets are in the pickle they should be moved continuously, so that all parts of the surfaces may be equally exposed to the action of the acid. The length of time for pickling will depend upon the temperature and strength of the acid and on the condition of the sheet surfaces. If the acid is fresh and the sheets have been close-annealed (that is, out of contact with the furnace gases), then the pickling may be done in about fifteen minutes; but if the acid is partly spent, or the sheets covered with heavy scale (as the result of open-annealing), thirty to forty-five minutes will be required. Heating the acid (done in the early days of galvanising by blowing steam into the tanks) will increase the speed of working, but the character of sheet-surface produced will not be so good as when pickled by the cooler acid. Any increase of temperature over and above that of the atmosphere required for the effective working of the pickle is soon obtained by the heat generated through the chemical action. If the pickling solution is too hot the action upon the sheets is not uniform, and the surfaces will be somewhat rougher. Occasionally a sheet will be found that contains a hard patch of scale or a scab, and this will have to be removed by a scraper or pick before attempting to pass through the galvanising bath. Sometimes a blister (a piece of double sheet which has not been properly welded in the manufacture) is found on a sheet, and great care should be taken to cut this away, as it will act as a receptacle for acid, which, when carried into the molten spelter, may cause a serious explosion. To obtain a good-looking surface after galvanising, the operator should be careful not to over-pickle, as this will cause the sheet to look "dead" and "dry." When properly cleaned the sheets are plunged into a water-tank for washing, and are then ready for the galvanising bath.

The quantity of acid used varies from 1½ to 4 carboys per ton of sheets, depending upon whether the sheets are close or open-annealed, or heavy or light. For economical working, the partly-spent acid from the large tank, when it becomes too slow for sheets, should be used as far as possible for small work, for which the time of pickling is not so important.

Where a large amount of work is done it is usual to test the acid with a Twaddel's hydrometer, the degree of reading, according to the density of the acid, varying from 24° to 30°. Without the acid is fairly pure the reading given on the hydrometer is not an exact indication of the strength of the acid from the galvaniser's point of view. A better and more effective test is to compare the relative amounts of zinc dissolved by equal quantities of acid taken from the different sample carboys. Thus, to give the result of one experiment: A certain quantity of 24° acid dissolves 5 oz. of zinc, whilst the same quantity of 30° acid dissolves 6 oz. of zinc. Their relative values, therefore, to the galvaniser are as 5 is to 6. In this way, by taking cost into account, it can be seen which is the most economical to use.

The amount of waste in pickling runs out to about 4 lb.

per 100 square ft. of open-annealed sheet iron to $2\frac{3}{4}$ lb. per 100 square ft. of close-annealed sheet. This gives, as near as possible, 33 lb. to the ton of 16 gauge and 82 lb. to the ton of 24 gauge of the former, galvanised, and 57 lb. to the ton of the latter.

Before proceeding to explain how sheets are passed through the galvanising bath, it is as well to call attention here to the fact that the quality of the galvanised sheet surface will very largely depend upon the kind of surface that is put upon the black sheet. If the iron is of an inferior quality with a coarse surface, or is over-pickled, no amount of care in galvanising will produce a good surface. This, indeed, is true of all surface treatment, whether tinning, painting, lacquering, or whatever it may be.

A sectional elevation of a galvanising bath, with the rolls in position, is shown in Fig. 338. A layer of flux, about 6 in. thick, of crude salammoniac (or muriate of ammonia, as it is called) is allowed to boil up in the flux-box, a bit of tallow being thrown in occasionally. The sheets are taken one by one and passed into the pot through the flux-box, down through the feeding-rolls, and up out between the surface, or leaving-rolls, and taken away, either by hand or travelling chains, boshed in a tank of warm water, and dried by passing through a drying stove.

To ensure a clean galvanised sheet the surface-rolls must be kept clear of all waste flux and scum, and the flux in the flux-box must not be allowed to get too dirty, or else some of it will be carried through to the leaving-rolls and mark the sheets.

Up to within a few years back all sheets were either drawn through the clear or through sand, the feeding-rolls alone being used, and these, of course, simply to carry the sheet through the molten metal. One object of the surface-rolls is to give a more uniform coating of zinc and impart a little better surface to the sheet. The primary object, however,

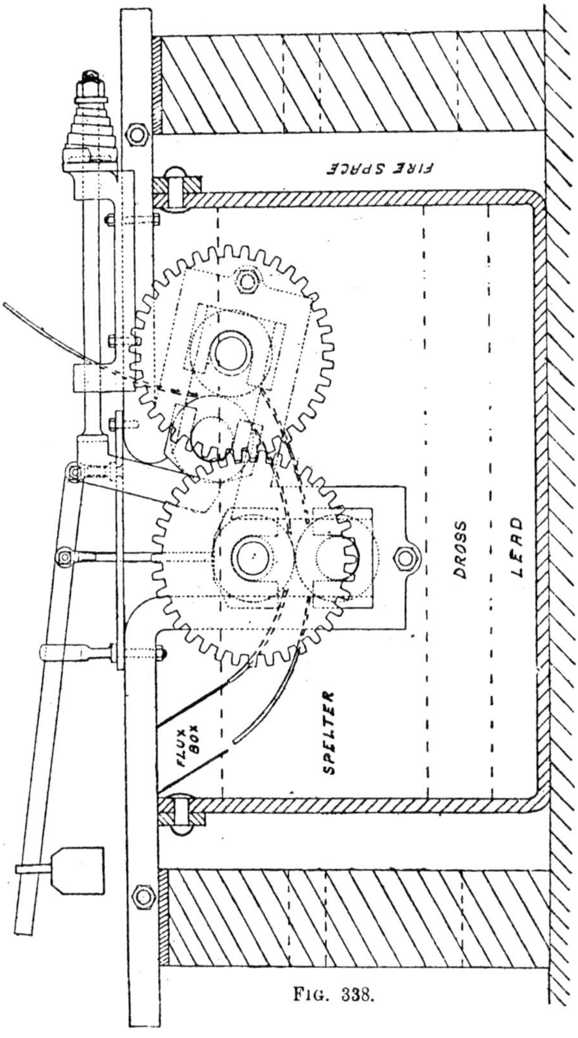

FIG. 338.

in the use of leaving-rolls is to squeeze as much zinc off the sheet as possible, and thus reduce the cost of manufacture. The result is that galvanised sheets of the present day are altogether inferior to what they were under the old system of manufacture. The following table, which has been compiled from experiments carried out by the writer, will give some indication of the altogether thinner coating of zinc which is now put upon sheets to that which was formerly the case:—

Kind of Sheets.	Spelter used per Ton of Galvanised Sheets.	
	Single Rolls.	Double Rolls.
28 gauge	504lb.	310lb.
26 „	420	254
24 „	367	216
22 „	308	168
20 „	287	126
18 „	216	104
16 „	173	86

It will be noticed that, generally, the amount of spelter used in the modern process is only about half that which was placed upon the sheets under the old method. In consequence of this much thinner coating of zinc, the "life" of galvanising sheets is not by any means as long as it was formerly In ordering large quantities of galvanised sheets, not only should the gauge of sheet or weight per square foot be specified, but also the thickness of zinc coating or weight of spelter per square foot to be added, if buyers desire to obtain sheets of lasting quality. By the old method it took from 15 lb. to 20 lb. of spelter to cover both sides of 100 square feet of sheet, whereas by the double-roll system only 9 lb. to 12 lb. of spelter are used in coating the same area.

Usually, when sheets are thicker than 16 gauge (that is, $\frac{1}{16}$in. thick) they are not run through the rolls, but are carefully dried and then plunged bodily into the molten spelter, being drawn up on the opposite side of the pot through a thick layer of sand, or sand and ashes, kept moistened by water being sprinkled upon it.

The purer the iron the better the coat, might be taken as a rule in galvanising, and that is the reason why mild steel never takes as firm a coat as the best iron. Any impurities, such as carbon, silicon, etc., offer resistance to the formation of a tenacious surface alloy of iron and zinc. The harder the steel the more tendency there is for the spelter to peel off when the sheet is bent.

The remarkable liking which zinc has for iron, and which is its chief advantage in forming a good coating on galvanised work, is also at the same time its greatest drawback in the ready formation of dross. As the sheet passes through the pot small particles of iron are detached from its surface, which combine with the zinc, forming an iron-zinc alloy which gradually precipitates at the bottom of the pot, and which has to be periodically removed. Also the molten zinc slowly dissolves away the walls of the wrought iron or mild steel pot, forming additional dross, so that its renewal is required every six to twelve months or so. This is one of the constant worries and expenses of the galvaniser, and will never be altered until a different material for the pot, or another system of carrying out the process, is adopted. In the present writer's experience a pot that lasted longer than any other was constructed of wrought iron, which gave the following analysis:—

Carbon	trace	
Silicon	·093	per cent.
Sulphur	trace	
Phosphorus	·357	"
Manganese	·057	"
Iron	99·493	"
	100·000	

For those who are interested in galvanising it may be useful to know the compositions of one or two samples of zinc spelter. The analyses of four specimens are given :—

	1	2	3	4
Zinc	99·900	98·80	98·526	98 34
Lead	—	1·036	1·002	1·535
Copper	trace	trace	·023	—
Cadmium	—	trace	—	·07
Tin	trace	—	·346	·015
Antimony	—	·087	·039	—
Iron	·084	·077	·014	·03
Arsenic	·010	—	·05	—
Sulphur	·006	—	—	·01
	100·000	100·000	100·000	100·000

Sample No. 1, it will be seen, is an exceptionally pure specimen of virgin spelter. The other three are samples of metal of the kind ordinarily in use. Any iron in spelter is most objectionable, as it all helps to form dross in the pot. Most of the lead present in a spelter falls to the bottom of the pot and there accumulates.

Lead and zinc have very little affinity: hence the former metal usually separates out, and on account of its being heavier than either spelter or dross, settles at the bottom of the galvanising bath, as shown in Fig. 338. When a bath has been in use several months, as much as a 6 in. depth of lead will sometimes have accumulated. In case the pot is shallow, and it is necessary to remove the lead, this can be readily done by standing a tube (one about 12 in. diameter, and a little longer than the depth of molten metal, made of $\frac{1}{8}$ in. plate will do) upright on the bottom of the bath, and lading out the spelter, when the molten lead will be forced up the tube, and can be laded out as required. The dross, on account of its greater density than spelter, and being lighter than lead, sinks through the former and floats on the latter, as seen by the layers in Fig. 338. It is usually removed by a perforated spoon or ladle. Its composition is generally something like the following:—

Zinc	92·554	per cent.
Lead	·050	"
Copper	·103	"
Cadmium	trace	
Tin	1·939	"
Antimony	·072	"
Iron	5·234	"
Arsenic	·012	"
Sulphur	·036	"
	100·000	

It will be observed that the pot is heated on the sides only; any bottom heating having a tendency to make the dross rise and mix with the spelter, thus causing the surface of articles which are being galvanised to become rough. Great care should be taken not to let the spelter become red-hot, as undue heating spoils the surface of the object to be galvanised (giving it a "dry" appearance), and at the same time accelerates the formation of dross.

As much as possible of the top of the molten metal should be covered with sand, or sand and fine ashes, to prevent the atmosphere from coming into contact with the spelter, and thus forming zinc-oxide. Also it will reduce the loss of heat by radiation if the top of the unused part of the bath is covered with a plate of iron or other suitable material. Small articles can be readily galvanised by fixing up a small iron pot on an ordinary fire, or by gas-heating. In this case, when the spelter has been in use some time, it will be found to have become mixed with dross. To remove this, the molten metal should be allowed to settle, being kept in a liquid state by the application of a gentle side heat, the dross being then laded out.

The amount of salammoniac used per ton of work done will usually be about 9 lb. to 12 lb.

One disadvantage in the galvanising of objects having

riveted or lap joints, is that these parts hold traces of acid or flux, which soon sets up corrosion, and causes the parts to decay. It is difficult to avoid this, except by galvanising both before and after manufacture. It is an advantage to wash the joints down with weak soda-water, thoroughly dry, and then force into the joints a little boiled oil. Stains or black spots on sheets are caused in the same way as above. When the surface is brocky or porous, acid is absorbed which shows up a day or so after galvanising, by forcing out black spots of chloride of zinc.

In galvanising odd work, if there are any parts like screw-threads, etc., that do not require coating with zinc, these should be covered with white lead before the article is put into the galvanising bath.

For cooking purposes zinc-coated articles are useless, on account of the readiness with which zinc is dissolved by various organic acids; but for temporary uses, such as with buckets, baths, etc., the zinc-coating is all right.

Tinning.

The process of tinning sheets follows very much the same lines as galvanising, the molten metal in this case being tin, and the flux generally a solution of chloride of zinc. The plates are run through several pairs of rolls, and ultimately up out through a "grease-pot" filled with palm oil.

Tinplate is of no use for outside purposes on account of the readiness with which it rusts. This is probably due to the fact that iron and tin have very little affinity to each other, causing the surface of the iron to be imperfectly coated. The miscroscopic points on the sheet left uncoated quickly rust when placed in a damp atmosphere, this being, no doubt, assisted by some electric action.

Terneplate is sheet iron or steel that has been coated with an alloy of tin and lead, the major portion of the alloy being composed of the latter metal.

Tin and copper have a very much greater liking for each other than tin and iron; therefore copper can be more readily and firmly tinned than iron. Its surface should be well cleaned, and then sprinkled with salammoniac, small pieces of tin placed on it, heated, and run over the surface, and finally wiped off with a wisp of tow. Any greasy parts on black iron or other metal that has to be tinned or galvanised, should first be either burned off or removed by a solution of soda. In tinning copper, if there are any parts that require to be left untinned, these should be first brushed over with whitening paste.

Tinned copper vessels make excellent cooking utensils where a quick heat is required, on account of the good conducting power for heat of copper, and also the cleanliness of a properly tinned copper vessel. Tin is not easily dissolved by vegetable or meat juices; but as copper quickly forms a poisonous verdigris, care should be taken to see that saucepans, etc., are kept properly tinned.

Lacquering, Colouring, etc.

To obtain various artistic effects, metals are sometimes coloured by dipping into different chemical solutions or by the combined action of air and heat to form tinted oxides, or by the application of coloured lacquers. Lacquers are practically varnishes, and when properly applied, preserve the surface of the metal from being acted upon by an inside atmosphere for a considerable length of time. In japanning or enamelling, the work is stoved in a suitable oven after the application of the enamel, to give it the necessary hardness. Lacquering and other solutions for every possible purpose can now be so cheaply obtained that it is not worth while attempting to make them up. White or other enamelled utensils, for culinary use, have their surfaces treated with a vitreous matter, afterwards being baked in an oven or furnace.

Protecting Plate Iron Work.

For plate work, other than boilers, there is no more effectual initial coating than boiled oil. To be lasting, the plates should first be cleaned of all scale that is likely to become detached. When bars have to be riveted to plates, and where the atmosphere is likely to get between, the inner surfaces of both plates and bars should first be oiled over. One of the most fruitful causes of rapid deterioration in plate and constructional iron work, and which may have serious results, is the oxidation of parts of the structure that when put together are not get-at-able to be scaled or painted. This should be guarded against as far as possible in the manner suggested above.

CHAPTER XXXVIII.

METALS AND THEIR PROPERTIES.

THERE are altogether in Nature between 50 and 60 different metals, but on account of the unfitness of many through difficulty of extraction from their ores, rarity, or rapid oxidation when exposed to the atmosphere, the numbers that can be used for general manufacturing purposes are only about a dozen.

The qualities possessed by metals which enable them to be used for so many purposes are:—

Metallic Lustre, or the property of reflecting rays of light.

Tenacity, or the strength with which the particles of which a metal is formed resist being pulled asunder.

Malleability. The property which many metals have of being hammered or rolled out into a large surface or thin sheet without fracture.

Ductility is the property which enables a metal to be drawn into a thin wire.

Specific Gravity, or relative weights of metals all compared to the weight of an equal volume of water.

Conducting Power for Heat. The property which metals possess in varying degree of transmitting heat along or through them.

Conducting Power for Electricity. The particular quality which metals have of becoming the medium for the passage of electricity.

Fusibility. The property which metals possess of becoming liquid when heated to a sufficiently high temperature.

Expansion and Contraction is the property which a metal has of increasing its length or volume when heated, or decreasing the same when cooled.

Specific Heat, or the relative quantities of heat absorbed by metals all compared to the heat absorbed by an equal weight of water when raised through the same temperature.

Metal	Specific Gravity.	Melting-Points (Fahr.).	Tenacity in Tons per Sq. In.
Platinum	21·53	3110	24
Gold...	19·34	1947	9
Mercury	13·59	−38	—
Lead	11·36	620	1·5
Silver	10·53	1715	18·2
Bismuth	9.79	507	1·5
Copper	8·95	1943	12 to 15
Nickel	8·82	2600	22
Iron	7·84	2737	18 to 24
Tin	7·29	449	2
Zinc	7·14	786	3
Antimony	6·71	1169	·5
Aluminium	2·67	1214	7 to 10

In the following table the metals are arranged in the order of their respective qualities, the first in the list being the best :—

Malleability	Ductility.	Conducting Power for Heat.	Conducting Power for Electricity.
Gold	Gold	Silver	Silver
Silver	Silver	Copper	Copper
Aluminium	Platinum	Gold	Gold
Copper	Aluminium	Aluminium	Aluminium
Tin	Iron	Zinc	Zinc
Platinum	Copper	Tin	Platinum
Lead	Zinc	Iron	Iron
Zinc	Tin	Lead	Tin
Iron	Lead	Platinum	Lead
—	—	Bismuth	Bismuth

Table of Weights, Expansion Multipliers, etc.

Metal.	Weight per Cubic Foot, in lbs.	Linear Expansion for 1° (Cent.).	Specific Heat.
Aluminium	166	·000023	·2143
Brass	520	·000018	·0939
Copper	550	·000017	·0951
Gold...	1150	·000015	·0324
Iron	480	·000013	·1138
Lead...	710	·000028	·0314
Platinum	1340	·000009	·0324
Silver	650	·000021	·0570
Tin	455	·000025	·0562
Zinc	440	·000029	·0955

Iron.

On account of the large supply, cheapness, and its many useful properties, iron ranks as the chief of metals. It is hardly, if ever, found in Nature in its pure state. It is extracted from its ores in the form of cast iron, which again when subjected to furnace treatment can either be converted into wrought iron or steel.

The main differences in the properties of cast and wrought iron and steel are chiefly due to the presence of carbon in the metal. Cast iron contains about 4·75 per cent. of carbon, steel up to about 1·75, and wrought iron from 0·01 up to about 0·2. The smallest variation in the amount of carbon present considerably alters the properties of the iron.

When iron contains an appreciable amount of sulphur it becomes brittle when heated and is called "hot short." If phosphorus is present the metal becomes "cold short."

The following table gives the percentage composition of several specimens of iron:—

	Sample of Pig Iron.	Sample of Good Iron.	Hot Short	Cold Short.	Low-moor.	Swedish
Carbon ...	3·302	0·080	trace	trace	trace	trace
Silicon ...	2·156	0·170	0·147	0·203	0·150	0·005
Phosphorus	1·258	0·246	0·456	0·480	0·145	0·007
Manganese	2·362	trace	trace	trace	trace	trace
Sulphur ...	0·036	0·010	0·091	0·030	0·010	trace
Iron ...	90·886	99·494	99·306	99·287	99·695	99·988

Copper.

This is the only metal which possesses a distinctly red colour, and of the ordinary manufacturing metals is, when pure, by far the toughest. It is most durable and an excellent conductor of heat; and on account of the facility with which it can be tinned is largely used in the making of the better-class cooking utensil.

Native copper is sometimes discovered in large masses, but the bulk of the copper of commerce is extracted from ores, cast into ingots and rolled into sheets or bars.

Zinc.

The chief use of this metal, which is known in the ingot form as "spelter," is in galvanising sheet iron. Its great advantage is, that while it quickly tarnishes or oxidises, the film so formed is indissoluble and forms a protective coating for the metal beneath.

For rolling into sheets it is essential that the metal should be pure, and also whilst passing through the rolls that its temperature should be between 200° and 250° F.

Aluminium.

This is a white malleable metal which is a little softer

than zinc. Its chief property is its remarkable lightness, being only about one-third the weight of iron. For sheet metal work its chief drawback is the difficulty with which it can be brazed or soldered.

It is never found in Nature in its pure state. Whilst it is the most widely diffused of all metals on the earth's surface, up to the present difficulty has been found in extracting it from its ores, which is the cause of its comparative dearness.

Tin.

This is a white metal with a slight yellow tinge, and is used principally as the coating metal for tinplates. It is not readily attacked by vegetable acids or meat juices, and this makes it of great value as a coating material for the interior of cooking utensils.

Lead.

Of the ordinary manufacturing metals lead is the softest and possesses the least strength. It is very heavy, being more than four times the weight of aluminium. Its softness renders it particularly suitable for working into the various shapes and forms required by the plumber. It resists the action of sulphuric or hydrochloric acid to a much greater degree than most of the ordinary metals. Nitric acid when strong scarcely attacks it, but when diluted rapidly dissolves the metal. Although lead is practically unacted upon by cold strong acids it is very soon corroded when exposed to an atmosphere containing much carbonic acid. The metal readily flows under pressure in the solid state, hence lead pipes can be formed by squirting from a hydraulic press.

Bismuth.

This is a white metal with a peculiar reddish tinge, and is very brittle. Its chief use is in being added to alloys of tin and lead, whose melting point it reduces in a remarkable manner.

Antimony.

This is a bluish-white feathery-looking metal, which is so crystalline that it may be broken and ground up to a powder. Its chief use is for hardening alloys of lead and tin, such as Britannia and Babbit metals.

Alloys.

Metals are often compounded with each other to obtain various properties not possessed by the metals themselves, such as: (1) Reduction of melting point to something lower than that of one or more of the constituent metals. (2) To increase the strength or toughness. (3) To obtain a different colour. (4) To resist oxidation, or corrosion of sea and other water. (5) To obtain a hardened metallic compound. (6) To facilitate the flow of metal in forming sound castings.

The following is a list of the more important alloys:—

Description.	Copper	Zinc.	Tin.	Properties.
Best Brass	72	28	—	Very malleable.
Ordinary Brass ...	66·6	33·3	—	
Muntz or Yellow Metal	60	40	—	Rolls hot, resists corrosion.
Dutch Metal ...	84	16	—	Highly malleable.
Bronze Coinage ...	95	1	4	
Gun Metal	90	—	10	Very tough.
Speculum Metal ...	66·6	—	33·3	Takes high polish.
Britannia Metal ...	1·8	—	92	and Antimony 6·2.
Babbit's Metal ...	1	—	50	" " 5.
Delta and Aich's Metal	55-60	38-44	with 1·5 to 4 of Iron.	
Pewter Plate ...	2	—	89 with Antimony 7 and Bismuth 2.	
German Silver ...	60	25	with Nickel 15.	

Common Pewter often consists of Lead and Tin alone.

CHAPTER XXXIX.

MENSURATION RULES.

Circumference of Circle.

Length of circumference = diameter multiplied by $3\frac{1}{7}$, or more accurately diameter × 3·1416.

Area of Triangle.

Multiply the base by half the perpendicular height.

Area of Circle.

Multiply radius by radius then by $3\frac{1}{7}$. Rules for the ellipse are given in Chapter XXI., and for the cylinder, cone, and sphere in Chapters XII., XXVII., and XXVIII.

Volume of Frustum of Cone or Pyramid.

Although this has been dealt with in Chapter XII. there is still another important method that can be applied in obtaining the volume of a bucket-shaped or other similar vessel, whether circular or not:—

Let a_1 = area of small end.
,, a_2 = ,, ,, large ,,
,, a = ,, ,, mid-section.

$$\text{Then volume} = \text{height} \times \frac{a_1 + 4a + a_2}{6}$$

Put in the form of a rule it becomes: "Add the areas of the ends to four times the area of the mid-section, multiply by the height, and divide by six."

Useful Data.

1 inch = 2·54 centimetres.
1 gallon = 277·274 (277¼ nearly) cubic inches.
1 cubic foot = 6¼ gallons.
1 cubic foot of fresh water weighs 62·3 lb.
(In ordinary calculations 62½ is used.)
1 cubic foot of sea water weighs 64 lb.
1 gallon of fresh water weighs 10 lb.
1 gallon of sea water weighs 10¼ lb.

Weights of Black Steel per Square Foot with Thicknesses in Inches and Millimetres.

Gauge.	Lbs. per Square Foot.	Thickness. Inches.	Thickness. Mm.
3/16	7·50	·1874	4·770
8	6·28	·1570	3·988
9	5·59	·1398	3·551
10	5·00	·1250	3·175
11	4·45	·1113	2·827
12	3·96	·0991	2·517
13	3·52	·0882	2·240
14	3·14	·0785	1·994
15	2·79	·0699	1·775
16	2·50	·0625	1·587
17	2·22	·0556	1·412
18	1·98	·0495	1·257
19	1·76	·0440	1·118
20	1·56	·0392	·996
21	1·39	·0349	·886
22	1·25	·0312	·794
23	1·11	·0278	·707
24	·99	·0247	·629
25	·88	·0220	·560
26	·78	·0196	·498
27	·69	·0174	·443
28	·62	·0156	·396
29	·55	·0139	·353
30	·49	·0124	·315

Approximate Gauges of Tinplates.

I.C.	is about	30	W.G.	easy
1 X	,,	28	,,	,,
1 XX	,,	27	,,	,,
1 XXX	,,	26	,,	,,
1 XXXX	,,	25	,,	,,
1 XXXXX	,,	24	,,	,,
1 XXXXXX	,,	23	,,	
S.D.C.	,,	27	,,	easy
S.D. X	,,	26	,,	
S.D. XX	,,	25	,,	
S.D. XXX	,,	24	,,	
S.D. XXXX	,,	24	,,	full
S.D. XXXXX...	,,	23	,,	
D.C. XXXXXX	,,	22	,,	full
D.C.	,,	28	,,	full
D X	,,	26	,,	
D XX	,,	25	,,	full
D XXX	,,	24	,,	
D XXXX	,,	22	,,	
D XXXXX	,,	21	,,	
D XXXXXX	,,	20	,,	

Steel Coke Tinplates.

	Size.	Plates in Box.		Size.	Plates in Box.
1 C ...	14 × 10	225	1 X ...	14 × 14	225
1 C ...	10 × 20	225	1 X ...	15 × 15	225
1 C ...	14 × 20	112	1 X ...	16 × 16	225
1 C ...	28 × 20	112	1 X ...	10 × 30	112
1 X ...	14 × 20	112	1 XX ...	10 × 30	112
1 X ...	28 × 20	112	1 XXX ...	10 × 30	112
1 XX ...	14 × 20	112	1 XX ...	$9\frac{3}{4} \times 19\frac{1}{2}$	225
1 XX ...	28 × 20	56	1 XX ...	$10\frac{1}{2} \times 21$	225
1 XXX ...	14 × 20	112	1 XXX ...	$12\frac{1}{2} \times 19\frac{1}{2}$	112
1 XXX ...	28 × 20	56	1 XXX ...	12 × 23	112
1 XXXX...	14 × 20	112	1 XXX ...	$12 \times 25\frac{3}{4}$	112
1 XXXX...	28 × 20	56	1 XXX ...	$12 \times 28\frac{3}{4}$	112
D C ...	17 × 25	50	1 XXX ...	12 × 32	112
DX ...	17 × 25	50	1 XXX ...	15 × 30	112
DXX ...	17 × 25	50	1 XXX ...	16 × 32	112
DXXX ...	17 × 25	50	1 XXX ...	17 × 34	112
DXXXX ...	17 × 25	50	1 XXX ...	$15\frac{1}{2} \times 19\frac{1}{2}$	112
DX ...	24 × 34	50	1 XXX ...	$17\frac{3}{4} \times 28\frac{3}{4}$	74
1 X ...	13 × 13	225	1 XXXX	12 × 32	112

Approximate Weights per Square Foot of Iron, Copper, and Brass.

W. G. Number.	Iron. In lbs. per sq foot.	Copper. In lbs. per sq. foot.	Brass. In lbs. per sq. foot.
1	$12\frac{1}{2}$	14	$13\frac{1}{2}$
2	11	13	$12\frac{1}{2}$
3	$10\frac{1}{2}$	12	$11\frac{1}{2}$
4	$9\frac{1}{2}$	11	$10\frac{1}{2}$
5	$8\frac{3}{4}$	$10\frac{1}{8}$	$9\frac{5}{8}$
6	$8\frac{1}{8}$	$9\frac{1}{2}$	9
7	$7\frac{1}{2}$	$8\frac{1}{2}$	8
8	$6\frac{7}{8}$	$7\frac{5}{8}$	$7\frac{1}{8}$
9	$6\frac{1}{4}$	7	$6\frac{1}{2}$
10	$5\frac{5}{8}$	$6\frac{1}{4}$	$5\frac{3}{4}$
11	5	$5\frac{1}{2}$	5
12	$4\frac{1}{2}$	5	$4\frac{5}{8}$
13	$3\frac{5}{8}$	$4\frac{1}{2}$	$4\frac{1}{4}$
14	$3\frac{5}{16}$	4	$3\frac{3}{4}$
15	3	$3\frac{1}{2}$	$3\frac{1}{4}$
16	$2\frac{1}{2}$	3	$2\frac{7}{8}$
17	$2\frac{1}{4}$	$2\frac{3}{4}$	$2\frac{5}{8}$
18	2	$2\frac{1}{4}$	$2\frac{1}{8}$
19	$1\frac{3}{4}$	2	$1\frac{7}{8}$
20	$1\frac{1}{2}$	$1\frac{3}{4}$	$1\frac{3}{4}$
21	$1\frac{3}{8}$	$1\frac{1}{2}$	$1\frac{1}{2}$
22	$1\frac{1}{4}$	$1\frac{3}{8}$	$1\frac{3}{8}$
23	$1\frac{1}{8}$	$1\frac{3}{16}$	$1\frac{3}{16}$
24	1	1	1
25	$0\frac{7}{8}$	$0\frac{7}{8}$	$0\frac{7}{8}$
26	$0\frac{3}{4}$	$0\frac{13}{16}$	$0\frac{13}{16}$
27	$0\frac{11}{16}$	$0\frac{3}{4}$	$0\frac{3}{4}$
28	$0\frac{5}{8}$	$0\frac{5}{8}$	$0\frac{5}{8}$
29	$0\frac{9}{16}$	$0\frac{9}{16}$	$0\frac{9}{16}$
30	...	$0\frac{1}{2}$	$0\frac{1}{2}$

Sheet Zinc.

Approximate Weights of light strengths in Sheets 8 ft. × 3 ft., showing the equivalent Wire Gauge to Zinc Gauge.

No. 4 Zinc	=	about	34 W.G.	...	=	$7\frac{1}{4}$	lb. per sheet.
5 ,,	=	,,	33 ,,		=	$8\frac{1}{2}$	,,
6 ,,	=	,,	31 ,,	full	=	$9\frac{3}{4}$	,,
7 ,,	=	,,	29 ,,	full	=	$11\frac{3}{4}$	,,
8 ,,	=	,,	28 ,,	full	=	$13\frac{1}{2}$	,,
9 ,,	=	,,	27 ,,		=	$15\frac{1}{2}$	,,
10 ,,	=	,,	26 ,,		=	$17\frac{1}{2}$	,,
11 ,,	=	,,	24 ,,		=	20	,,
12 ,,	=	,,	23 ,,		=	$22\frac{3}{4}$	,,
13 ,,	=	,,	22 ,,	full	=	$25\frac{1}{2}$	,,
14 ,,	=	,,	21 ,,		=	$28\frac{1}{4}$	,,
15 ,,	=	,,	20 ,,		=	$32\frac{3}{4}$	,,
16 ,,	=	,,	19 ,,		=	$37\frac{1}{4}$	,,
17 ,,	=	,,	17 ,,		=	$41\frac{3}{4}$	,,

Sheet Copper.

Equivalent Gauges and Weights for Sheets 4 ft. × 2 ft.

4 ft. × 2 ft. × 4 lb. = 30 W.G.		
4 ft. × 2 ft. × 5 lb. = 28 ,,		
4 ft. × 2 ft. × 6 lb. = 27 ,,	easy	
4 ft. × 2 ft. × 7 lb. = 25 ,,		
4 ft. × 2 ft. × 8 lb. = 24 ,,		
4 ft. × 2 ft. × 9 lb. = 23 ,,	easy	
4 ft. × 2 ft. × 10 lb. = 22 ,,	,,	
4 ft. × 2 ft. × 12 lb. = 21 ,,		
4 ft. × 2 ft. × 14 lb. = 20 ,,		
4 ft. × 2 ft. × 16 lb. = 19 ,,		

4 ft. × 2 ft. × 18 lb = 18 W.G.	
4 ft. × 2 ft. × 22 lb. = 17 ,,	
4 ft. × 2 ft. × 24 lb. = 16 ,,	
4 ft. × 2 ft. × 28 lb. = 15 ,,	
4 ft. × 2 ft. × 32 lb. = 14 ,,	
4 ft. × 2 ft. × 36 lb. = 13 ,,	
4 ft. × 2 ft. × 40 lb. = 12 ,,	
4 ft. × 2 ft. × 44 lb. = 11 ,,	
4 ft. × 2 ft. × 50 lb. = 10 ,,	

CHAPTER XL.

ANNEALING, WELDING, ETC.

Annealing.

IN the operation of rolling, hammering, or drawing, metals become hard and brittle; and to avoid fracture in further working the sheets or plates, it is essential that these should be softened, or annealed, as it is called.

Sheet iron or steel is made by passing pieces of the metal, in almost a white-hot condition, backwards and forwards through powerful rolls and rolling down to the required thickness. After rolling, the sheets are very hard, and have to be kept in an annealing furnace for several hours to bring them back to the soft state. The length of time they are in the furnace, and the slowness of cooling, more or less determine the degree of softness of the sheet. When the sheets are placed in batches in the furnace and heated in an uncovered state, they are said to be **open-annealed.** For some purposes, however, batches of sheets are placed in iron boxes and annealed without coming in contact with the atmosphere or the furnace flames; these are called **close-annealed** sheets. The open-annealed sheets have more scale on them than the close-annealed sheets, the latter, of course, having a much smoother surface. To obtain a good smooth surface, sheets are sometimes run, when cold, through smooth rolls after they are close-annealed; and this quality of iron is called **cold-rolled-close-annealed.**

In stretching the edge of an article, throwing off a flange, or in raising, hollowing, stamping, or spinning, some judgment must be exercised as to the suitable times for annealing. One kind of a job may only require to be softened once, whilst others may have to be annealed several times before the sheet metal can be worked with safety up to the required shape. In any case, care ought always to be taken against working a metal up to the splitting or cracking point for the want of annealing.

In annealing iron or steel the highest degree of softness is obtained when the sheet or plate is allowed to remain red hot as long as possible and to cool out very slowly. In thin sheet metal care should be taken that the edge of the sheet is not "burnt" or over-annealed. Even if a piece is not burnt out the edge may be got to a white heat, and this part will break away when being hammered.

Copper becomes soft when made red hot and allowed to cool out slowly in the air or plunged into water. When cooled out in water there is the additional advantage that the surface of the sheet is cleaned in the process by the removal of the scale in the water. This is especially the case if the surface of the copper is sprinkled with common salt before the sheet is made red hot.

Brass is annealed by gradually heating, and then being allowed to cool out slowly.

Zinc gets rather brittle at low temperatures. This is well known to those who work sheet zinc during winter in a cold workshop. For safe working during cold weather, sheet zinc should be warmed so that it can just be handled, and this is especially so if any sharp bends or edges have to be made.

In working upon any part of a sheet or plate that is to be used in a pipe or vessel that is to be subjected to a pressure, the greatest care must be taken that no part of the metal is left in a stressed condition, either through hammering or local heating. Serious results sometimes happen through want of thought in this direction. The metal can generally be brought to a proper condition by careful heating with the blowpipe or furnace, not only the parts that have been worked, but also the surrounding metal that may have been affected.

Every time a piece of metal is made red hot, whilst in contact with the atmosphere, fresh scale forms on its surface. This is due to the oxygen in the air combining with the metal to form an oxide. It is therefore evident if we require a metal not to scale or waste during annealing, it must be kept out of contact with the atmosphere, and this is in many cases an exceedingly difficult thing to do. Small articles in iron may be covered with rust or oxide, and copper may be buried in ash dust. Furnaces for large work are now being constructed in which metals can be heated out of contact with the atmosphere.

Theory of Annealing. Why metals become hard when worked, or why they become soft under heat treatment, are difficult questions to answer. Or, again, why a metal like steel becomes hard when plunged into water, or copper under the same treatment becomes soft, is no easy task to solve. Sufficient to say that these matters are now being carefully investigated, and at the no distant future a full scientific explanation will be forthcoming. We can, however, imagine that under hammering or rolling, the particles of the metal become pushed or crushed into unnatural positions, and then the metal is strained or hard. When heated, and whilst the metal is in the soft state, we may suppose that the particles then assume their natural position, and the metal comes back to its normal condition of softness.

Autogenous Welding.

The welding together of two pieces of the same metal without the introduction of a solder is termed "autogenous soldering." The term, therefore, has a very restricted meaning, and up to the present can only apply to the welding of iron or steel, "lead burning," and, to a slight extent, cast iron and copper. It is true that cast iron, aluminium, and some alloys can be fused together, but these require the aid of a flux.

Autogenous welding permits the fusing together of bars or plates of iron or steel, from the thinnest sheet up to almost any thickness of plate that is used in the boilershop. With great care it is possible to make the welded portion as strong as the solid bar or plate, but generally it is the safest plan to assume that the welded part will be slightly weaker, say to the extent of about 5 per cent. of the solid plate.

Of recent years the process of autogenous welding by the oxy-acetylene method has come rapidly to the front, and has now a very wide application.

There are two systems in operation—namely, the "high pressure" and the "low pressure." In the former case, where dissolved acetylene is used, the two gases are under high pressure in cylinders; and for use the acetylene is reduced to a pressure of about $5\frac{1}{2}$ lb. per square inch before it passes to the blowpipe, the oxygen being regulated to give the correct flame. This latter is judged by the appearance of the flame, and with a little experience the proper condition of the blowpipe flame for welding can be arrived at.

To ensure success, the work should be properly prepared before attempting to weld. When it is desired to make a butt joint on plates over $\frac{1}{8}$ in. in thickness, the two edges of the plates should be bevelled so that when they come together a "V" is formed, that should be as wide as the plates are in

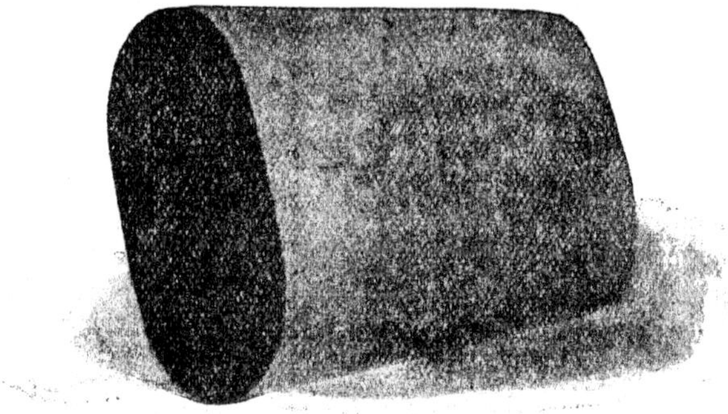

Shows a welded Steel Petrol Tank.

thickness. The complete removal of any rust or scale is also essential. The feeding metal, with wrought iron or steel, should be, if possible, shearings from material similar to the

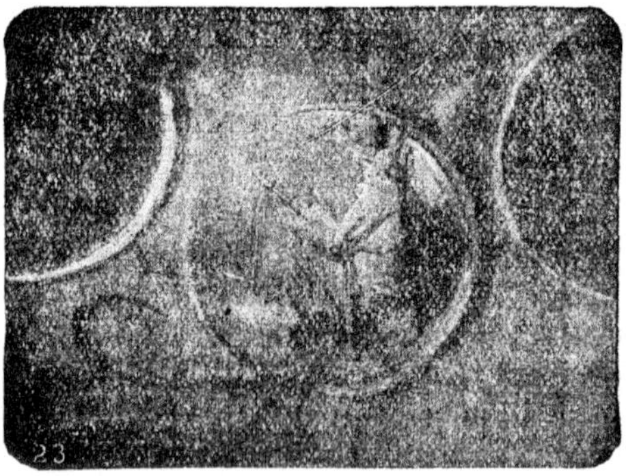

Repairs to Marine Boiler Furnace by Acetylene Process.

metal worked upon, and should never be of greater section or diameter than the thickness of the work; and in case of heavy work, should not be greater than $\frac{1}{4}$ in. diameter or $\frac{3}{16}$ in. square. In practice, Swedish iron wire is very commonly used as a filling-in medium, which, in such cases, accounts for the loss of strength when dealing with boiler plates.

The following table will give some idea of the gas consumption per foot run, on different gauge iron or steel, and the rate at which work can be carried out:

Thickness of Plate.			Consumption of Acetylene, cubic feet.	Consumption of Oxygen, cubic feet.	Speed of work in foot run of weld per hour.
Wire Gauge.	Inches.	mm.			
20	·04	1·0	1·8	2·25	50
16	·06	1·5	2·7	3·50	40
14	·08	2·0	3·6	4·50	35
12	·10	2·5	5·4	7·75	30
10	·12	3·0	8·0	10·00	24
8	·16	4·0	12·5	15·70	18
	$\frac{1}{4}$	5-6	18·0	22·00	14
	$\frac{5}{16}$	7-8	27·0	33·00	10
	$\frac{3}{8}$	9-10	36·0	44·00	7

When the metal is over $\frac{3}{8}$ in. in thickness, the gas consumption, per foot run, increases rapidly, and the speed of work falls similarly.

When acetylene, in either the high or low pressure system, is used with oxygen, in a properly designed blowpipe, it splits up into its component parts—hydrogen and carbon—at the base of the flame, carbon only taking part in the burning, due to the fact that hydrogen will not combine with oxygen at the temperature carbon will; consequently the hydrogen remains free and forms a protecting zone at the blowpipe tip, where the carbon is burning. The high flame temperature obtained, combined with the fact that there is a zone of free hydrogen, renders the flame very reducing and extremely suitable for many operations, which would otherwise have to be carried out by a more costly and probably less efficient method, and which would, in many cases, be altogether impracticable. The temperature of the oxy-acetylene flame is very high, being about 6,350° F.

To ensure the complete combustion of acetylene, theoretically 2·45 volumes of oxygen are required to one of acetylene, but in actual practice it is found that the proportions vary

Repairs to a Steamer by Oxy-Acetylene Welding.

between 1·6 to 1·0 and 1·0 to 1·0, the lower figure being found to be correct when heavy work is done with the high-pressure system, and the higher when working with the low-pressure system or the high-pressure system on light materials.

The autogenous process can be used for the welding of

bicycle, motor-car, and aeroplane frames, and in many cases act as a substitute for rivets; the welding and repairing of boilers, tanks, ships, etc.; the filling in of parts that have been worn or corroded away; and the repairing of all kinds of cracks.

In the repairing of cracks in riveted joints or other work that is held tight in position, it is important that part of the work around the crack should be loosened by unriveting, so as to allow for the expansion and contraction that takes place when the broken portion is passing through the welding

Wrought-iron Scroll Work, built up by Oxy-Acetylene Process.

process. It is also important that the parts all around the repaired portion should be carefully heated up so that as far as possible no undue strain shall be placed upon, or any stresses left in, the plate or bar.

The illustrations shown will give some idea of the range of work that can be carried out with the oxy-acetylene welding blowpipe.

Cutting Metal with Oxygen.

In addition to welding, as explained above, acetylene can also be used as the heating agent in a special blowpipe for oxygen cutting. The blowpipe is so arranged that a separate jet of oxygen may be discharged through the centre of the blowpipe flame when the metal is heated up to melting-point. This immediately produces combustion of the metal with the resulting formation of oxide. The jet of oxygen is made sufficiently strong to blow away this iron oxide in front of it, with the result that a clean, narrow cut is effected through the metal at a speed of travel which is comparable with hot sawing. The metal on each side of the cut is neither melted nor injured in any way, as the action proceeds too rapidly for the heat to spread. The cutting may be made to follow any desired line or curve as required.

Some idea of the rate of cutting can be gleaned from the fact that ½-in. plate can be cut at the rate of 1 ft. per minute, and boiler plate, of 1½ in. thickness, being cut at about one-half this speed.

The proportionate consumption of acetylene to that of oxygen varies from 25 per cent. for the thinnest section of plate to 10 per cent. for the thickest section of plate.

CHAPTER XLI.

MISCELLANEOUS PATTERNS.

Gear Case for Mitre Wheels.

THE making of a sheet metal covering to act as a guard for bevel wheels is not by any means a difficult matter, the chief consideration being the setting out of the patterns to work up

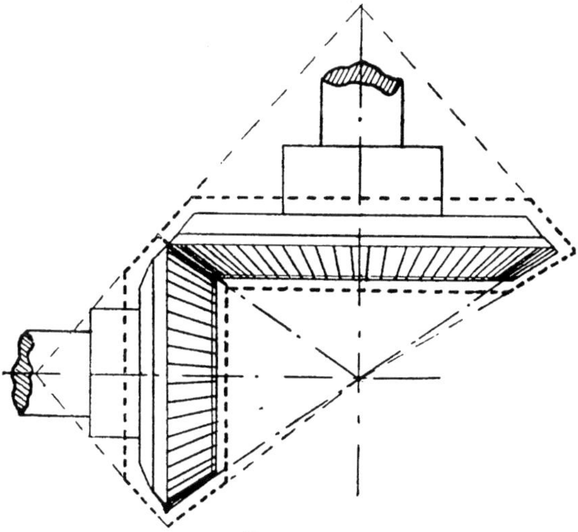

FIG. 339.

exactly to the required shape. A view of a pair of wheels is shown in Fig. 339, the thick dotted lines representing the

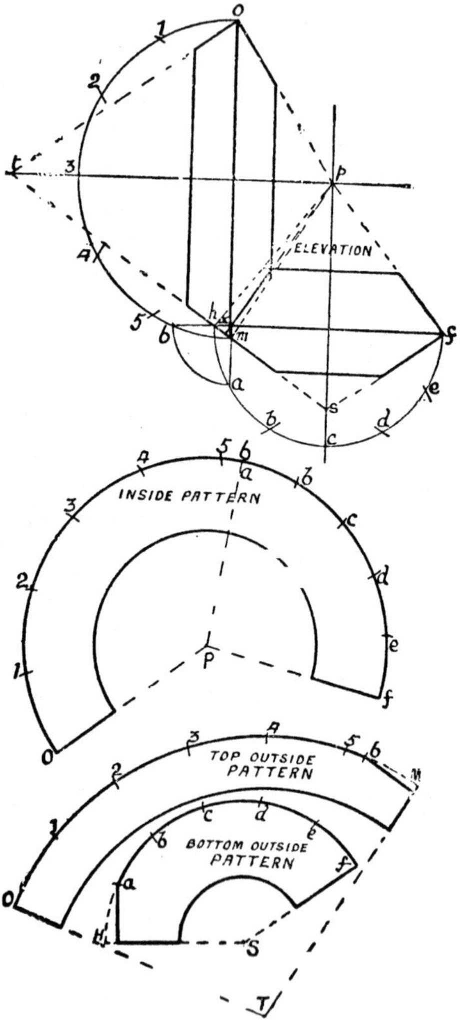
ELEVATION
INSIDE PATTERN
TOP OUTSIDE PATTERN
BOTTOM OUTSIDE PATTERN

FIG. 340.

gear case. A little consideration will show that the guard is formed from the surfaces of four cones, arranged in such a way to intersect or cut into each other so as to give the necessary opening for the two wheels to gear together. The setting out of the patterns is fully explained by the diagrams on Fig. 340.

The case is made in two equal halves, these, when put over the wheels, being fastened together with slip wire hasps at *f* and *o*. It will thus be seen that each half of the cover is formed of four half frustums of cones.

The inside cones have a common apex at *p*, and overlap or intersect each other so as to give the opening for the two wheels to come together. Thus, the base of one cone is *f h* and the other *o m*, these crossing at *x*. In setting down the bases of the cones the sizes of these must be arranged so as to bring the lines *x* 6 and *x a* equal in length, these giving half the width of the opening of the cones at the bases.

In marking out the pattern for the insides of the case, P *o* on the pattern is made equal to *p o* on the elevation; the lengths 0 1, 1 2, etc., up to 6, being set off from the top semicircle, and the lengths *a b*, *b c*, etc., from the bottom semicircle.

The patterns for the outside are marked out in a similar manner, T *o* and S *f* on the patterns being equal, respectively, to *t o* and *s f* on the elevation. The girth around each one will be the same length as that of the corresponding semicircle in the elevation. The straight lines at the ends are obtained by seeing that M 6 and H *a* on the patterns are respectively the same lengths as the arcs *m* 6 and *h a* on the elevation, the lines then being drawn square to the outside lines of the patterns. It should be observed that these two lines come the same length if the patterns are marked out correctly.

Half discs to put into the backs and fronts of the casing

will be required, but patterns for these are not shown. Allowances for jointing must be added to the patterns to suit the method of jointing adopted.

Square Cover or Dome of Semicircular Section.

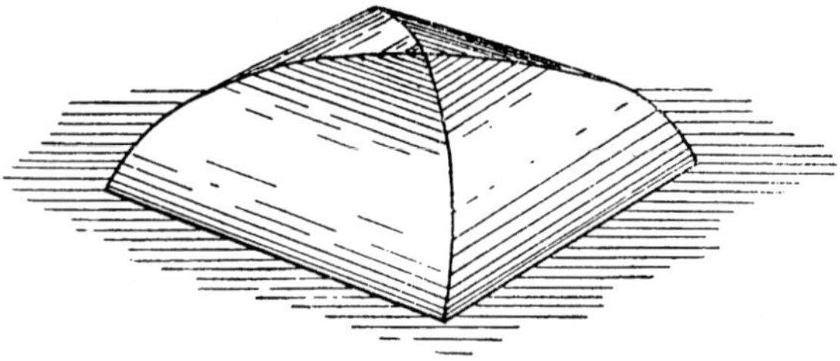

FIG. 341.

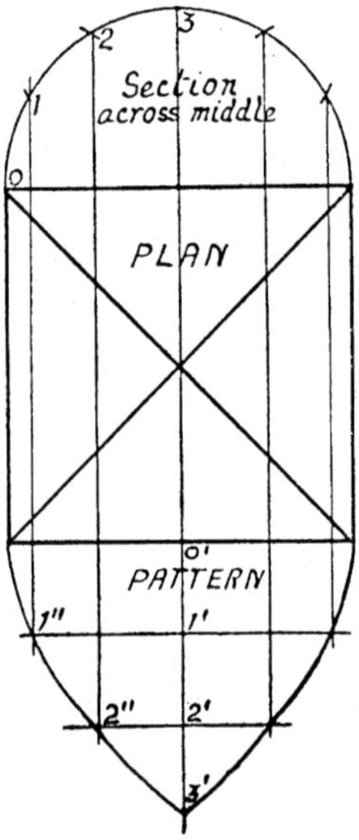

FIG. 342.

A very effective-looking cover can be made up out of four pieces of sheet metal to the shape as shown in Fig. 341. The shape is semicircular in both directions.

The pattern is marked out as seen in Fig. 342. A semicircle is described on one side of the plan, one-half being divided into three equal parts 0, 1, 2, and 3. For the pattern, the lengths 0′ 1′, 1′ 2′, and 2′ 3′ are made equal to the lengths of the correspondingly numbered arcs on the section; lines drawn through each point across and, on to these, lines run down from 1 and 2; thus the points 1″ and 2″ are obtained. The points so found are carefully joined up with a free flowing curve; the other side of the pattern being obtained in the same manner.

It will readily be seen that all the lines required for the setting out of the pattern can be obtained by drawing only one-eighth of the plan and half of the section. The full plan has been drawn in here to better explain the method. The joints can be made either by soldering, knocking-up, or any other way as required.

Rectangular Cover of Circular and Elliptical Sections.

The pattern for this is shown marked out on Fig. 343. The side pieces are quarter circle in section, and to join on to these, if the joints are to be diagonal as in plan, it will be necessary for the end pieces to be a quarter of an ellipse in shape. The pattern for the side is set out exactly as for the square cover. Before the end pattern can be struck out it is necessary to draw the shape of the half section as shown. This is done by running lines down from 3, 2, and 1 on the semicircle to the diagonal line on the plan; then drawing across and marking up M A to equal D 3, N *b* to equal E 2,

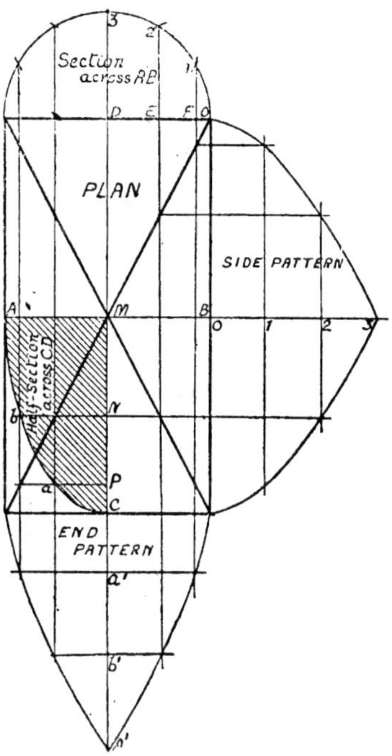

FIG. 343.

and P *a* to equal F 1. Joining these points up will give a quarter of an ellipse. For the end pattern the girth line C *a'* *b'* A' must be set out to equal in length the parts with the corresponding letters on the quarter ellipse. Lines are drawn across from the points, and others to meet these from the

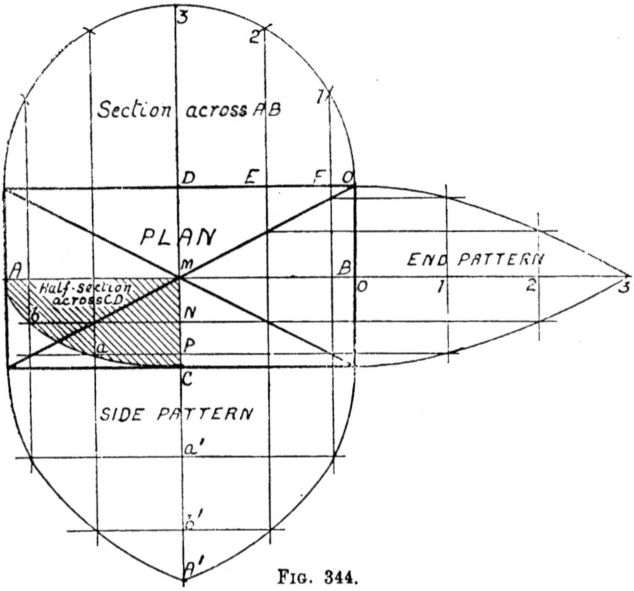

FIG. 344.

semicircle, and thus points on the pattern curve obtained. These being carefully joined will give the pattern as required.

In Fig. 344 it will be seen that the semicircular section runs lengthways of the cover; the setting out of the two patterns is, however, exactly the same as in the last case.

Conical Pipe on Spherical Dome.

If the conical pipe has to fit on the middle of the spherical dome, it will at once be seen that the cut on the end of the cone must be square to its centre line. But if the conical pipe fits on the side of the dome, as in Fig. 345, then the setting out of the pattern becomes a much more difficult matter.

To set out the pattern for the latter case, it will be necessary to first of all obtain points on the elevation of the joint line. To do this the principle adopted is to imagine horizontal cuts taken through the cone and sphere. These sections would of course be circles, and where they intersect each other would give points on the joint line. The arcs shown on Fig. 345 represent parts of the section circles. Thus, to obtain one point: With centre *c* (on the centre line of sphere) and radius *c b*, the arc *b e* is drawn; then with centre *d* (on the centre line of the cone) and radius *d a*, the arc *a e* is drawn; from the point *e*, where the two arcs intersect, a perpendicular is dropped on to the line *a c*, giving *f*, this being one point on the joint line. In the same manner, as many other points as are required can be obtained. Through each point so found, a line from the apex of the cone is drawn down to the base.

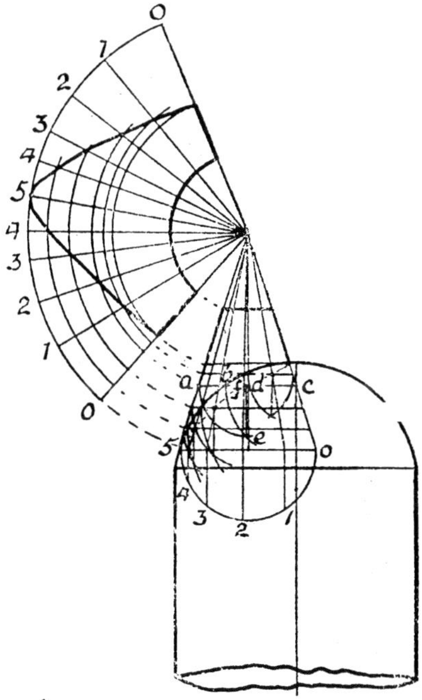

Fig. 345.

and then from the base on to the semicircle as shown. The lengths of arcs on the semicircle are then set around for the girth of the pattern curve, as 0, 1, 2, etc., and the radial lines drawn; these latter are then cut by swinging the lengths

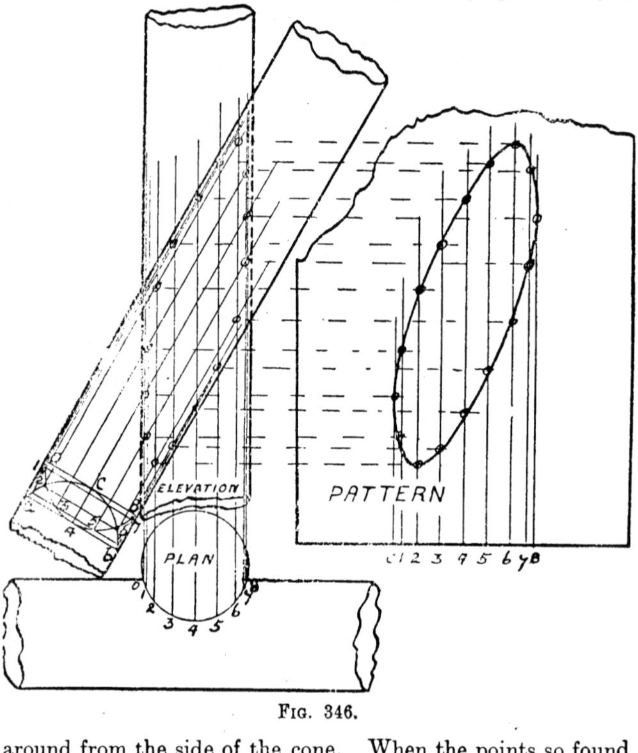

FIG. 346.

around from the side of the cone. When the points so found are joined up, the pattern is complete.

Cylindrical Crossed Tubes.

A somewhat interesting case of pattern-cutting is that shown in Fig. 346. It will be seen that the tubes cross and cut part way into each other, and as both pipes are the same size, the shape of the hole in each pipe will be the same.

The setting out of the pattern is now explained. The part of the circle, 0 to 8 in plan, which shows the distance the pipes cut into each other, is divided up into any convenient number of parts; the same being done with an exactly similar arc on the pipe in the elevation. An arc (*a c b*) of the same size and shape is drawn in the reverse direction as shown. Lines 7 1, 6 2, and 5 3, are drawn across to cut this arc, and from the points of intersection lines are drawn along the pipe; where these meet the lines on the vertical pipe, will give points on the elevation of the joint line. The girth on the pattern is laid out from the numbered arc on the plan; lines run up, and others drawn across from the elevation as shown, and thus points on the hole obtained.

The holes on the two pipe patterns will, of course, be the same shape if the pipes are of equal diameter; but if one pipe is larger than the other, the holes will be of different shapes and require to be set out separately. The same method, however, as shown for equal pipes, can also be applied to those of unequal diameter.

Curved Square Hood.

A hood with curved back, throat, and sides, as shown in Fig. 347, can have its patterns marked out with very little trouble if the method as shown on the diagram is followed.

For the cheek pattern, the lengths 0 1, 1 2, etc., are made equal in length to the arcs with the same numbers on the end elevation, and the width projected down from the side elevation, these points then being joined up. The curve *a* 4 at the top of the pattern will be the same as that on the elevation above. The back pattern lengths are taken from the side elevation, lines drawn across, and these cut off to the required lengths by projecting down from the end elevation. The throat pattern is obtained by taking the lengths from the throat in the side elevation, and the widths across from the end elevation.

The four parts can be joined together by either knocking-

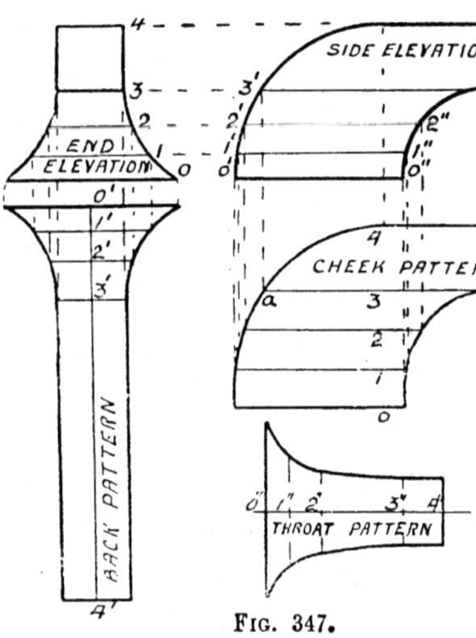

Fig. 347.

up, riveting, or any other method as required, the necessary allowances being put on the patterns.

Offside Conical Crosspipe in Conical Tube.

Perhaps one of the most peculiar and difficult patterns to mark out is that for a conical pipe which fits inside another conical pipe, and whose centre lines do not meet and are also inclined to each other. The position of the pipes can be best understood by reference to the plan and elevation in Fig. 348.

In this case, and in most others, the main difficulty lies in

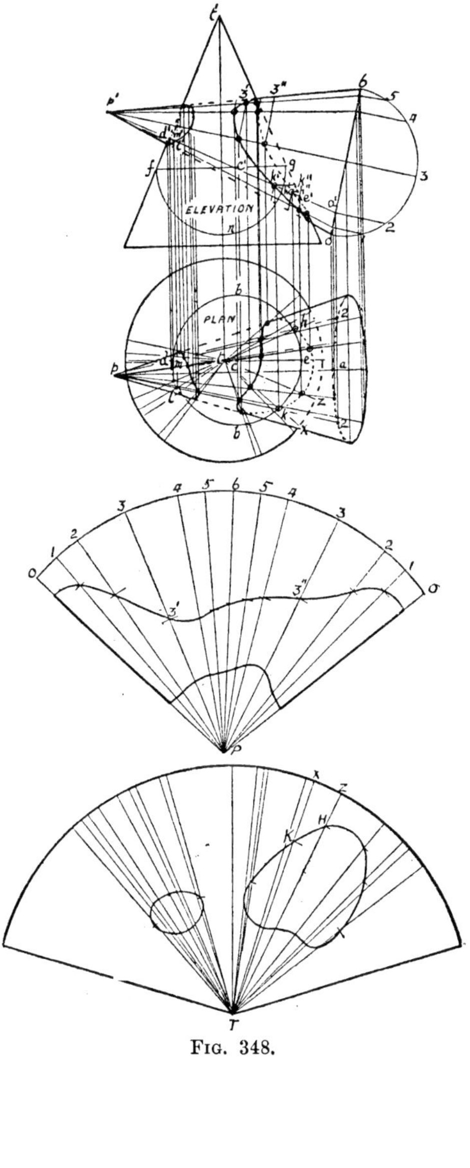

FIG. 348.

obtaining points on the curves of intersection of the two pipes; as when this is accomplished very little trouble is experienced in afterwards laying out the patterns.

To obtain points on the joint curve several methods can be used. In this case the idea is to take cutting planes which all pass through the apex p' of one cone, and thus give triangular sections; the sections of the other cone being elliptical. Where the pairs of triangles and ellipses intersect will give points on the curve of interpenetration. It will be as well to explain how to obtain one set of points. A semicircle is described on the base of the inclined cone and divided into six parts. Take the point marked 2. A line is drawn square to the base, giving a', and this point joined to p'. A projector is drawn from a' to the centre line of the cone in plan, giving a; then a 2 in the plan is made equal to a' 2 in the elevation, the points marked 2 being joined to p. The line d' e' is bisected and a horizontal line (f g) drawn through its middle point c'. On f g a semicircle is described, and a perpendicular (c' n) dropped from c' on to it; the length, c b in the plan, being made equal to c' n, and d and e obtained by projecting down from the corresponding points in the elevation. Through the points d b e b an ellipse is drawn, or such parts of it as are required to cut the lines marked p 2. Thus four points, h k l m, are found, and these projected up to the elevation. In the same manner any other number of points on the joint curve can be obtained.

To strike out the pattern for the outside cone: Lines are first drawn from t, in the plan, through each point on the curves—thus, to show two, t x and t z—and from the lengths of arcs obtained, the girth curve on the pattern is laid out and radial lines drawn, as shown by T X and T Z. The points on the pattern holes are found by running lines from the points on the elevation of the joint curve to the outside of the cone; thus, the lengths T K and T H on the pattern are respectively equal to $t' k''$ and $t' h''$ on the elevation. In a similar way all the other points required are obtained.

The pattern for the inclined cone or inside tube is set out in the usual way, lines being run out to the outside line of the cone and lengths taken off. Thus, P 3′ and P 3″ on the pattern are made equal in length to the corresponding lines p' 3′ and p' 3″ on the elevation

In complicated work of the above description, the setting out must be done as accurately as possible if it is desired that the parts shall fit neatly together.

CHAPTER XLII.

SHEET AND PLATE METAL WORKING MACHINES AND TOOLS.

THE machines and tools used in sheet and plate metal work are of almost infinite variety, especially in the former class of work.

For plate work the machines usually used are those for straightening, punching, shearing, planing, rolling, drilling, riveting, together with hydraulic presses for flanging, and a variety of pneumatic and electrically driven tools.

In sheet metal work the chief machine is perhaps that of the press, in which blanks of almost any shape and size can be cut out, and objects pressed or drawn into any form as required.

The spinning lathe, too, is used for a great number of operations in light work, either in trimming, flanging, bulging, burnishing, beading, curling, or wiring of stamped or drawn articles. Also, with suitable tools, hollow circular objects can be spun right up into shape from the flat disc.

A few of the more important machines used by sheet metal workers are shown in the following pages, also several bench tools, in addition to those already given in the former chapters.

Inclined Power Press for Cutting, Piercing and Forming.

Sections as formed in the above Press.

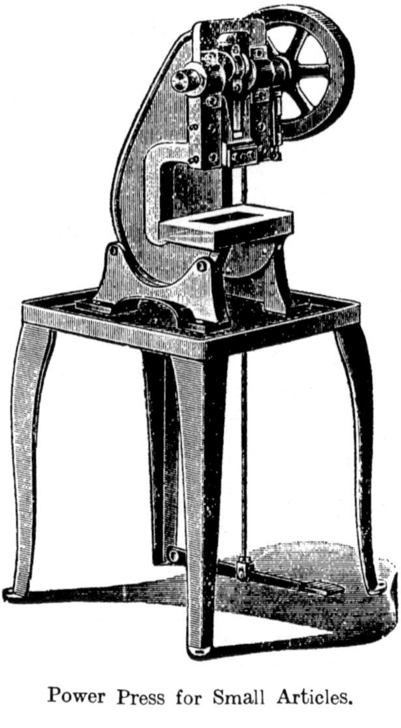

Power Press for Small Articles.

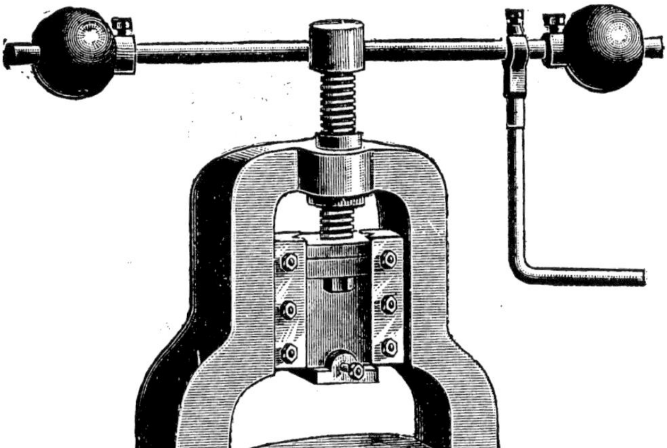

Double-Arm Screw Press.

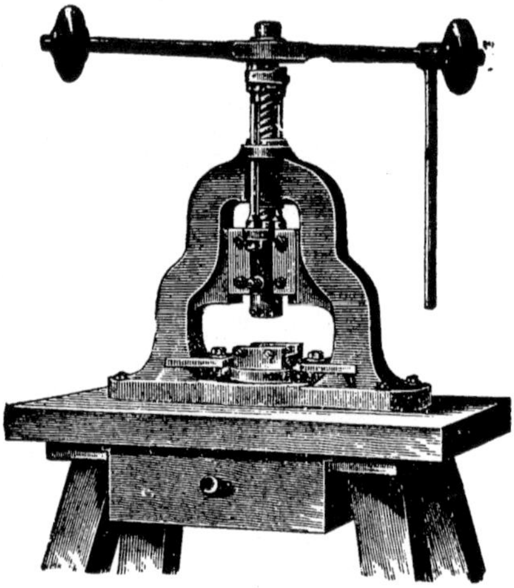

Small Double-Arm Bench Press.

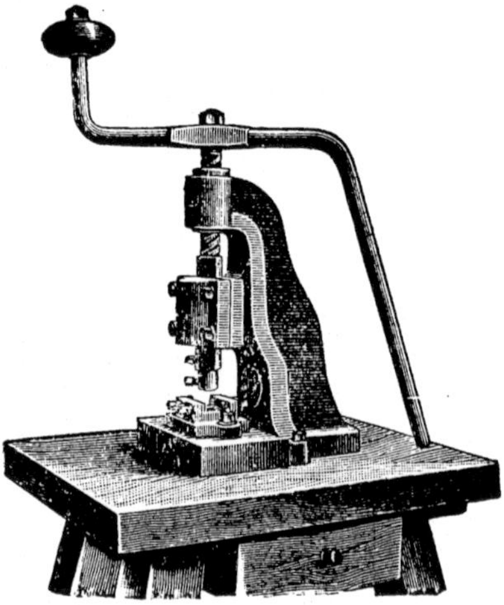

Small Fly Press.

Multiple Punching Press for Sheet Metal.

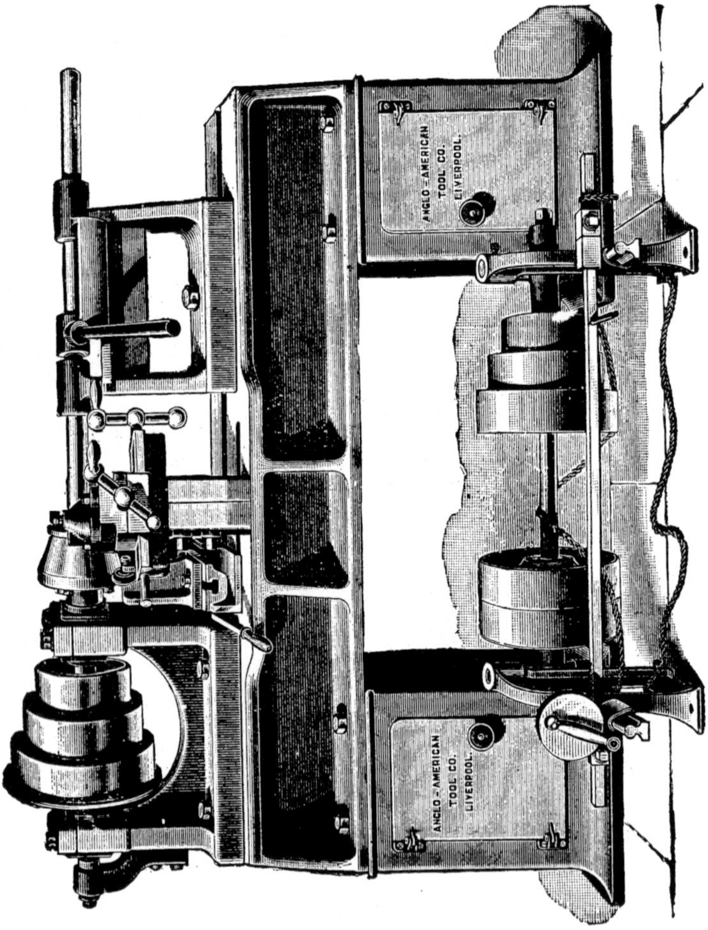

Sheet Metal Spinning Lathe.

Trimming and Beading Lathe.

Seaming Machine.

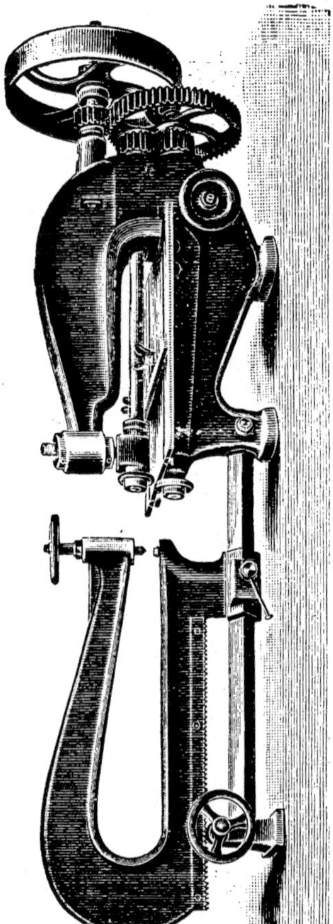

Circle and Strip Shearing Machine.

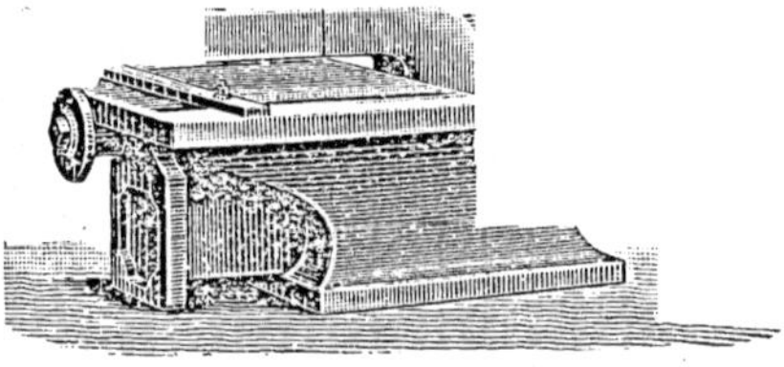

Hand Strip Cutting Machine.

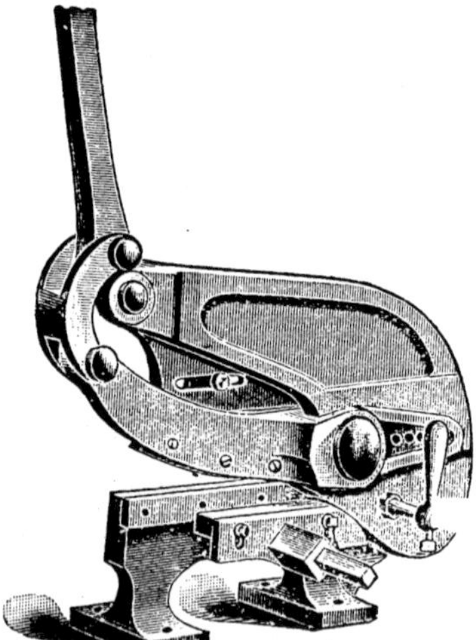

Hand Lever Shears.

Treadle Guillotine Squaring Shears.

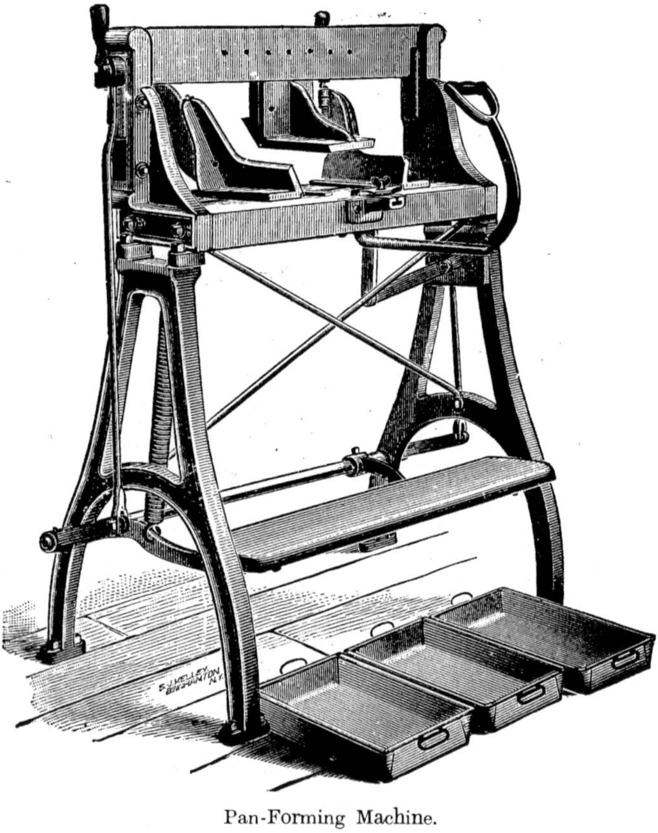

Pan-Forming Machine.

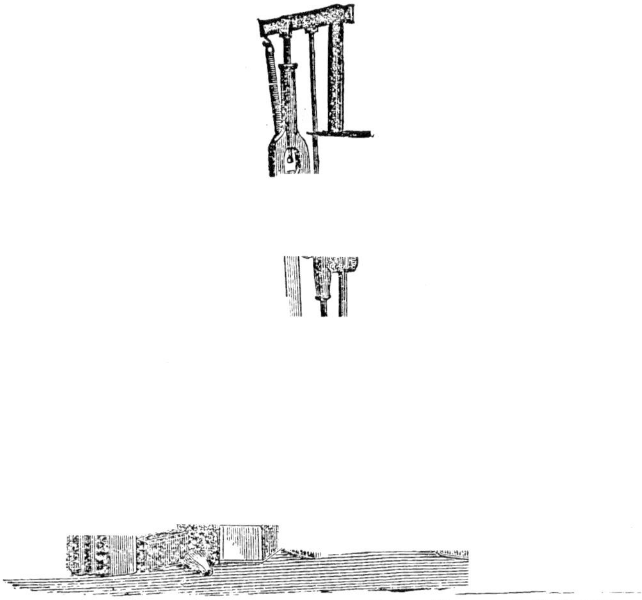

Crimping Machine for Fixing Tops and Bottoms.

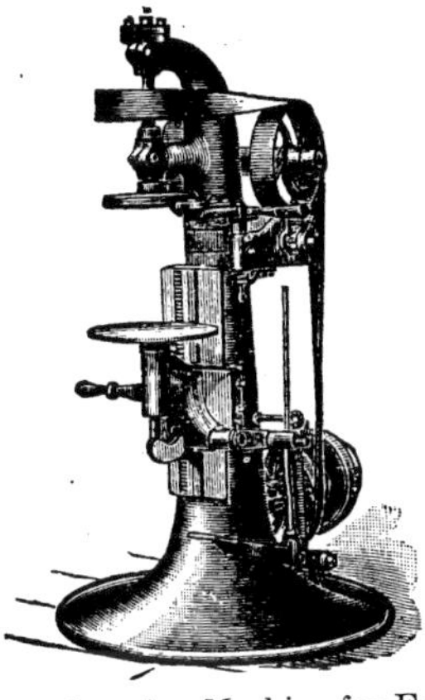

Double Seaming Machine for Ends.

Treadle Grooving Machine

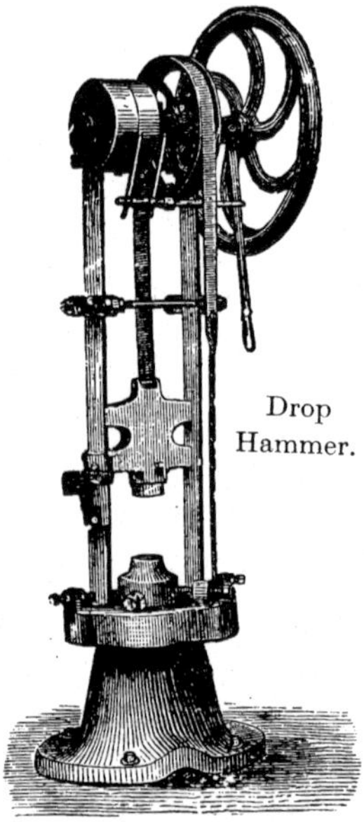

Drop Hammer.

Spring Hammer.

Drawing Machine for Mouldings.

Hand Rolls for Sheet Metal Work.

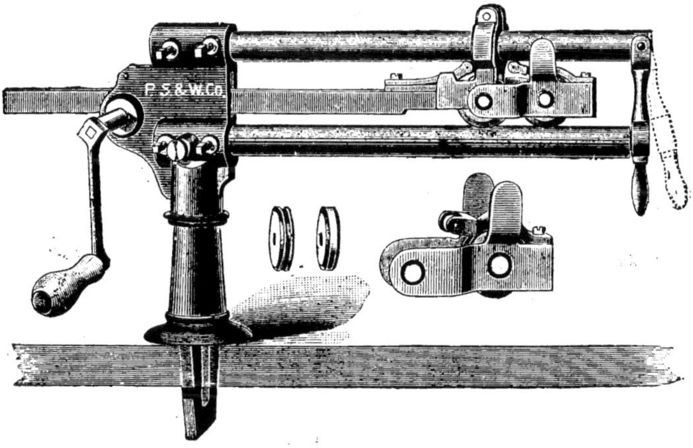

Combined Grooving and Closing Machine.

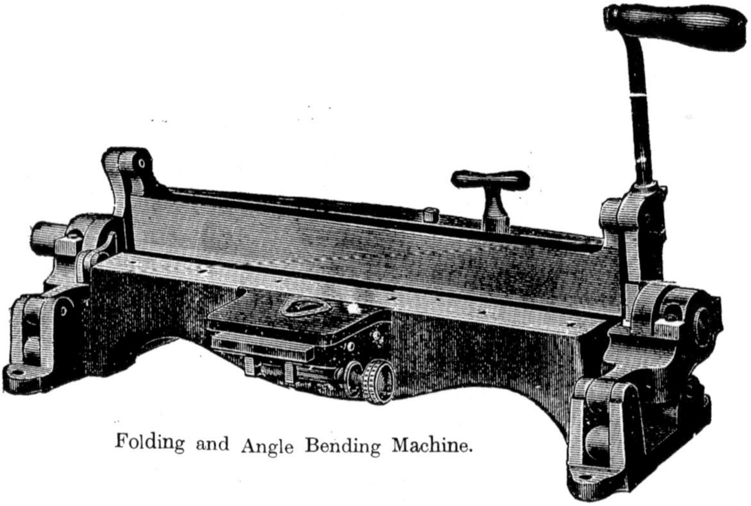

Folding and Angle Bending Machine.

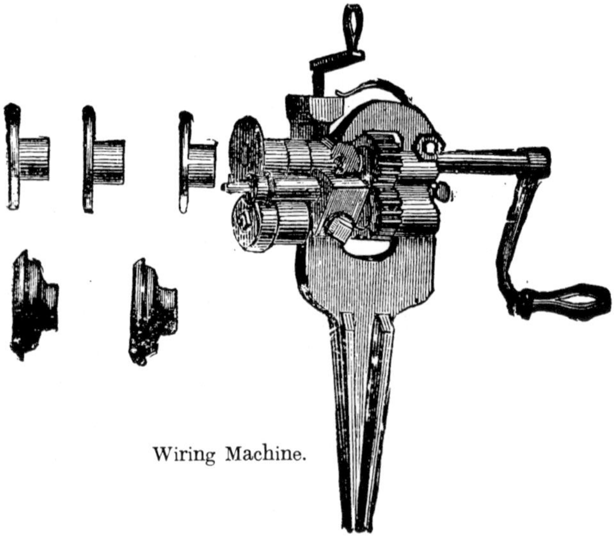

Wiring Machine.

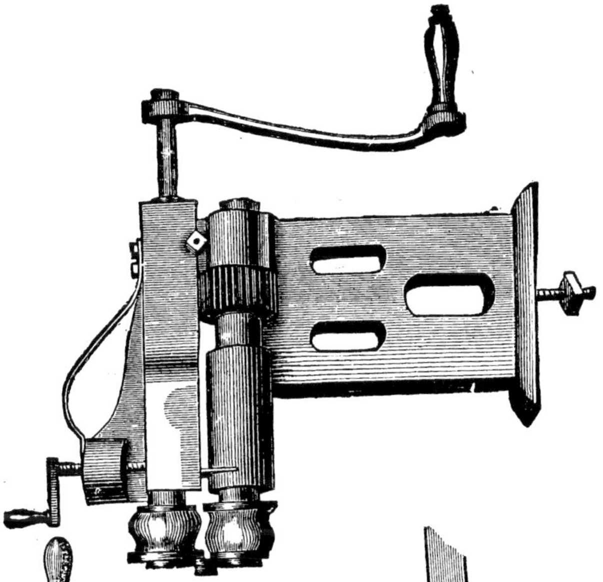

Swaging Machine.

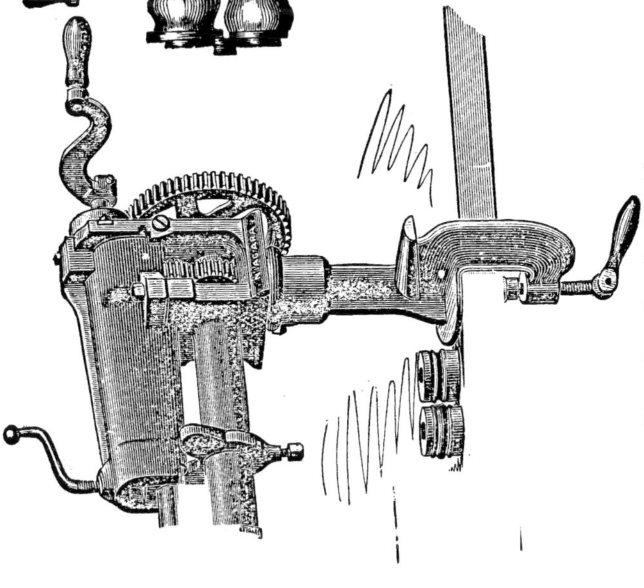

Long Arm Swaging Machine.

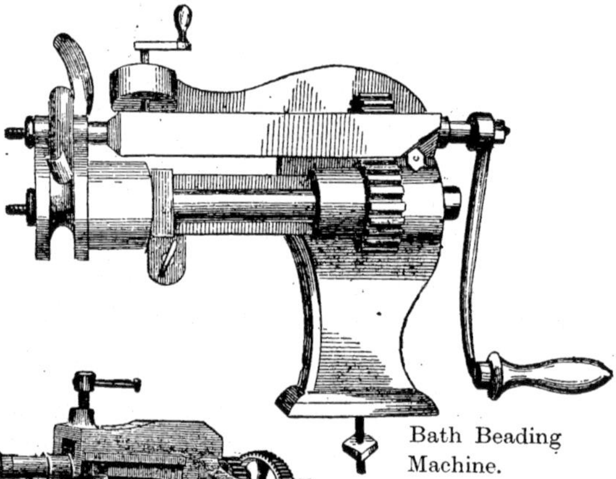

Bath Beading Machine.

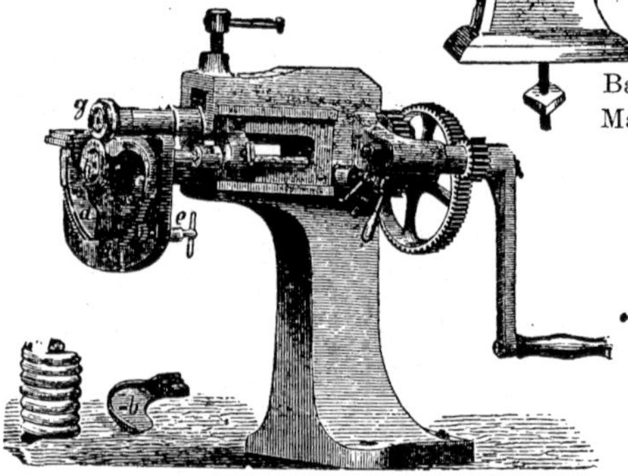

Astragal Forming Machine.

Cone Rolling Machine.

Knocking-up Machine.

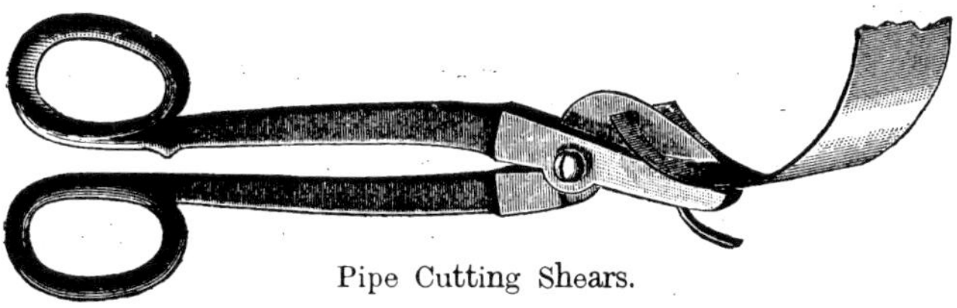

Pipe Cutting Shears.

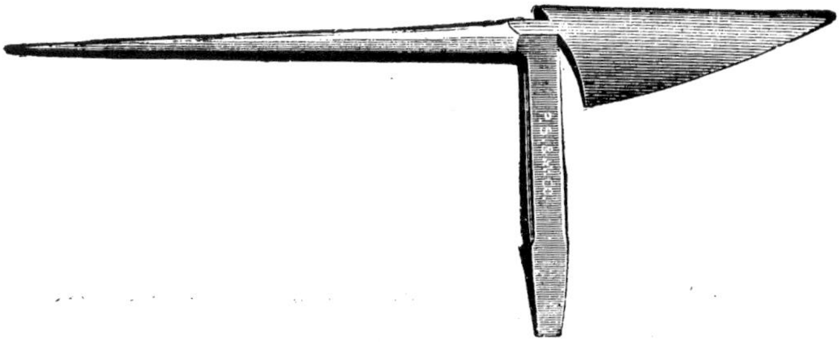

Funnel Stake with Beck.

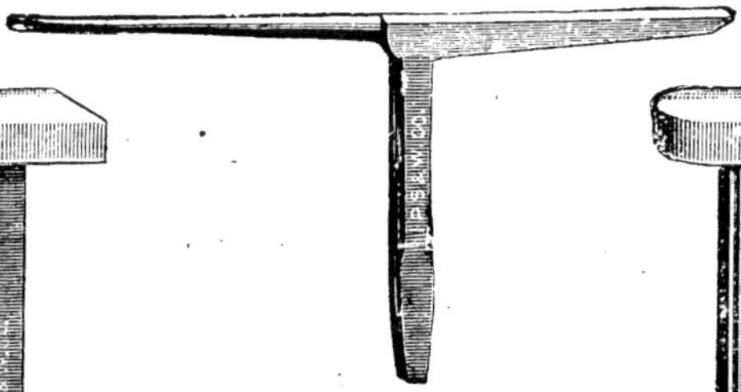

Small Beck Stake.

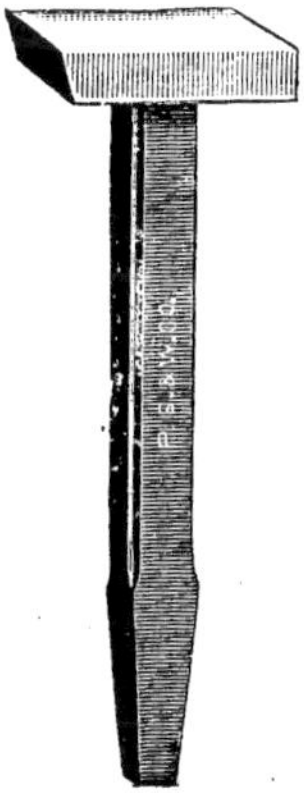

Pan Stake (Square).

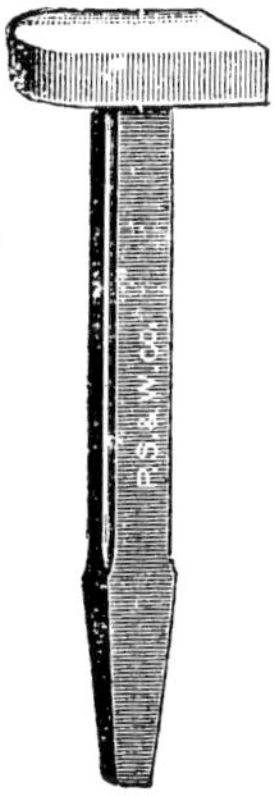

Pan Stake (Round-ended)

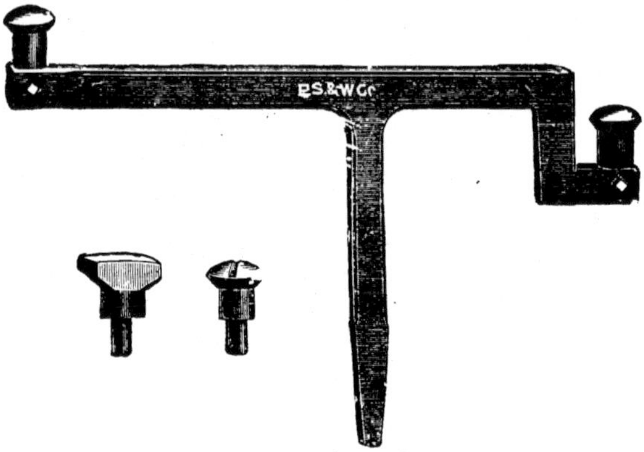

Horse and Heads.

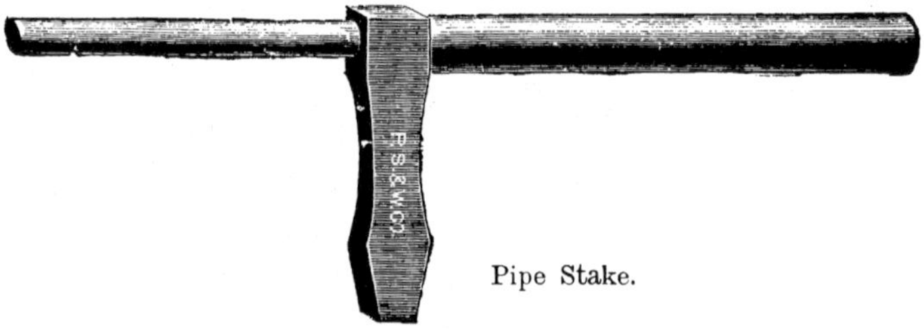

Pipe Stake.

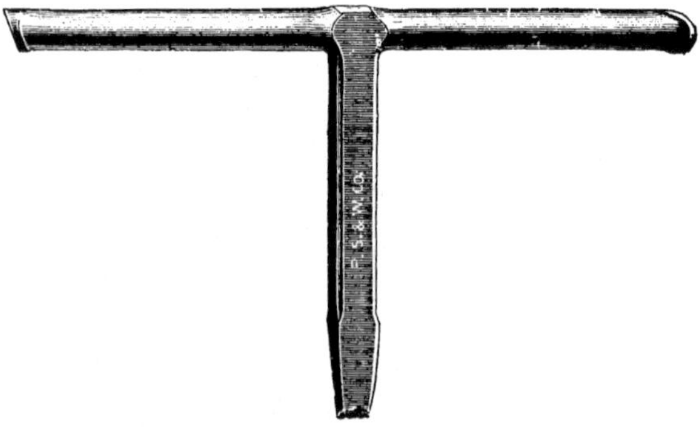

Small Pipe Stake.

Creasing Iron.

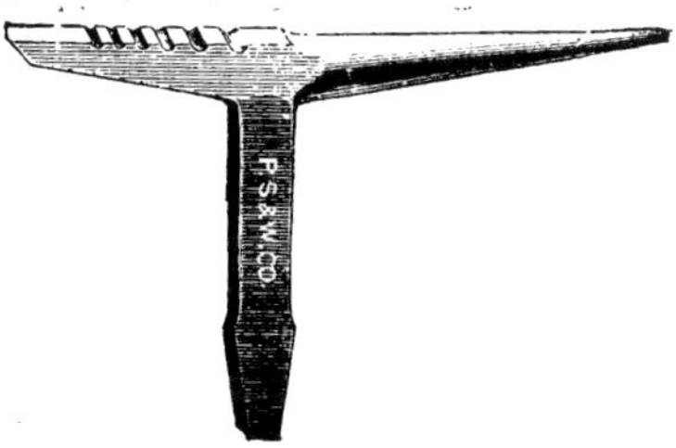

Creasing Stake and Horn.

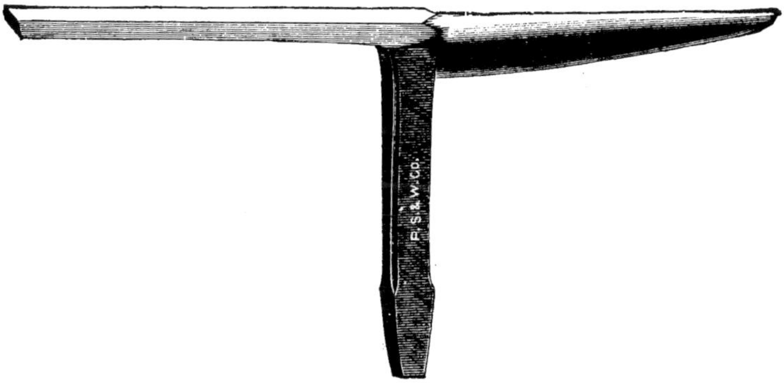

Double-ended Round and Square Side Stake.

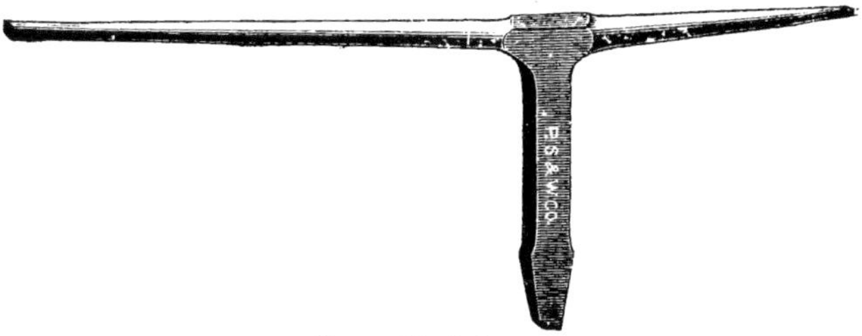

Large Beck-Iron.

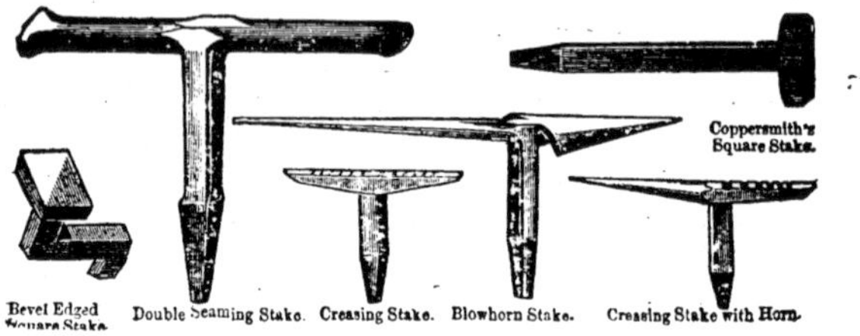
Coppersmith's Square Stake.
Bevel Edged
Double Seaming Stake.
Creasing Stake.
Blowhorn Stake.
Creasing Stake with Horn.

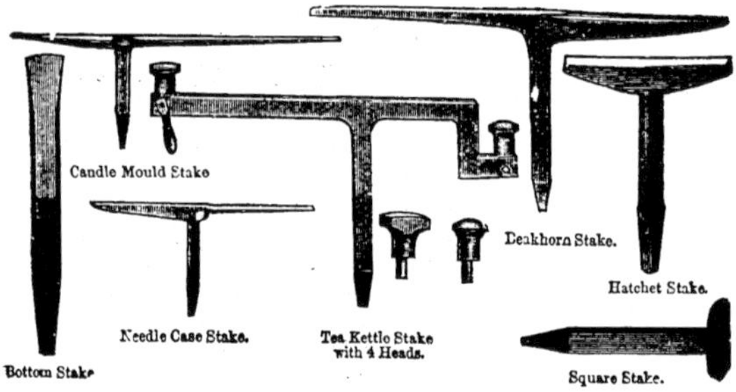
Candle Mould Stake
Beakhorn Stake.
Hatchet Stake.
Needle Case Stake.
Tea Kettle Stake with 4 Heads.
Bottom Stake
Square Stake.

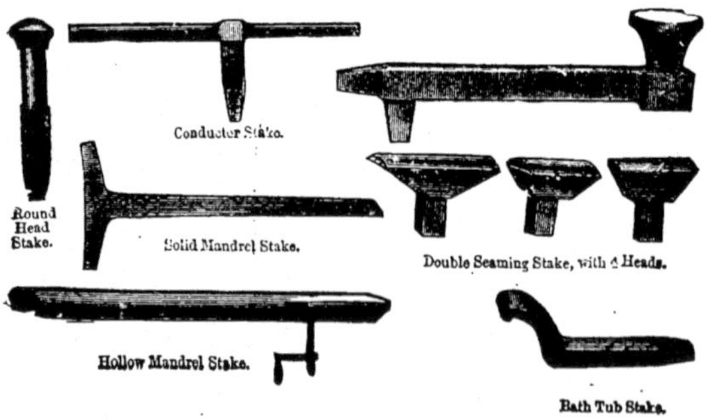
Conductor Stake.
Round Head Stake.
Solid Mandrel Stake.
Double Seaming Stake, with 4 Heads.
Hollow Mandrel Stake.
Bath Tub Stake.

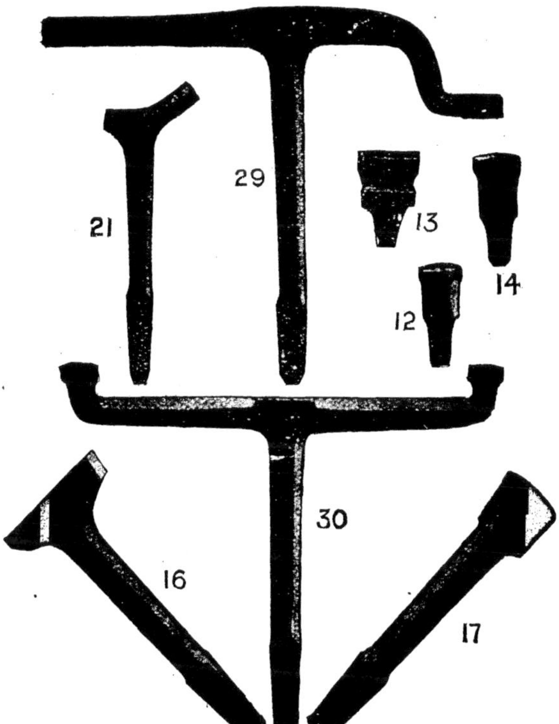

No. 21. Teapot Neck Tool.
No. 29. Tinsmith's Horse.
No. 12. Round Head for Horse.
No. 13. Long Head for Horse.
No. 14. Oval Head for Horse.
No. 30. Saucepan Belly Stake.
No. 16. Tea-Kettle Bottom Stake.
No. 17. Round-Bottom Stake.

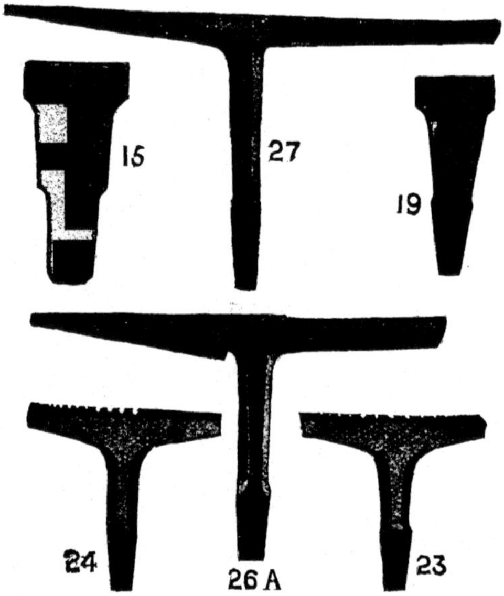

No. 15. Tinsmith's Anvil.
No. 27. Beck-Iron.
No. 19. Anvil Stake.
No. 24. Crease Iron.
No. 26a. Funnel and Side Stake.
No. 23. Grooving Stake.

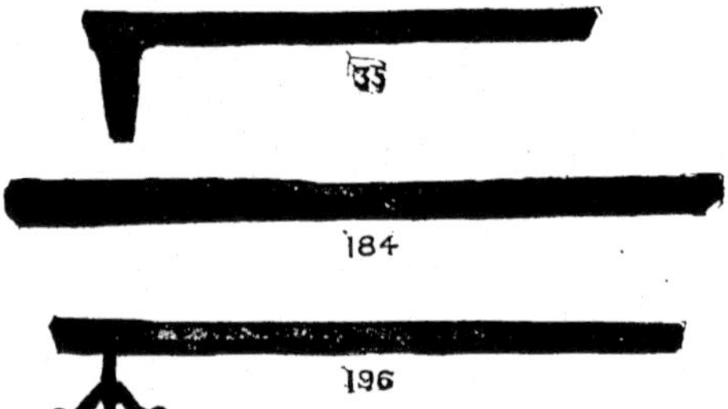

No. 35. Pipe Stake. No. 184. Cast Mandril.
No. 186. Hollow Mandril Stake.

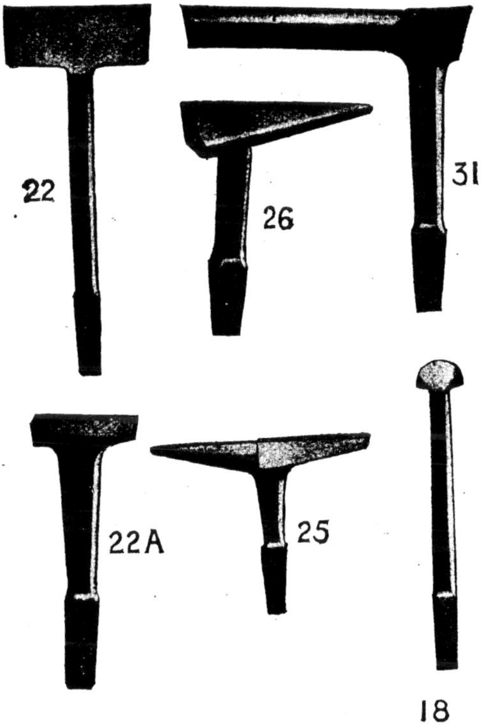

No. 22. Hatchet Stake.
No. 26. Funnel Stake.
No. 31. Side Stake.
No. 22a. Drip Pan Stake.
No. 25. Extinguisher Stake.
No. 18. Half-moon Stake.

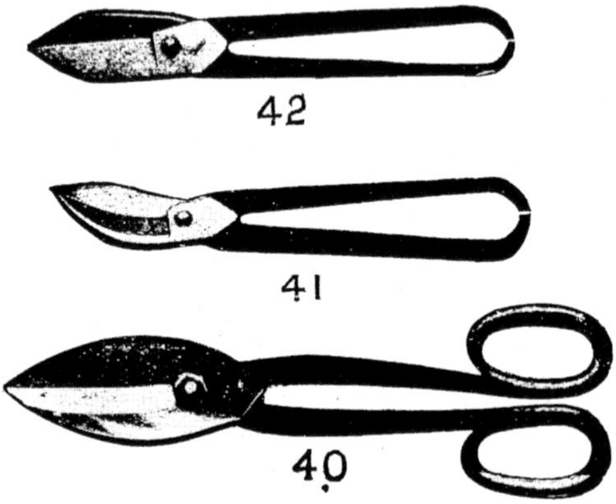

No. 42. Straight Snips. No. 41. Bent Snips.
No. 40. Scotch Shears.

Stock Shears (Right-Handed).

Block Shears (Left-Handed).

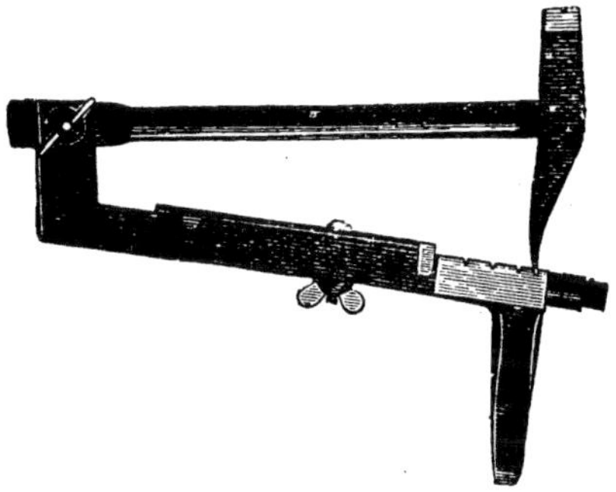

Hand Swage.

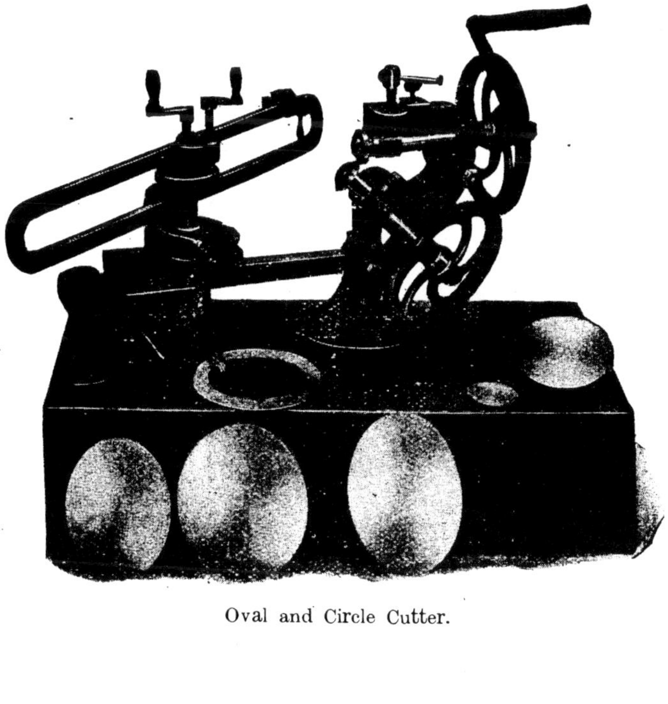

Oval and Circle Cutter.

Hand Beading Machines for tinplates.

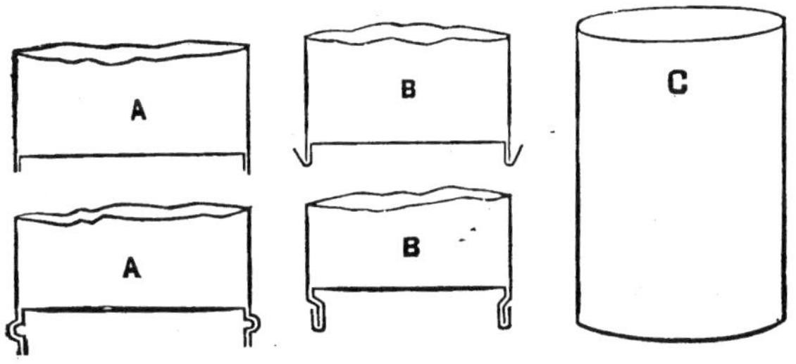

Beading or Swaging Machine, fitted with clamping Device.

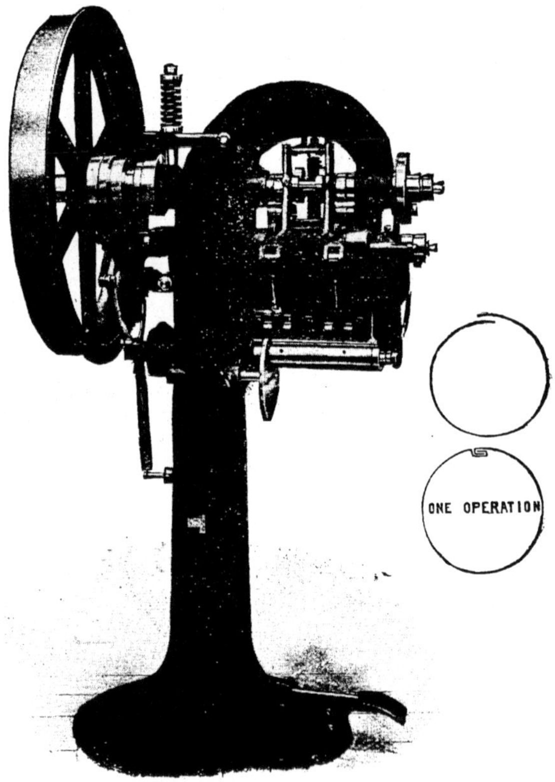

Folding and Grooving Press.

Inclinable Power Press, with Double Roller-feed and Automatic Release Motion.

Drawing Press.

Power Stamping Press.

Sheet Iron Worker's Guillotine Shears.

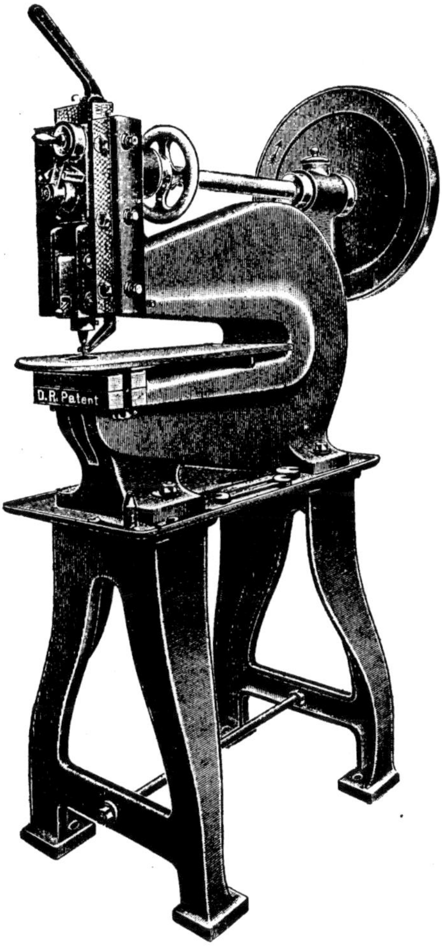

Jig Punching Machine.

Specimens of Work, Patent Jig Punching Machine.

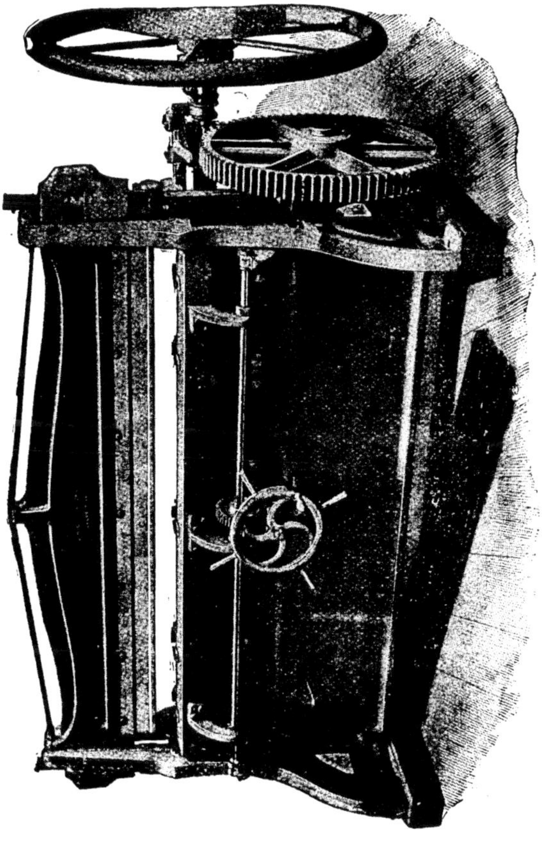

Guttering Machine.

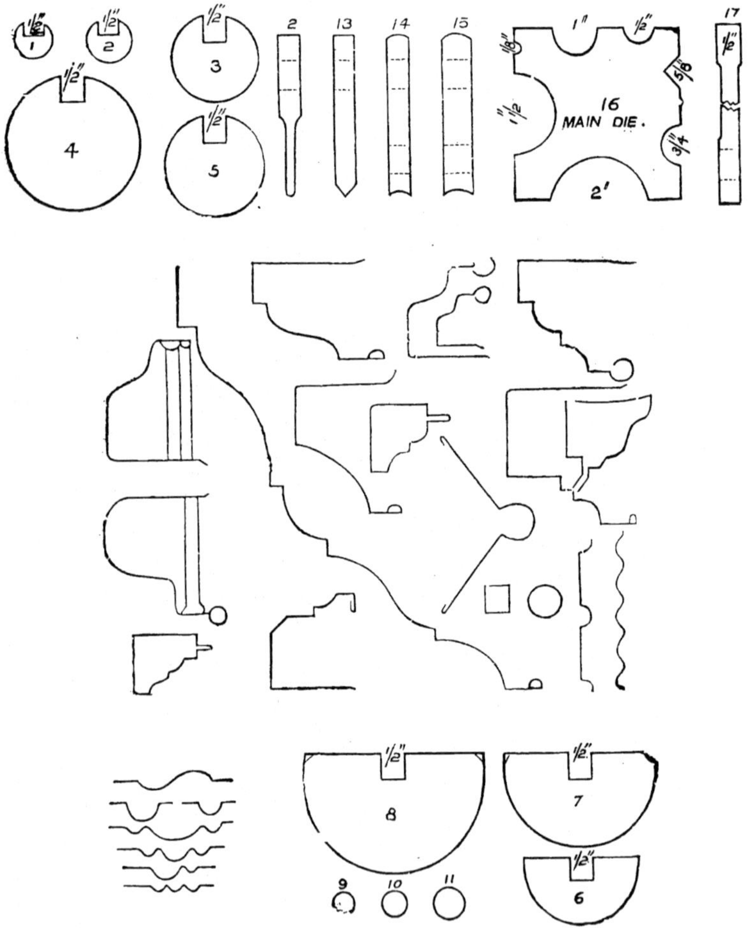

Sections of Work which can be produced with Guttering Machine, but expert workmen can produce almost any form desired.

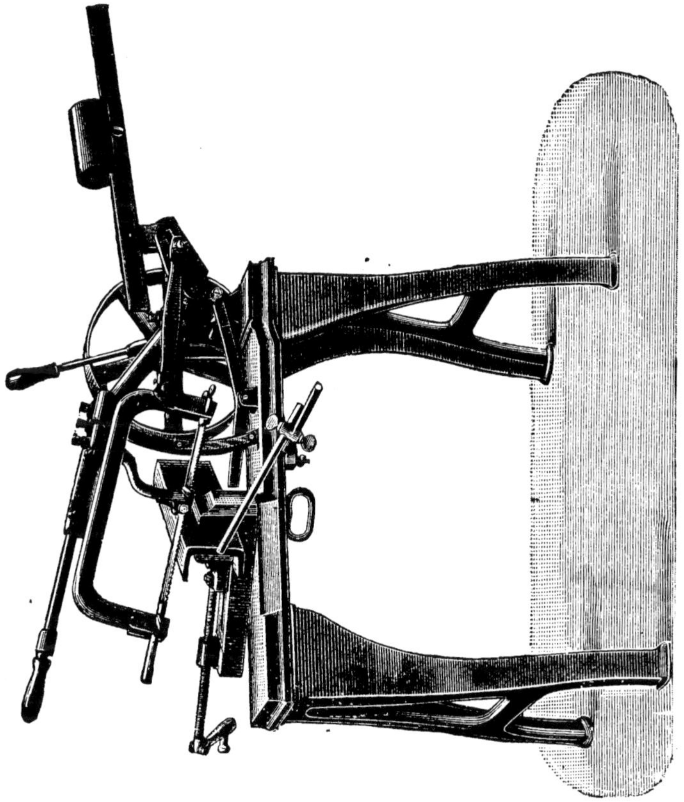

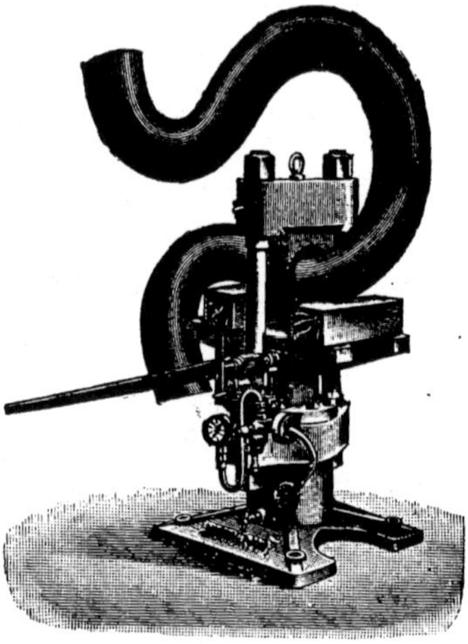

Hydraulic Pipe-Bending Machine.
For Bending Copper, Iron, and Steel Pipes up to 14 in. diameter.

Punching Bear.

Fan Forge, with Geared Blower for Hand Power.

Power Driven Shearing, Punching, and Bar-Cutting Machine.

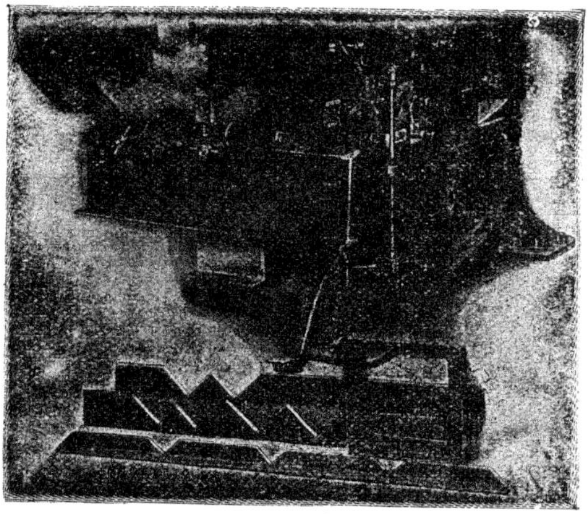

Cutting Mitre out of Flange of Joist.

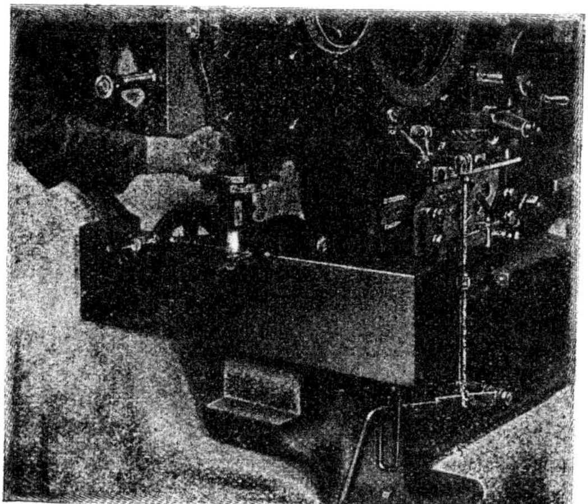

Cutting Mitre out of Flange of Channel.

Cutting Tee-Bar at 90° and Punching Angle Bar Inside.

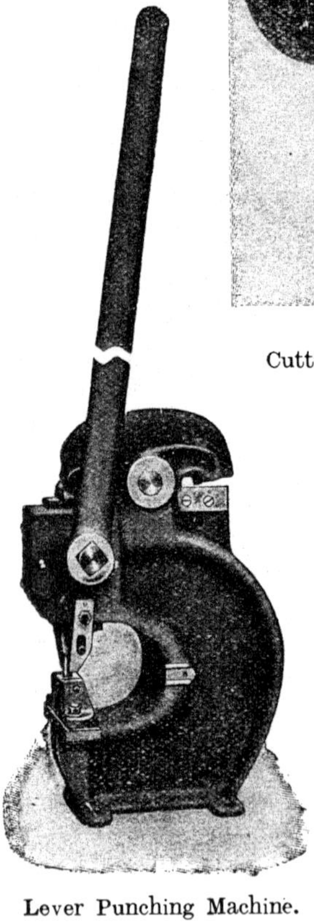

Lever Punching Machine.

Circular Cold Sawing Machine.

INDEX

A LIST OF BOOKS

PUBLISHED BY

Sir Isaac Pitman & Sons, Ltd.

(*Incorporating WHITTAKER & CO.*)

1 AMEN CORNER, LONDON, E.C. 4

A complete Catalogue giving full details of the following books will be sent post free on application.

ALL PRICES ARE NET.

	s.	d.
Alternating-Current Work. W. Perren Maycock .	7	6
Arithmetic of Electrical Engineering. Whittaker's	3	6
Arithmetic of Alternating Currents. E. H. Crapper .	3	0
Architectural Hygiene, or Sanitary Science as applied to Buildings. B. F. and H. P. Fletcher .	6	0
Art and Craft of Cabinet Making. D. Denning . .	6	0
Astronomy, for General Readers. G. F. Chambers .	4	0
Atlantic Ferry: its Ships, Men and Working, The. A. J. Maginnis	3	0
Baudôt Printing Telegraphic System. H. W. Pendry	3	0
Calculus for Engineering Students. J. Stoney .	3	6
Carpentry and Joinery: a Practical Handbook for Craftsmen and Students. B. F. and H. P. Fletcher	7	6
Central Station Electricity Supply. A. Gay and C. H. Yeaman	12	6
Colour in Woven Design: a Treatise on Textile Colouring. R. Beaumont	21	0
Common and Technical Terms in the English and Spanish Languages. R. D. Monteverde . . .	2	6
Conversion of Heat into Work. Sir W. Anderson .	6	0
Concrete Steel Buildings: being a Continuation of the Treatise on Concrete-Steel. W. N. Twelvetrees .	12	0

	s.	d.
Continuous-Current Dynamo Design, Elementary Principles of. H. M. Hobart	9	0
Concrete-Steel: a Treatise on Reinforced Concrete Construction. W. N. Twelvetrees	7	6
Continuous Current Motors and Control Apparatus. W. Perren Maycock	7	6
Design of Alternating Current Machinery. J. R. Barr and R. D. Archibald	15	0
Direct Current Electrical Engineering. J. R. Barr	12	0
Dissections Illustrated. C. G. Brodie	21	0
Drawing and Designing for Marine Engineering. C. W. Roberts	6	0
Drawing and Designing. C. G. Leland	2	6
Dynamo: its Theory, Design and Manufacture, The. C. C. Hawkins and F. Wallis. In two vols. . Each	12	6
Electric Light Fitting: a Treatise on Wiring for Lighting, Heating, &c. S. C. Batstone	6	0
Electro Plater's Handbook. G. E. Bonney . . .	3	6
Electrical Experiments. ,, . . .	3	0
Electrical Instrument Making for Amateurs. S. R. Bottone	3	6
Electric Motors: how made and how used. S. R. Bottone	3	0
Electric Bells and all about Them. S. R. Bottone .	2	6
Electric Traction. A. T. Dover	21	0
Electrical Engineers' Pocket Book. K. Edgcumbe .	6	0
Electric Motors and Control Systems. A. T. Dover.	16	0
Electric Motors—Continuous, Polyphase and Single-Phase Motors. H. M. Hobart	21	0
Electric Lighting and Power Distribution. Vol. I. W. Perren Maycock	7	6
Electric Lighting and Power Distribution. Vol. II. W. Perren Maycock	10	6
Electric Wiring, Fittings, Switches and Lamps. W. Perren Maycock	9	0
Electric Wiring Diagrams. W. Perren Maycock .	3	0
Electric Wiring Tables. W. Perren Maycock .	4	0
Electric Circuit Theory and Calculations. W. Perren Maycock	6	0
Electrical Measuring Instruments. Murdoch and Oschwald	12	6
Electric Traction. J. H. Rider . . .	12	6
Electric Light Cables. S. A. Russell	10	6

	s.	d.
ELECTRICITY IN HOMES AND WORKSHOPS. S. F. Walker .	6	0
ELECTRIC LIGHTING FOR MARINE ENGINEERS. ,, .	5	6
ELECTRICAL ENGINEERS' POCKET BOOK. Whittaker's .	6	0
ELEMENTARY GEOLOGY. A. J. Jukes-Browne . . .	3	0
ELEMENTARY TELEGRAPHY. H. W. Pendry . . .	3	6
ELEMENTARY AERONAUTICS, OR THE SCIENCE AND PRACTICE OF AERIAL MACHINES. A. P. Thurston	4	0
ELEMENTARY GRAPHIC STATICS. J. T. Wight . . .	5	0
ENGINEER DRAUGHTSMEN'S WORK: HINTS TO BEGINNERS IN DRAWING OFFICES	2	6
ENGINEERING WORKSHOP EXERCISES. E. Pull . .	2	6
ENGINEERS' AND ERECTORS' POCKET DICTIONARY: ENGLISH, GERMAN, DUTCH. W. H. Steenbeek . . .	2	6
ENGLISH FOR TECHNICAL STUDENTS. F. F. Potter . .	2	0
EXPERIMENTAL MATHEMATICS. G. R. Vine		
Book I, with Answers		9
,, II, with Answers	1	0
EXPLOSIVES INDUSTRY, RISE AND PROGRESS OF THE BRITISH	18	0
FIELD WORK AND INSTRUMENTS. A. T. Walmisley .	6	0
FIRST BOOK OF ELECTRICITY AND MAGNETISM. W. Perren Maycock	5	0
GALVANIC BATTERIES: THEIR THEORY, CONSTRUCTION AND USE. S. R. Bottone	7	6
GAS, OIL AND PETROL ENGINES: INCLUDING SUCTION GAS PLANT AND HUMPHREY PUMPS. A. Garrard . .	6	0
GAS AND GAS FITTINGS. H. F. Hills	6	0
GAS SUPPLY, IN PRINCIPLES AND PRACTICE: A GUIDE FOR THE GAS FITTER, GAS ENGINEER AND GAS CONSUMER. W. H. Y. Webber	4	0
GEOMETRICAL OPTICS. T. H. Blakesley	3	0
GERMAN GRAMMAR FOR SCIENCE STUDENTS. W. A. Osborne	3	0
HANDRAILING FOR GEOMETRICAL STAIRCASES. W. A. Scott	2	6
HIGH-SPEED INTERNAL COMBUSTION ENGINES. A. W. Judge	18	0
HISTORICAL PAPERS ON MODERN EXPLOSIVES. G. W. MacDonald	9	0
HOW TO MANAGE THE DYNAMO. S. R. Bottone . .	1	0
HYDRAULIC MOTORS AND TURBINES. G. R. Bodmer .	15	0
INDUCTION COILS. G. E. Bonney	6	0
INSPECTION OF RAILWAY MATERIAL. G. R. Bodmer .	5	6
INSULATION OF ELECTRIC MACHINES. H. W. Turner and H. M. Hobart	15	0

	s.	d.
LAND SURVEYING AND LEVELLING. A. T. Walmisley .	7	6
LEATHER WORK. C. G. Leladd	5	0
LEKTRIC LIGHTING CONNECTIONS. W. Perren Maycock .		9
LENS WORK FOR AMATEURS. H. Orford	3	6
LIGHTNING CONDUCTORS AND LIGHTNING GUARDS. Sir O. Lodge	15	0
LOGARITHMS FOR BEGINNERS	1	6
MAGNETO AND ELECTRIC IGNITION. W. Hibbert . .	3	0
MANAGEMENT OF ACCUMULATORS. Sir D. Salomons . .	7	6
MANUAL INSTRUCTION—WOODWORK. Barter, S. . .	7	6
,, ,, DRAWING ,, . .	4	0
MANUFACTURE OF EXPLOSIVES. 2 Vols. O. Guttmann .	50	0
MECHANICAL TABLES, SHOWING THE DIAMETERS AND CIRCUMFERENCES OF IRON BARS, ETC. J. Foden . .	2	0
MECHANICAL ENGINEERS' POCKET BOOK. Whittaker's .	5	0
MECHANICS' AND DRAUGHTSMEN'S POCKET BOOK. W. E. Dommett	—	
METAL TURNING. J. G. Horner	4	0
METAL WORK—REPOUSSÉ. C. G. Leland	5	0
METRIC AND BRITISH SYSTEMS OF WEIGHTS AND MEASURES. F. M. Perkin	2	0
MINERALOGY: THE CHARACTERS OF MINERALS, THEIR CLASSIFICATION AND DESCRIPTION. F. H. Hatch .	6	0
MINING MATHEMATICS (PRELIMINARY). G. W. Stringfellow	1	0
MODERN ILLUMINANTS AND ILLUMINATING ENGINEERING. Dow and Gaster	15	0
MODERN PRACTICE OF COAL MINING. Kerr and Burns. Parts . . . . each		0
MODERN OPTICAL INSTRUMENTS. H. Orford		0
MODERN MILLING. E. Pull		0
MOVING LOADS ON RAILWAY UNDER BRIDGES. H. Bamford	5	6
OPTICAL ACTIVITY AND CHEMICAL COMPOSITION. H. Lanbolt	5	6
OPTICS OF PHOTOGRAPHY AND PHOTOGRAPHIC LENSES. J. T. Taylor	4	0
PIPES AND TUBES: THEIR CONSTRUCTION AND JOINTING. P. R. Bjorling	4	0
PLANT WORLD: ITS PAST, PRESENT AND FUTURE, THE. G. Massee	3	0
POLYPHASE CURRENTS. A. Still	7	6
POWER WIRING DIAGRAMS. A. T. Dover	7	6
PRACTICAL EXERCISES IN HEAT, LIGHT AND SOUND. J. R. Ashworth	2	6

	s.	d.
PRACTICAL ELECTRIC LIGHT FITTING. F. C. Allsop .	6	0
PRACTICAL EXERCISES IN MAGNETISM AND ELECTRICITY. J. R. Ashworth	2	6
PRACTICAL SHEET AND PLATE METAL WORK. E. A. Atkins	7	6
PRACTICAL IRONFOUNDING. J. G. Horner . . .	6	0
PRACTICAL EDUCATION. C. G. Leland.	5	0
PRACTICAL TESTING OF ELECTRIC MACHINES. L. Oulton and N. J. Wilson	6	0
PRACTICAL TELEPHONE HANDBOOK AND GUIDE TO THE TELEPHONIC EXCHANGE. J. Poole	—	
PRACTICAL ADVICE FOR MARINE ENGINEERS. C. W. Roberts	3	6
PRACTICAL DESIGN OF REINFORCED CONCRETE BEAMS AND COLUMNS. W. N. Twelvetrees	7	6
PRINCIPLES OF FITTING. J. G. Horner . . .	6	0
PRINCIPLES OF PATTERN-MAKING ,, . . .	4	0
QUANTITIES AND QUANTITY TAKING. W. E. Davis . .	4	0
RADIO-TELEGRAPHIST'S GUIDE AND LOG BOOK. W. H. Marchant	5	6
RADIUM AND ALL ABOUT IT. S. R. Bottone . . .	1	6
RAILWAY TECHNICAL VOCABULARY. L. Serraillier . .	7	6
RESEARCHES IN PLANT PHYSIOLOGY. W. R. G. Atkins .	9	0
ROSES AND ROSE GROWING. Kingsley, R. G. . .	7	6
ROSES, NEW		9
RUSSIAN WEIGHTS AND MEASURES, TABLES OF. Redvers Elder	—	
SANITARY FITTINGS AND PLUMBING. G. L. Sutcliffe . .	6	0
SIMPLIFIED METHODS OF CALCULATING REINFORCED CONCRETE BEAMS. W. N. Twelvetrees		9
SLIDE RULE. A. L. Higgins		6
SMALL BOOK ON ELECTIC MOTORS, A. C. C. AND A. C. W. Perren Maycock.	5	0
SPANISH IDIOMS WITH THEIR ENGLISH EQUIVALENTS. R. D. Monteverde	3	0
SPECIFICATIONS FOR BUILDING WORKS AND HOW TO WRITE THEM. F. R. Farrow	4	0
STEEL WORK ANALYSIS. J. O. Arnold and F. Ibbotson .	12	6
STRESSES AND STRAINS: THEIR CALCULATION, ETC. F. R. Farrow	6	0
STRUCTURAL IRON AND STEEL. W. N. Twelvetrees. .	7	6
SUBMARINES, TORPEDOES AND MINES. W. E. Dommett .	3	6
SURVEYING AND SURVEYING INSTRUMENTS. G. A. T. Middleton	6	0
TABLES FOR MEASURING AND MANURING LAND. J. Cullyer	3	0

This page intentionally left blank

Except of course for that line above. And this one.
And whilst we have your attention, perhaps you'd
like to follow us on Instagram @sothis.press. Go on.

Made in United States
North Haven, CT
04 November 2026

21780191R00323